T0256089

Annals of Mathematics Studies

Number 49

ANNALS OF MATHEMATICS STUDIES

Edited by Robert C. Gunning, John C. Moore, and Marston Morse

1. Algebraic Theory of Numbers, *by* HERMANN WEYL
3. Consistency of the Continuum Hypothesis, *by* KURT GÖDEL
11. Introduction to Nonlinear Mechanics, *by* N. KRYLOFF *and* N. BOGOLIUBOFF
16. Transcendental Numbers, *by* CARL LUDWIG SIEGEL
17. Problème Général de la Stabilité du Mouvement, *by* M. A. LIAPOUNOFF
19. Fourier Transforms, *by* S. BOCHNER *and* K. CHANDRASEKHARAN
20. Contributions to the Theory of Nonlinear Oscillations, Vol. I, *edited by* S. LEFSCHETZ
21. Functional Operators, Vol. I, *by* JOHN VON NEUMANN
22. Functional Operators, Vol. II, *by* JOHN VON NEUMANN
24. Contributions to the Theory of Games, Vol. I, *edited by* H. W. KUHN *and* A. W. TUCKER
25. Contributions to Fourier Analysis, *edited by* A. ZYGMUND, W. TRANSUE, M. MORSE, A. P. CALDERON, *and* S. BOCHNER
27. Isoperimetric Inequalities in Mathematical Physics, *by* G. POLYA *and* G. SZEGO
28. Contributions to the Theory of Games, Vol. II, *edited by* H. W. KUHN *and* A. W. TUCKER
29. Contributions to the Theory of Nonlinear Oscillations, Vol. II, *edited by* S. LEFSCHETZ
30. Contributions to the Theory of Riemann Surfaces, *edited by* L. AHLFORS *et al.*
31. Order-Preserving Maps and Integration Processes, *by* EDWARD J. MCSHANE
33. Contributions to the Theory of Partial Differential Equations, *edited by* L. BERS, S. BOCHNER, *and* F. JOHN
34. Automata Studies, *edited by* C. E. SHANNON *and* J. MCCARTHY
36. Contributions to the Theory of Nonlinear Oscillations, Vol. III, *edited by* S. LEFSCHETZ
38. Linear Inequalities and Related Systems, *edited by* H. W. KUHN *and* A. W. TUCKER
39. Contributions to the Theory of Games, Vol. III, *edited by* M. DRESHER, A. W. TUCKER *and* P. WOLFE
40. Contributions to the Theory of Games, Vol. IV, *edited by* R. DUNCAN LUCE *and* A. W. TUCKER
41. Contributions to the Theory of Nonlinear Oscillations, Vol. IV, *edited by* S. LEFSCHETZ
42. Lectures on Fourier Integrals, *by* S. BOCHNER. In preparation
43. Ramification Theoretic Methods in Algebraic Geometry, *by* S. ABHYANKAR
44. Stationary Processes and Prediction Theory, *by* H. FURSTENBERG
45. Contributions to the Theory of Nonlinear Oscillations, Vol. V, CESARI, LASALLE, *and* LEFSCHETZ
46. Seminar on Transformation Groups, *by* A. BOREL *et al.*
47. Theory of Formal Systems, *by* R. SMULLYAN
48. Lectures on Modular Forms, *by* R. C. GUNNING
49. Composition Methods in Homotopy Groups of Spheres, *by* H. TODA
50. Cohomology Operations, *lectures by* N. E. STEENROD, *written and revised by* D. B. A. EPSTEIN
51. Lectures on Morse Theory, *by* J. W. MILNOR
52. Advances in Game Theory, *edited by* M. DRESHER, L. SHAPLEY, *and* A. W. TUCKER. In preparation

COMPOSITION METHODS IN HOMOTOPY GROUPS OF SPHERES

BY

Hirosi Toda

PRINCETON, NEW JERSEY
PRINCETON UNIVERSITY PRESS
1962

L. C. Card 62-12617

Printed in the United States of America

CONTENTS

COMPOSITION METHODS IN HOMOTOPY GROUPS OF SPHERES

INTRODUCTION

The i-th homotopy group $\pi_i(X)$ of a topological space X is considered as the set of the homotopy classes of the mappings from i-sphere S^i into X preserving base points. One of the main problems in homotopy theory is to determine the homotopy groups $\pi_i(S^n)$ of spheres, since this is the first fundamental difficulty in the computations of the homotopy groups of polyhedra and topological spaces.

The group $\pi_i(S^n)$ is trivial if $i<n$ or $i>1$, $n = 1$. The first sequence of non-trivial groups is $\pi_n(S^n)$, $n = 1,2,\dots$. The group $\pi_n(S^n)$ is infinite cyclic and the homotopy class of a mapping of S^n into itself is characterized by the Brouwer degree of the mapping. The second example of non-trivial groups appeared in Hopf's work [8], in which he gave a homomorphism H of $\pi_{2n-1}(S^n)$ into the group Z of integers. If n is even, then the image $H(\pi_{2n-1}(S^n))$ of Hopf's homomorphism contains $2Z$, in particular if $n = 2,4$ or 8 then H is onto and the so-called Hopf fibre map $h : S^{2n-1} \longrightarrow S^n$ has Hopf invariant 1, i.e $H(h) = 1$. It is remarkable that the groups $\pi_n(S^n)$ and $\pi_{4n-1}(S^{2n})$ are the only examples of infinite groups [13].

The first example of a non-trivial finite group $\pi_i(S^n)$ was presented by Freudenthal in [7] as the result: $\pi_{n+1}(S^n) \approx Z_2 = Z/2Z$ for $n \geqq 3$. In the paper [7], Freudenthal's suspension homomorphism

$$E : \pi_i(S^n) \longrightarrow \pi_{i+1}(S^{n+1})$$

was defined and played an important role. The group $\pi_3(S^2)$ is infinite cyclic and generated by the class η_2 of the Hopf map : $S^3 \longrightarrow S^2$. The generator of the group $\pi_{n+1}(S^n)$ is $\eta_n = E^{n-2}\eta_2$, where E^{n-2} is the (n-2)-fold interation of E. The above suspension homomorphism E is an isomorphism if $i < 2n-1$, and this provides the concept of stable group

$$G_k = \lim_{n\to\infty} \pi_{n+k}(S^n)$$

which is isomorphic to $\pi_{n+k}(S^n)$ if $n > k+1$.

The second sequence of the groups $\pi_{n+2}(S^n) \approx Z_2$, $n \geqq 2$, was determined by G. Whitehead [27], and the generator of $\pi_{n+2}(S^n)$ is the composition $\eta_n \circ \eta_{n+1}$. In general, the composition operator

$$\circ : \pi_j(S^i) \times \pi_i(S^n) \longrightarrow \pi_j(S^n)$$

is defined simply by taking the homotopy class of the composition $g \circ f$: $S^j \longrightarrow S^i \longrightarrow S^n$ of mappings f and g.

We see in these examples, together with the beautiful isomorphism

$$\pi_{i-1}(S^{n-1}) + \pi_i(S^{2n-1}) \longrightarrow \pi_i(S^n), \qquad n = 2,4,8$$

given by the correspondence $(\alpha,\beta) \longrightarrow E\alpha + \gamma \circ \beta$ for the class γ of Hopf fibre map h (see [17]), that the suspension homomorphism and the composition operator are fundamental tools for calculating the generators of the homotopy groups of spheres.

The purpose of this book is to compute the groups $\pi_{n+k}(S^n)$ for $k \leq 19$, by means of the suspension homomorphisms, the compositions and secondary operators derived from compositions. The book will be divided into two parts. In the first part, we shall compute the 2-primary components of $\pi_{n+k}(S^n)$ for $k \leq 13$ by use of purely composition methods and without use of cohomological operations, topology of Lie groups and other methods. It seems that our composition methods are not sufficient to determine the groups $\pi_{n+14}(S^n)$, and also it was seen that purely cohomological methods as in [23] are not sufficient to determine the groups of the stable case. But, combining these two sorts of methods the groups $\pi_{n+14}(S^n)$ can be determined. Thus, in the second part, we shall no more insist on using only composition methods but apply cohomological methods. Some of the results on the stable groups implied from the algebraic structure of the cohomological operations will be used without proofs, because the theory of the cohomological operations and its applications is quite different from our geometrical situation and is still in the process of development.

The readers are assumed to know familiar concepts of algebraic topology, and the recent book "Homotopy Theory" of Hu will give some good information for readers who are not familiar with homotopy problems.

The first chapter is a general consideration of the set $\pi(X \to Y)$ of the homotopy classes of maps between topological spaces X and Y, related with suspensions, reduced joins, compositions and secondary compositions. For a triple (α,β,γ) of elements of $\pi_{k+n}(S^m) \times \pi_j(S^k) \times \pi_i(S^j)$ having vanishing compositions; $\alpha \circ E^n\beta = 0$ and $\beta \circ \gamma = 0$, the secondary composition

$$\{\alpha,\ E^n\beta,\ E^n\gamma\}_n \subset \pi_{i+n+1}(S^m)$$

is defined; it is a coset of the subgroup $\alpha \circ E^n \pi_{i+1}(S^k) + \pi_{j+n+1}(S^m) \circ E^{n+1}\gamma$ of $\pi_{i+n+1}(S^m)$. Roughly speaking, the secondary composition is also a sort of composition, namely, it is represented by a composition $f \circ E^n g$: $S^{i+n+1} \longrightarrow E^nK \longrightarrow S^m$ of the n-fold suspension $E^n g$ of a mapping g : $S^{i+1} \longrightarrow K$ and a mapping $f : E^nK \longrightarrow S^n$ such that $K = S^n \cup e^{j+1}$ is a cell complex having β as the class of the attaching map of e^{j+1}, the restriction of f on $E^nS^k = S^{k+n}$ represents α and such that g maps the upper-hemisphere of S^{i+1} into K like the suspension of a representative of γ and lower-hemisphere of S^{i+1} into S^k.

Chapter II contains the sequence

$$\cdots \longrightarrow \pi_i(S^n) \xrightarrow{E} \pi_{i+1}(S^{n+1}) \xrightarrow{H} \pi_{i+1}(S^{2n+1}) \xrightarrow{\Delta} \pi_{i-1}(S^n) \xrightarrow{E} \pi_i(S^{n+1}) \cdots$$

of James [10], which is exact if n is odd and if $i < 3n-3$, and which is an exact sequence of the 2-primary components if $i > 2n$. The above H is a generalization of the Hopf homomorphism, first given by G. Whitehead in [26] for $i < 3n-3$. Several propositions on the relations of secondary compositions with H and Δ are proved in this chapter.

The anti-commutativity : $\alpha \circ \beta = (-1)^{ij} \beta \circ \alpha$, $\alpha \in G_i$, $\beta \in G_j$, of the composition operator was proved in [5] by use of reduced joins. Chapter III is an application of results about the reduced join to secondary compositions. The stable secondary composition

$$< \alpha, \beta, \gamma > \in G_{i+j+k+1}/(\alpha \circ G_{j+k+1} + \beta \circ G_{i+j+1})$$

for the triple $(\alpha, \beta, \gamma) \in G_i \times G_j \times G_k$ of $\alpha \circ \beta = \beta \circ \gamma = 0$ is defined as the suspension limit of the usual secondary composition. For this stable operation we derived in [25] anti-commutativity

$$< \alpha, \beta, \gamma > = (-1)^{ij+jk+ki+1} < \gamma, \beta, \alpha >$$

and the Jacobi identity

$$(-1)^{ik} < \alpha, \beta, \gamma > + (-1)^{ji} < \beta \;\; \gamma, \alpha > + (-1)^{kj} < \gamma, \alpha, \beta > \equiv 0.$$

Chapter IV is a preliminary for the computations of the 2-primary components of $\pi_{n+k}(S^n)$ and it derives an exact sequence

$$\cdots \longrightarrow \pi^n_i \xrightarrow{E} \pi^{n+1}_{i+1} \xrightarrow{H} \pi^{2n+1}_{i+1} \xrightarrow{\Delta} \pi^n_{i-1} \longrightarrow \cdots$$

which is adapted from James' sequence, and where π^n_i coincides with the 2-primary component of $\pi_i(S^n)$ if $i \neq n, 2n-1$.

Chapters V, VI, and VII are the computations of π^n_{n+k} for $k \leq 13$. The computations are done by induction both on k and n. The first example of a homotopy group of spheres which differs from Z or Z_2 is $\pi_6(S^3) \approx Z_{12} (\pi^3_6 \approx Z_4)$. From the above exact sequence we obtain the following group extension:

$$0 \longrightarrow Z_2 \xrightarrow{E} \pi^3_6 \xrightarrow{H} Z_2 \longrightarrow 0.$$

Then a generator of π^3_6 is given by the secondary composition $\{\eta_3, 2\iota_4, \eta_4\}_1$ (ι_4 generates $\pi_4(S^4)$) and the structure of the group extension is determined by properties of the secondary composition (cf [20]).

Chapter VIII is a discussion on Steenrod's squaring operations $\{Sq^i\}$, in particular the functional squaring operations ([18]) for Sq^2, Sq^4, and Sq^8 give generators of π^n_{n+1}, π^n_{n+3} and π^n_{n+7}. Furthermore some discussions on Sq^{16} are contained in this chapter.

Chapter IX is devoted to proving a lemma useful for obtaining a generator of G_{11} and it will be applied in Chapter X to compute the group π^n_{n+14}.

In Chapter XI, we shall define some modifications of G. Whitehead's J-homomorphism [26]; these will give much information about iterated suspension homomorphisms $E^k : \pi_i(S^n) \rightarrow \pi_{i+k}(S^{n+k})$.

Chapters X and XII are the computation of the groups π^n_{n+k} for $14 \leq k \leq 19$.

To complete the table of $\pi_{n+k}(S^n)$ in Chapter XIV, we shall use Chapter XIII about the computation of the odd primary components of π^n_{n+k} (S^n) for $k \leq 19$, and in which we shall not bother to compute the odd components further. The computation of the odd components of $\pi_{n+k}(S^n)$ for larger k will be carried out elsewhere.

PART 1.

CHAPTER I

General Compositions and Secondary Compositions

Denote by I^n the unit n-cube in euclidean n-space, i.e.,

$I^n = \{(t_1,\ldots,t_n) \mid t_1,\ldots,t_n$ real numbers, $0 \leqq t_i \leqq 1$ for $i = 1,\ldots,n\}$.

Denote by S^n the unit n-sphere in euclidean (n+1)-space, i. e.,

$$S^n = \{(t_1,\ldots,t_{n+1}) \mid t_1,\ldots,t_{n+1} \text{ real numbers, } t_1^2 + \cdots + t_{n+1}^2 = 1\}.$$

Define a continuous mapping

$$(1.1) \qquad \psi_n : I^n \longrightarrow S^n$$

as follows. The center $c_0 = (\frac{1}{2},\ldots,\frac{1}{2})$ of I^n is mapped by ψ_n to the point $e_0^* = (-1,0,\ldots,0)$ of S^n. Each point x of the boundary $\dot{I}^n$ of I^n is mapped by ψ_n to the point $e_0 = (1,0,\ldots,0)$ of S^n. Then, each segment $\overrightarrow{c_0x}$ is mapped by a uniform velocity onto a great semi-circle $\overset{\frown}{e_0^* e_0}$ whose tangent at the point e_0^* is parallel to the vector $(-1) \times \overrightarrow{c_0x} \subset (-1) \times I^n$. ψ_n has the following properties.

(1.2). $\psi_n(\dot{I}^n) = e_0$ and ψ_n is a homeomorphism of $I^n - \dot{I}^n$ onto $S^n - e_0$. $\psi_n(t) = (\cos 2\pi t, \sin 2\pi t)$. $\psi_{n+1}(t_1,\ldots,t_n,\frac{1}{2}) = i(\psi_n(t_1,\ldots,t_n))$ for an injection $i : S^n \longrightarrow S^{n+1}$ given by $i(s_1,\ldots,s_{n+1}) = (s_1,\ldots,s_{n+1},0)$. Let $\psi_n(t_1,\ldots,t_n) = (s_1,\ldots,s_{n+1})$. If $t_i \geqq \frac{1}{2}$ (resp. $t_i \leqq \frac{1}{2}$), then $s_{i+1} \geqq 0$ (resp. $s_{i+1} \leqq 0$). If t_i is replaced by $1-t_i$, then s_{i+1} is replaced by $-s_{i+1}$. If t_i and t_j are interchanged then so are s_{i+1} and s_{j+1}.

Throughout this book, we associate with each topological space X a point x_0 of X, which is called as the base point of X. For the sphere S^n, we take the point e_0 as the base point.

Every mapping and homotopy of topological spaces will have to <u>preserve the base points</u>, i.e., a mapping $f : X \longrightarrow Y$ and a homotopy $H : X \times I^1 \longrightarrow Y$ will satisfy the conditions $f(x_0) = y_0$ and $H(x_0,t) = y_0$ for all $t \in I^1$.

Denote by $\pi(X \longrightarrow Y)$ the set of the homotopy classes of the mappings $f : (X,x_0) \longrightarrow (Y,y_0)$. $\{f\} \in \pi(X \longrightarrow Y)$ indicates the homotopy class of f.

In particular, if $X = S^n$, then $\pi(X \longrightarrow Y)$ coincides with the n-dimensional homotopy group $\pi_n(Y)$ of Y, with respect to the base point y_0.

Consider two elements $\alpha\in\pi(X \longrightarrow Y)$ and $\beta\in\pi(Y \longrightarrow Z)$ and let $f : X \longrightarrow Y$ and $g : Y \longrightarrow Z$ be representatives of α and β respectively. Then the composition $g\circ f : X \longrightarrow Z$ of f and g represents an element of $\pi(X \longrightarrow Z)$ which is independent of the choice of the representatives f and g, and the class of $g\circ f$ is denoted by $\beta\circ\alpha\in\pi(X \longrightarrow Z)$ and called as the composition of α and β.

The formula $\beta\circ\alpha = f^*(\beta) = g_*(\alpha)$ defines mappings

$$f^* : \pi(Y \longrightarrow Z) \longrightarrow \pi(X \longrightarrow Z)$$

and

$$g_* : \pi(X \longrightarrow Y) \longrightarrow \pi(X \longrightarrow Z)$$

induced by f and g respectively.

Obviously the composition operator is associative:

$$(\gamma\circ\beta)\circ\alpha = \gamma\circ(\beta\circ\alpha)$$

and thus $(f_2\circ f_1)^* = f_1^*\circ f_2^*$ and $(g_2\circ g_1)_* = g_{2*}\circ g_{1*}$.

The reduced join $A \# B$ of two spaces A and B, with the base points a_0 and b_0, is obtained from the product $A \times B$ by shrinking the subset $A \vee B = A \times b_0 \cup a_0 \times B$ to a single point which is taken as the base point of $A \# B$. Denote by $\phi_{A,B} : A \times B \longrightarrow A \# B$ the shrinking map which defines $A \# B$.

Two reduced joins $(A \# B) \# C$ and $A \#(B \# C)$ are identified with the correspondence $\phi_{A \# B,C}(\phi_{A,B}(a,b),c) \longleftrightarrow \phi_{A,B \# C}(a, \phi_{B,C}(b,c))$, and denoted simply by $A \# B \# C$.

For two spheres S^m and S^n, we identify $S^m \# S^n$ with S^{m+n} by a mapping

(1.3) $$\phi_{m,n} : S^m \times S^n \longrightarrow S^{m+n}$$

given by the formula

$$\phi_{m,n}(\psi_m(x), \psi_n(y)) = \psi_{m+n}(x,y) \quad \text{for} \quad (x,y) \in I^m \times I^n = I^{m+n} .$$

This identification allows us to identify $(S^m \# S^n) \# S^p$ with $S^m \# (S^n \# S^p)$ as above, since the equality $\phi_{m+n,p}(\phi_{m,n}(a,b),c) = \phi_{m,n+p}(a, \phi_{n,p}(b,c))$, $(a,b,c) \in S^m\times S^n\times S^p$, holds. For two mappings $f : A \longrightarrow A'$ and $g : B \longrightarrow B'$ (of course $f(a_0) = a_0'$ and $g(b_0) = b_0'$), their reduced join $f \# g : A \# B \longrightarrow A' \# B'$ is naturally defined from the product $f \times g : A \times B \longrightarrow A' \times B'$ of f and g by $(f\#g)\circ\phi_{A,B} = \phi_{A',B'}\circ(f\times g)$. The following properties are checked easily.

(1.4)
o). If $f \simeq f' : (A, a_0) \longrightarrow (A', a_0')$ and $g \simeq g' : (B, b_0) \longrightarrow (B', b_0')$, then $f \# g \simeq f' \# g' : (A\#B, a_0\#b_0) \longrightarrow (A'\#B', a_0'\#b_0')$.
i). $(f\#g) \# h = f \# (g\#h)$.
ii). $(f_2\circ f_1) \# (g_2\circ g_1) = (f_2\#g_2) \circ (f_1\#g_1)$.

Let f and g be representatives of $\alpha\in\pi(A \longrightarrow A')$ and $\beta\in\pi(B \longrightarrow B')$ respectively, then the class of $f \# g$ is independent of the choice of f and g, by o) of (1.4), and the class is denoted by $\alpha \# \beta\in\pi(A\#B \longrightarrow A'\#B')$ and called as the reduced join of α and β. It follows from (1.4)

(1.5) i). $(\alpha\#\beta)\#\gamma = \alpha\#(\beta\#\gamma)$.
ii). $(\alpha'\circ\alpha)\#(\beta'\circ\beta) = (\alpha'\#\beta')\circ(\alpha\#\beta)$.

The reduced join $X \# S^n$ of a space X and the n-sphere S^n is denoted by $E^nX = X \# S^n$ and called as the n-fold suspension of X. E^1X is denoted simply by $EX = X \# S^1$. By identification $X \# S^{m+n} = X \# S^m \# S^n$, we have $E^n(E^mX) = E^{m+n}X$.

For a given mapping $f : X \longrightarrow Y$, its reduced join with the identity of S^n is called as the <u>n-fold suspension of f</u> and denoted by $E^nf : E^nX \longrightarrow E^nY$. ($Ef = E^1f$).

The equality $E^n(E^mf) = E^{m+n}f$ holds. It follows from (1.4) and (1.5)

(1.6)
0). $f \simeq g : X \longrightarrow Y$ <u>implies</u> $E^nf \simeq E^ng$, <u>and the correspondence</u> $\{f\} \longrightarrow \{E^nf\}$ <u>defines a mapping</u> $E^n : \pi(X \longrightarrow Y) \longrightarrow \pi(E^nX \longrightarrow E^nY)$.
i). $E^n(E^m\alpha) = E^{n+m}\alpha$ <u>for</u> $\alpha \in \pi(X \longrightarrow Y)$.
ii). $E^n(g\circ f) = E^ng \circ E^nf$ <u>and</u> $E^n(\beta\circ\alpha) = E^n\beta \circ E^n\alpha$.

The suspension EX of X is also defined by a mapping $d_X : X \times I^1 \longrightarrow EX$ given by setting $d_X = \emptyset_{X,S^n} \circ (1_X \times \psi_1)$, where 1_X denotes the identity of X. d_X shrinks the subset $X \times \dot{I}^1 \cup x_0 \times I^1$ to the base point of EX, and this property characterizes d_X. The suspension $Ef : EX \longrightarrow EY$ of a mapping $f : X \longrightarrow Y$ is also given by the formula

$$Ef(d_X(x,t)) = d_Y(f(x),t), \quad x\in X, \quad t\in I^1.$$

Consider the set $\pi(EX \longrightarrow Y)$ of homotopy classes. For given mappings $f,g : (EX, x_0) \longrightarrow (Y,y_0)$, define mappings $f + g$ and $-f : (EX, x_0) \longrightarrow (Y,y_0)$ by the formulas

$$(f+g)(d_X(x,t)) = \begin{cases} f(d_X(x,2t)), & 0 \leq t \leq \frac{1}{2} \\ g(d_X(x,2t-1)), & \frac{1}{2} \leq t \leq 1, \end{cases} \quad \text{and } (-f)(d_X(x,t)) = f(d_X(x,1-t)).$$

Then these operations + and - are compatible with the homotopy and thus induces those in $\pi(EX \longrightarrow Y)$. By the operations, $\pi(EX \longrightarrow Y)$ forms a group, similar to the fundamental group $\pi_1(Y_0^X)$ of the mapping space Y_0^X of the mappings : $(X,x_0) \longrightarrow (Y,y_0)$. In fact, $\pi(EX \longrightarrow Y)$ and $\pi_1(Y_0^X)$ are canonically isomorphic if X is locally compact. We have easily

(1.7). <u>For mappings</u> $f : Y \longrightarrow Z$, <u>and</u> $g : W \longrightarrow X$, $f_* : \pi(EX \longrightarrow Y) \longrightarrow \pi(EX \longrightarrow Z)$ <u>and</u> $Eg^* : \pi(EX \longrightarrow Y) \longrightarrow \pi(EW \longrightarrow Y)$ <u>are homomorphisms</u>. <u>Thus</u> $\{f\}\circ(\alpha+\alpha') = \{f\}\circ\alpha + \{f\}\circ\alpha'$ <u>and</u> $(\alpha+\alpha')\circ E\{g\} = \alpha\circ E\{g\} + \alpha'\circ E\{g\}$.

The group $\pi(E^nX \longrightarrow Y)$ is abelian if $n \geq 2$.

In the case $X = S^n$, we denote d_X by

(1.8) $$d_n : S^n \times I^1 \longrightarrow S^{n+1}$$

which is defined by the formula $d_n(\psi_n(t_1,\ldots,t_n),t) = \psi_{n+1}(t_1,\ldots,t_n,t)$.

The group operations of $\pi(EX \to Y)$ and $\pi_{n+1}(Y)$ coincide in this case.

Denote by $\Omega Y = \{\ell : (I^1, \dot{I}^1) \to (Y, y_0)\}$ the space of the loops in Y with the base point y_0. For a given mapping $f : X \to Y$, define a mapping $\Omega f : \Omega X \to \Omega Y$ by the formula

$$(\Omega f(\ell))(t) = f(\ell(t)), \quad \ell \in \Omega X, \quad t \in I^1.$$

If $f \simeq g$, then $\Omega f \simeq \Omega g$ and we obtain a correspondence $\Omega : \pi(X \to Y) \to \pi(\Omega X \to \Omega Y)$ by $\Omega\{f\} = \{\Omega f\}$. Obviously $\Omega(g \circ f) = \Omega g \circ \Omega f$ and $\Omega(\beta \circ \alpha) = \Omega(\beta) \circ \Omega(\alpha)$.

Consider the space $\Omega(EX)$ of loops in EX. For a point x of X, define a loop $i(x)$ in EX by the formula $i(x)(t) = d_X(x,t)$. Then we obtain an injection $i : X \to \Omega(EX)$, which will be called as the canonical injection. The following diagram is commutative

(1.9)
$$\begin{array}{ccc} X & \xrightarrow{\quad f \quad} & Y \\ \downarrow i & & \downarrow i \\ \Omega(EX) & \xrightarrow{\Omega(Ef)} & \Omega(EY) \end{array}$$

For a mapping $F : EX \to Y$, denote that

$$\Omega_0 F = \Omega F \circ i : X \to \Omega(EX) \to \Omega Y.$$

$\Omega_0 F$ is defined directly by the formula $\Omega_0 F(x)(t) = F(d_X(x,t))$. The correspondence $F \to \Omega_0 F$ is a homeomorphism between compact subsets of $\{EX \to Y\}$ and $\{X \to \Omega Y\}$ and thus induces a one-to-one correspondence

(1.10) $$\Omega_0 : \pi(EX \to Y) \cong \pi(X \to \Omega Y).$$

This Ω_0 is an isomorphism if we give a multiplication in $\pi(X \to \Omega Y)$ which is induced by the loop-multiplication of ΩY. This multiplication coincides with the above "+" if X is a suspension EX'. The following diagram is commutative.

(1.11)
$$\begin{array}{ccc} \pi(X \to Y) & \xrightarrow{\quad E \quad} & \pi(EX \to EY) \\ & {\searrow}^{i_*} \quad {\swarrow}^{\Omega_0} & \\ & \pi(X \to \Omega(EY)). & \end{array}$$

For mappings $f : X \to Y$, $g : EY \to Z$ and $h : Z \to W$ and for their homotopy classes $\alpha = \{f\}$, $\beta = \{g\}$ and $\gamma = \{h\}$, the following relations hold.

(1.12)
$$\Omega_0(g \circ Ef) = \Omega_0 g \circ f, \quad \Omega_0(h \circ g) = \Omega h \circ \Omega_0 g,$$
$$\Omega_0(\beta \circ E\alpha) = \Omega_0\beta \circ \alpha, \quad \Omega_0(\gamma \circ \beta) = \Omega\gamma \circ \Omega_0\beta.$$

Denote by $0 \in \pi(X \to Y)$ the class of the trivial map $X \to y_0 \in Y$.

Let $n \geq 0$ be an integer. Consider elements $\alpha \in \pi(E^nY \to Z)$, $\beta \in \pi(X \to Y)$ and $\gamma \in \pi(W \to X)$ which satisfy

(1.13) $$\alpha \circ E^n\beta = 0 \quad \text{and} \quad \beta \circ \gamma = 0.$$

Let $a : (E^nY, y_0) \to (Z, z_0)$, $b : (X, x_0) \to (Y, y_0)$ and $c : (W, w_0) \to (X, x_0)$ be representatives of α, β and γ respectively.

Then by the assumption (1.13), there exist homotopies $A_t : (E^nX, x_0) \to (Z, z_0)$ and $B_t : (W, w_0) \to (Y, y_0)$, $0 \leq t \leq 1$, such that $A_0 = a \circ E^n b$, $B_0 = b \circ c$, $A_1(E^nX) = z_0$ and $B_1(W) = y_0$.

Construct a mapping $H : (E^{n+1}W, w_0) \to (Z, z_0)$ by the formula

$$(1.14) \qquad H(d(w,t)) = \begin{cases} a(E^n B_{2t-1}(w)) & \text{for } \tfrac{1}{2} \leq t \leq 1, \\ A_{1-2t}(E^n c(w)) & \text{for } 0 \leq t \leq \tfrac{1}{2}, \end{cases}$$

where $w \in E^nW$ and $d : E^nW \times I^1 \to E^{n+1}W$ is an identification defining $E^{n+1}W = E(E^nW)$. H is well-defined, continuous and it depends on the choices of a, b, c, A_t and B_t.

Define a <u>secondary composition</u> $\{\alpha, E^n\beta, E^n\gamma\}_n \subset \pi(E^{n+1}W \to Z)$ as the set of all the homotopy classes of the mappings H given as above.

<u>Lemma 1.1</u>: <u>$\{\alpha, E^n\beta, E^n\gamma\}_n$ is a double coset of the subgroups $\pi(E^{n+1}X \to Z) \circ E^{n+1}\gamma$ and $\alpha \circ E^n(\pi(EW \to Y))$ in $\pi(E^{n+1}W \to Z)$. If $\pi(E^{n+1}W \to Z)$ is abelian, in particular if $n > 0$, $W = EW'$ or $Z = \Omega Z'$, then $\{\alpha, E^n\beta, E^n\gamma\}_n$ is a coset of the subgroup $\pi(E^{n+1} \to Z) \circ E^{n+1}\gamma + \alpha \circ E^n(\pi(EW \to Y))$.</u>

Here we use the following notations. Let A and B be subsets of $\pi(Y \to Z)$ and $\pi(X \to Y)$ respectively, then $A \circ B$ denotes the set of all $\alpha \circ \beta$ for $\alpha \in A$, $\beta \in B$. Let A_1 and A_2 be subsets of $\pi(EX \to Y)$. Then $A_1 + A_2$ denotes the set of all $\alpha_1 + \alpha_2$ for $\alpha_1 \in A_1$, $\alpha_2 \in A_2$.

<u>Proof</u>: It is easily verified that any mapping of (1.14) is homotopic to a mapping of (1.14) for fixed representatives a,b and c. Let H be given as above and let H' be given by use of other null-homotopies A'_t and B'_t of $a \circ E^n b$ and $b \circ c$, respectively. Consider mappings $F : (E^{n+1}X, x_0) \to (Z, z_0)$ and $G : (EW, w_0) \to (Y, y_0)$ given by

$$F(d_R(x,t)) = \begin{cases} A_{2t-1}(x), & \tfrac{1}{2} \leq t \leq 1, \\ A'_{1-2t}(x), & 0 \leq t \leq \tfrac{1}{2}, \end{cases} \qquad G(d_W(w,t)) = \begin{cases} B'_{2t-1}(w), & \tfrac{1}{2} \leq t \leq 1, \\ B_{1-2t}(w), & 0 \leq t \leq \tfrac{1}{2}, \end{cases}$$

where $R = E^nX$, $x \in R$ and $w \in W$. Then it is verified that H' is homotopic to $F \circ E^{n+1}c + (H + a \circ E^nG \circ \sigma)$, where σ is a homeomorphism of $E^{n+1}W = W \# S^{n+1}$ on itself given by the formula $\sigma(\phi(w, \psi_{n+1}(t_1, \ldots, t_n, t))) = \phi(w, \psi_{n+1}(t, t_1, \ldots, t_n))$. Since the correspondence $\psi_{n+1}(t_1, \ldots, t_n, t) \longleftrightarrow \psi_{n+1}(t, t_1, \ldots, t_n)$ is a homeomorphism of S^n with the degree $(-1)^n$, then σ is homotopic to the identity or a reflection of $E^{n+1}W$. Therefore $\{H'\} = F \circ E^{n+1}\gamma + (\{H\} \pm \alpha \circ E^n\{G\})$ and H and H' belong to the same double coset.

Conversely, any element of $\pi(E^{n+1}W \to Z)$, which belongs to the same double coset of $\{H\}$, is represented by a mapping $F' \circ E^{n+1}c + (H + a \circ E^nG' \circ \sigma)$ for some mappings $F' : (E^{n+1}X, x_0) \to (Z, z_0)$ and $G' : (EW, w_0) \to (Y, y_0)$. By setting

$$A'_t(x) = \begin{cases} F'(d_R(x, 2-2t)), & \frac{1}{2} \leq t \leq 1, \\ A_{2t}(x), & 0 \leq t \leq \frac{1}{2}, \end{cases} \qquad B'_t(w) = \begin{cases} G'(d_W(w, 2t-1)), & \frac{1}{2} \leq t \leq 1, \\ B_{2t}(w), & 0 \leq t \leq \frac{1}{2}, \end{cases}$$

we have null-homotopies A'_t and B'_t of $a \circ E^n b$ and $b \circ c$. Let H' be constructed by use of A'_t and B'_t then H' is homotopic to $F' \circ E^{n+1}c + (H + a \circ E^n G')$. So, any element belonging to the same double coset of $\{H\}$ is represented by a mapping of (1.14). Consequently we have proved that $\{\alpha, E^n\beta, E^n\gamma\}_n$ is a double coset of $\pi(E^{n+1}X \longrightarrow Z) \circ E^{n+1}\gamma$ and $\alpha \circ E^n(\pi(EW \longrightarrow Y))$. The second part of the lemma follows easily.

Let $n \geq m \geq 0$. The mapping of (1.14) may be regarded that it is constructed from the mappings a, $E^{n-m}b$, $E^{n-m}c$ and the homotopies A_t, $E^{n-m}B_t$. It follows then

$$(1.15) \qquad \{\alpha, E^n\beta, E^n\gamma\}_n \subset \{\alpha, E^m(E^{n-m}\beta), E^m(E^{n-m}\gamma)\}_m .$$

For the case $n = 0$, we write $\{\alpha, \beta, \gamma\}_0$ simply as $\{\alpha, \beta, \gamma\} \subset \pi(EW \longrightarrow X)$.

<u>Proposition 1.2</u>: 0). <u>If one of</u> α, β <u>or</u> γ <u>is</u> 0, <u>then</u> $\{\alpha, E^n\beta, E^n\gamma\}_n \equiv 0$.

i). <u>If</u> $\alpha \circ E^n\beta = \beta \circ \gamma = 0$, <u>then</u> $\{\alpha, E^n\beta, E^n\gamma\}_n \circ E^{n+1}\delta \subset \{\alpha, E^n\beta, E^n(\gamma \circ \delta)\}_n$.

ii). <u>If</u> $\alpha \circ E^n\beta = \beta \circ \gamma \circ \delta = 0$, <u>then</u> $\{\alpha, E^n\beta, E^n(\gamma \circ \delta)\}_n \subset \{\alpha, E^n(\beta \circ \gamma), E^n\delta\}_n$.

III). <u>If</u> $\alpha \circ E^n\beta \circ E^n\gamma = \gamma \circ \delta = 0$, <u>then</u> $\{\alpha \circ E^n\beta, E^n\gamma, E^n\delta\}_n \subset \{\alpha, E^n(\beta \circ \gamma), E^n\delta\}_n$.

IV). <u>If</u> $\beta \circ E^n\gamma = \gamma \circ \delta = 0$, <u>then</u> $\alpha \circ \{\beta, E^n\gamma, E^n\delta\}_n \subset \{\alpha \circ \beta, E^n\gamma, E^n\delta\}_n$.

<u>Proof</u>: 0). If α or β is 0, then we may chose a or b as the trivial map and A_t as the trivial homotopy. It follows that the class of the mapping H of (1.14) belongs to $\alpha \circ E^n(\pi(EW \longrightarrow Y))$. Thus $\{\alpha, E^n\beta, E^n\gamma\}_n \equiv 0$. The case $\gamma = 0$ is proved similarly.

i). According to the definition, consider the mapping H of (1.14), whose class belongs to $\{\alpha, E^n\beta, E^n\gamma\}_n$, given from a, b, c, A_t and B_t. Let d be a representative of δ. Similarly, by use of the mappings $a, b, c \circ d$ and the homotopies A_t and $B_t \circ d$, we obtain a mapping H' whose class belongs to $\{\alpha, E^n\beta, E^n(\gamma \circ \delta)\}_n$ such that the equality $H' = H \circ E^{n+1}d$ holds. Thus i) is proved. The proof of ii), iii), and iv) is similar, and left to the reader.

<u>Proposition 1.3</u>: $-\Omega_0\{\alpha, E^{n+1}\beta, E^{n+1}\gamma\}_{n+1} = \{\Omega_0\alpha, E^n\beta, E^n\gamma\}_n$ <u>and</u> $-E\{\alpha, E^n\beta, E^n\gamma\}_n \subset \{E\alpha, E^{n+1}\beta, E^{n+1}\gamma\}_{n+1}$ $(n \geq 0)$.

<u>Proof</u>: Let a, b and c be representatives of $\alpha \in \pi(E^{n+1}Y \longrightarrow Z)$, $\beta \in \pi(X \longrightarrow Y)$ and $\gamma \in \pi(W \longrightarrow Z)$ respectively. An arbitrary element of $\{\alpha, E^{n+1}\beta, E^{n+1}\gamma\}_{n+1}$ is represented by a mapping $H : E^{n+2}W \longrightarrow Z$ which is given by

$$H(d_{ER}(d_R(w, s), t)) = \begin{cases} a(E^{n+1}B_{2t-1}(d_R(w, s))), & \frac{1}{2} \leq t \leq 1, \\ A_{1-2t}(E^{n+1}c(d_R(w, s))), & 0 \leq t \leq 1, \end{cases}$$

where $w \in R = E^n W$ and A_t and B_t are null-homotopies of $a \circ E^{n+1}b$ and $b \circ c$ respectively. $d_{ER}(d_R(w,s),t) = \emptyset(w, \psi_2(s,t))$ for an identification $\emptyset : R \times S^2 \longrightarrow R \# S^2 = E^2R$ defining E^2R. Let $\sigma : E^{n+2}W \longrightarrow E^{n+2}W$ be a homeomorphism given by $\sigma(d_{ER}(d_R(w,s),t)) = d_{ER}(d_R(w,t),s)$. Then, as is seen in the proof of the lemma 1.1, $\{H \circ \sigma\} = -\{H\}$. By (1.12) and by the definition of Ω_0, we have that

$$\Omega_0(H \circ \sigma)(d_R(w,t)) = \begin{cases} \Omega_0 a(E^n B_{2t-1}(w)), & \frac{1}{2} \leq t \leq 1, \\ \Omega_0 A_{1-2t}(E^n c(w)). & 0 \leq t \leq \frac{1}{2}. \end{cases}$$

Thus $\Omega_0\{H \circ \sigma\}$ belongs to $\{\Omega_0\alpha, E^n\beta, E^n\gamma\}_n$. Conversely, an arbitrary ele- of $\{\Omega_0\alpha, E^n\beta, E^n\gamma\}_n$ is represented by a mapping $\Omega_0(H \circ \sigma)$ as above, since Ω_0 is a homeomorphism. Then the equality of this proposition is proved.

Let ι be the class of the canonical injection $i : Z \longrightarrow \Omega(EZ)$. By (1.11) and by iv) of Proposition 1.2,

$$-E\{\alpha, E^n\beta, E^n\gamma\}_n = -\Omega_0^{-1}(\iota \circ \{\alpha, E^n\beta, E^n\gamma\}_n) \subset -\Omega_0^{-1}\{\iota \circ \alpha, E^n\beta, E^n\gamma\}_n$$

$$= \{\Omega_0^{-1} \circ \iota \circ \alpha, E^{n+1}\beta, E^{n+1}\gamma\}_{n+1} = \{E\alpha, E^{n+1}\beta, E^{n+1}\gamma\}_{n+1} .$$

Then the proof of Proposition 1.3 is established.

<u>Proposition 1.4</u>: <u>If</u> $\alpha \circ E^n\beta = \beta \circ \gamma = \gamma \circ \delta = 0$, <u>then</u>

$$\{\alpha, E^n\beta, E^n\gamma\}_n \circ E^{n+1}\delta = (-1)^{n+1}(\alpha \circ E^n\{\beta, \gamma, \delta\}) .$$

<u>Proof</u>: Let a,b,c and d be representatives of $\alpha \in \pi(E^nY \longrightarrow Z)$, $\beta \in \pi(X \longrightarrow Y)$, $\gamma \in \pi(W \longrightarrow X)$ and $\delta \in \pi(V \longrightarrow W)$, respectively. An arbitrary element of $\{\alpha, E^n\beta, E^n\gamma\}_n \circ E^{n+1}\delta$ is represented by a mapping $H : E^{n+1}V \longrightarrow Z$ given by

$$H(d_S(s,t)) = \begin{cases} a(E^n B_{2t-1}(E^n d(s))), & \frac{1}{2} \leq t \leq 1, \\ A_{1-2t}(E^n c(E^n d(s))), & 0 \leq t \leq \frac{1}{2}, \end{cases}$$

where A_t and B_t are null-homotopies of $a \circ E^n b$ and $b \circ c$, respectively, and $s \in S = E^n V$. Let C_t be a null-homotopy of $c \circ d$. Consider a homotopy $H_u : (E^{n+1}V, v_0) \longrightarrow (Z, z_0)$ given by the formula

$$H_u(d_S(s,t)) = \begin{cases} a(B_{2t-1}(E^n d(s))) & \text{for } \frac{1}{2} \leq t \leq 1, \\ A_{1-2t}(E^n C_{2u(1-2t)}(s)) & \text{for } 0 \leq t \leq \frac{1}{2} \text{ and } 0 \leq u \leq \frac{1}{2}, \\ A_{(2-2u)(1-2t)}(E^n C_{1-2t}(s)) & \text{for } 0 \leq t \leq \frac{1}{2} \text{ and } \frac{1}{2} \leq u \leq 1. \end{cases}$$

Then it is verified directly that $H_0 = H$ and $H_1 = a \circ E^n K \circ \sigma$ for a mapping $K : EV \rightarrow Y$ given by

$$K(d_V(v,t)) = \begin{cases} b(C_{2t-1}(v)), & \frac{1}{2} \leq t \leq 1 , \\ B_{1-2t}(d(v)), & 0 \leq t \leq \frac{1}{2} . \end{cases}$$

and for a homeomorphism σ of $E^{n+1}V$ on itself given by the formula

$$\sigma(\emptyset(v, \psi_{n+1}(t_1, \ldots, t_n, t))) = \emptyset(v, \psi_{n+1}(1-t, t_1, \ldots, t_n))$$

where $\emptyset : V \times S^{n+1} \rightarrow E^{n+1}V$ is an identification defining $E^{n+1}V$. The mapping K represents an element of $\{\beta, \gamma, \delta\}$. Since the correspondence ψ_{n+1} $(t_1, \ldots, t_n, t) \longleftrightarrow \psi_{n+1}(1-t, t_1, \ldots, t_n)$ is a homeomorphism of S^{n+1} with the degree $(-1)^{n+1}$, then $\{H\} = \{a \circ E^n K \circ \sigma\}$ is an element of $(-1)^{n+1}(\alpha \circ E^n$

$\{\beta, \gamma, \delta\})$. Conversely an arbitrary element of $(-1)^{n+1}(\alpha \circ E^n\{\beta, \gamma, \delta\})$ is represented by the above composition $a \circ E^n K \circ \sigma$ for suitable homotopies B_t and C_t. By use of the homotopy H_{1-u}, it is proved that $\{a \circ E^n K \circ \sigma\} \in \{\alpha, E^n\beta, E^n\gamma\}_n \circ E^{n+1}\delta$. Then the proof of Proposition 1.4 is completed.

Proposition 1.5: *Assume that* $\alpha \circ \beta = \beta \circ \gamma = \gamma \circ \delta = \delta \circ \varepsilon = 0$, $0 \in \alpha \circ \{\beta, \gamma, \delta\}$ *and* $0 \in \{\beta, \gamma, \delta\} \circ E\varepsilon$. *Then there exist elements* $\lambda \in \{\alpha, \beta, \gamma\}$, $\mu \in \{\beta, \gamma, \delta\}$ *and* $\nu \in \{\gamma, \delta, \varepsilon\}$ *such that* $\lambda \circ E\delta = \alpha \circ \mu = 0$, $\beta \circ \nu = \mu \circ E\varepsilon = 0$ *and the sum* $\{\lambda, E\delta, E\varepsilon\} + \{\alpha, \mu, E\varepsilon\} + \{\alpha, \beta, \nu\}$ *contains the zero element*.

Briefly says,

$$\{\{\alpha, \beta, \gamma\}, E\delta, E\varepsilon\} + \{\alpha, \{\beta, \gamma, \delta\}, E\varepsilon\} + \{\alpha, \beta, \{\gamma, \delta, \varepsilon\}\} \equiv 0 .$$

This proposition is a generalization of ii) of Theorem 4.3 in [25], obtained by changing spheres by topological spaces, and the proofs are same.

Proposition 1.6: *Let* $\alpha, \alpha' \in \pi(E^nY \longrightarrow Z)$, $\beta, \beta' \in \pi(X \longrightarrow Y)$ *and* $\gamma, \gamma' \in \pi(W \longrightarrow X)$. *Then*

$\{\alpha, E^n\beta, E^n\gamma\}_n + \{\alpha, E^n\beta, E^n\gamma'\}_n \supset \{\alpha, E^n\beta, E^n(\gamma + \gamma')\}_n$ *if* $n \geq 1$ *or* $W = EW'$,

$\{\alpha, E^n\beta, E^n\gamma\}_n + \{\alpha, E^n\beta', E^n\gamma\}_n = \{\alpha, E^n(\beta + \beta'), E^n\gamma\}_n$ *if* $n \geq 1$ *or* $\gamma = E\gamma'(X = EX', W = EW')$,

$\{\alpha, E^n\beta, E^n\gamma\}_n + \{\alpha', E^n\beta, E^n\gamma\}_n \supset \{\alpha + \alpha', E^n\beta, E^n\gamma\}_n$ *if* $n \geq 1$ *or* $\beta = E\beta'$ *and* $\gamma = E\gamma'(Y = EY', X = EX', W = EW')$.

Proof: First remark that the group $\pi(E^{n+1}W \longrightarrow Z)$ is abelian, since $n \geq 1$ *or* $W = EW'$.

We shall prove the last relation. $\{\alpha + \alpha', E^n\beta, E^n\gamma\}_n$ is a coset of the subgroup $(\alpha + \alpha') \circ E^{n+1}\pi(W \longrightarrow X) + \pi(E^{n+1}X \longrightarrow Z) \circ E^{n+1}\gamma$. $\{\alpha, E^n\beta. E^n\gamma\}_n + \{\alpha', E^n\beta\ E^n\gamma\}_n$ is a coset of the subgroup $\alpha \circ E^{n+1}\pi(W \longrightarrow X) + \alpha' \circ E^{n+1}\pi(W \longrightarrow X) + \pi(E^{n+1}X \longrightarrow Z) \circ E^{n+1}\gamma$ which contains $(\alpha + \alpha') \circ E^{n+1}(W \longrightarrow X) + \pi(E^{n+1}X \longrightarrow Z) \circ E^{n+1}\gamma$. Thus it is sufficient to prove that $\{\alpha, E^n\beta, E^n\gamma\}_n + \{\alpha', E^n\beta, E^n\gamma\}_n$ and $\{\alpha + \alpha', E^n\beta, E^n\gamma\}_n$ have a common element.

According to the definition of the secondary composition, we construct mappings H and $H' : (E^{n+1}W, w_0) \longrightarrow (Z, z_0)$ by (1.14), which represent elements of $\{\alpha, E^n\beta, E^n\gamma\}_n$ and $\{\alpha', E^n\beta, E^n\gamma\}_n$, respectively from representatives a, a', b and c of α, α', β and γ, respectively, and by use of null-homotopies A_t of $a \circ E^nb$, A'_t of $a' \circ E^nb$ and B_t of $b \circ c$. Let $\phi : E^{n-1}W \times S^2 \longrightarrow E^{n+1}W = E^{n-1} \# S^2$ be a mapping defining $E^{n+1}W = E^2E^{n-1}W$, where $E^{n-1}W = W'$ if $n = 0$. Let $\sigma : E^{n+1}W \longrightarrow E^{n+1}W$ be a homeomorphism given by $\sigma(\phi(w, \psi_2(s,t))) = \phi(w, \psi_2(t,s))$, then σ is homotopic to the reflection of $E^{n+1}W$ and thus $\{H_0 \circ \sigma\} = -\{H_0\}$ for arbitrary $H_0 : (E^{n+1}W, w_0) \longrightarrow (Z, z_0)$. Define a mapping $H'' : (E^{n+1}W, w_0) \longrightarrow (Z, z_0)$ by

$$H''(\phi(w, \psi_2(s,t))) = \begin{cases} H(\phi(w, \psi_2(2s,t))) & \text{for } 0 \leq s \leq \tfrac{1}{2}, \\ H'(\phi(w, \psi_2(2s-1,t))) & \text{for } \tfrac{1}{2} \leq s \leq 1. \end{cases}$$

Then it is verified directly that $H'' \circ \sigma = (H \circ \sigma) + (H' \circ \sigma)$ and that H'' is constructed by (1.14) by use of the mappings $a + a'$, b, c and homotopies $A_t + A_t'$, B_t. It follows that

$$\{H\} + \{H'\} = \{H''\} \in \{\alpha + \alpha', E^n\beta, E^n\gamma\}_n$$

and the class $\{H''\}$ is a required common element. Thus the last relation of Proposition 1.6 is proved. The proof of the first two relations is similar and omitted.

A <u>cone</u> CX over a base space X will mean a space given from the product $X \times I^1$ by shrinking its subset $X \times (1) \cup x_0 \times I^1$ to a single point. Denote by $d_X' : X \times I^1 \longrightarrow CX$ the identification defining CX. We shall identify the space X with the base $d_X'(X \times (0))$ by the correspondence $x \longleftrightarrow d_X'(x, 0)$.

Consider an element β of $\pi(X \longrightarrow Y)$ and let $f : (X, x_0) \longrightarrow (Y, y_0)$ be a representative of β. A space which is obtained from the disjoint union of CX and Y by identifying X with its image under f, will be denoted by $Y \cup_f CX$ or $Y \cup_\beta CX$.

The homotopy type of $Y \cup_f CX$ is independent of the choice of representatives f of β.

An element $\bar{\alpha}$ of $\pi(Y \cup_\beta CX \longrightarrow Z)$ will be called as an <u>extension</u> of $\alpha \in \pi(Y \longrightarrow Z)$, if the restriction $g|Y$ of a representative g of $\bar{\alpha}$ represents α. An element $\tilde{\gamma}$ of $\pi(EW \longrightarrow Y \cup_\beta CX)$ is called a <u>coextension</u> of $\gamma \in \pi(W \longrightarrow X)$, if $\tilde{\gamma}$ is represented by a mapping $h : EW \longrightarrow Y \cup_\beta CX$ which satisfies the condition

$$h(d_W(w,t)) = \begin{cases} d_X'(C(w), 1-2t) & \text{if } 0 \leq t \leq \frac{1}{2}, \\ \in Y & \text{if } \frac{1}{2} \leq t \leq 1, \end{cases}$$

where $C : W \longrightarrow X$ is a representative of γ.

An extension $\bar{\alpha}$ of α exists if and only if $\alpha \circ \beta = 0$. A coextension $\tilde{\gamma}$ of γ exists if and only if $\beta \circ \gamma = 0$.

We shall use the following identification

(1.16) $$E^n(Y \cup_\beta CX) = E^nY \cup_{\beta'} C(E^nX), \qquad \beta' = E^n\beta,$$

given by the correspondence $\emptyset(d_X'(x, t), \psi_n(t_1, \ldots, t_n)) \longleftrightarrow d_Z'(\emptyset'(x, \psi_n(t_1, \ldots, t_n)), t)$, where $Z = E^nX$ and $\emptyset$ and $\emptyset'$ are mappings which define $E^n(Y \cup_\beta CX) = (Y \cup_\beta CX) \# S^n$ and $E^nX = X \# S^n$ respectively.

<u>Proposition 1.7</u>: <u>Assume that</u> $\alpha \circ E^n\beta = \beta \circ \gamma = 0$ <u>for</u> $\alpha \in \pi(E^nY \longrightarrow Z)$, $\beta \in \pi(X \longrightarrow Y)$ <u>and</u> $\gamma \in \pi(W \longrightarrow X)$. <u>Consider an extension</u> $\bar{\alpha} \in \pi(E^n(Y \cup_\beta CX) \longrightarrow Z)$ <u>of</u> α <u>and a coextension</u> $\tilde{\gamma} \in \pi(EW \longrightarrow Y \cup_\beta CX)$ <u>of</u> γ. <u>Then the set</u> $\{\bar{\alpha} \circ E^n\tilde{\gamma}\}$ <u>of the compositions</u> $\bar{\alpha} \circ E^n\tilde{\gamma}$ <u>coincides with the secondary composition</u> $(-1)^n\{\alpha, E^n\beta, E^n\gamma\}_n$.

<u>Proof</u>: For the above mapping h, we set $h(d_W(w, t)) = B_{2t-1}(w)$ for $\frac{1}{2} \leq t \leq 1$, then B_t is a null-homotopy of $b \circ c$, where $b = f$ is a representative of β. We consider also that a representative g of $\bar{\alpha}$ satisfies the formula $g(d_X'(x, t)) = A_t(x)$ for a null-homotopy A_t of

$A_0 = a \circ E^n b = (g | E^n Y) \circ E^n f$. Then it is verified directly that $g \circ E^n h = H \circ \sigma^{-1}$ for a mapping H given by (1.14) and for a homeomorphism σ of $E^{n+1}W$ in the proof of Lemma 1.1. Then the class $\bar{\alpha} \circ E^n \tilde{\gamma}$ of $g \circ E^n h$ belongs to $(-1)^n \{\alpha, E^n\beta, E^n\gamma\}_n$. Conversely, any element of $(-1)^n\{\alpha, E^n\beta, E^n\gamma\}_n$ is represented by a mapping of the form $g \circ E^n h = H \circ \sigma^{-1}$. Thus Proposition 1.7 is proved.

<u>Proposition 1.8</u>: *α, β and γ are same as the above proposition. Let $\tilde{\beta} \in \pi(E^{n+1}X \longrightarrow Z \cup_\alpha CE^n Y)$ be a coextension of $E^n\beta$, Then the set of all the compositions $\tilde{\beta} \circ E^{n+1}\gamma$ coincides with $-i_* \{\alpha, E^n\beta, E^n\gamma\}_n$, where i is the injection of Z into $Z \cup_\alpha CE^n Y$.*

<u>Proof</u>: Consider a homotopy $H_s : (E^{n+1}W, w_0) \longrightarrow (Z \cup_\alpha CE^n Y, z_0)$ given by the formula (by the same notations of (1.14)).

$$H_s(d(w, t)) = \begin{cases} A_{2t-1}(E^n c(w)) & \text{for } \tfrac{1}{2} \leq t \leq 1, \\ d'_V(E^n B_{s(1-2t)}(w),(1-s)(1-2t)) & \text{for } 0 \leq t \leq \tfrac{1}{2}, \end{cases}$$

where $V = E^n Y$. Then it is verified directly that $H_1 = -H$ and H_0 represents $\tilde{\beta} \circ E^{n+1}\gamma$, where H is given by (1.14). This proves Proposition 1.8.

Remark that $-i_* \{\alpha, E^n\beta, E^n\gamma\}_n$ is a coset of $i_* \pi(E^{n+1}X \longrightarrow Z) \circ E^{n+1}\gamma$, since Lemma 1.1 and since $i_*(\alpha) = 0$.

Denote by

$$p : X \cup_\gamma CW \longrightarrow EW \tag{1.17}$$

a mapping given by the formulas $p(d'_W(w, t)) = d_W(w, t)$ and $p(X) = w_0$. Then p is a mapping which shrinks X to a point. We have that

(1.18) $p_*(\tilde{\delta}) = E\delta$ *for a coextension* $\tilde{\delta} \in \pi(EV \longrightarrow X \cup_\gamma CW)$ *of* $\delta \in \pi(V \longrightarrow W)$.

<u>Proposition 1.9</u>: *α, β and γ are same as Proposition 1.7. Let $\bar{\beta} \in \pi(X \cup_\gamma CW \longrightarrow Y)$ be an extension of β. Then there exists an element λ of $\pi(E^{n+1}W \longrightarrow Z)$ such that $(E^n p)^*\lambda = \alpha \circ E^n\bar{\beta}$. The set $\{\lambda\}$ of such elements forms a coset of $\pi(E^{n+1}X \longrightarrow Z) \circ E^{n+1}\gamma$ which is a subset of $\{\alpha, E^n\beta, E^n\gamma\}_n$. Furthermore, any element λ of $\{\alpha, E^n\beta, E^n\gamma\}_n$ satisfies the relation $(E^n p)^* \lambda = \alpha \circ E^n\bar{\beta}$ for some choice of $\bar{\beta}$.*

<u>Proof</u>: The element $\alpha \circ E^n\bar{\beta}$ is represented by a mapping $G : E^n(X \cup_\gamma CW) \longrightarrow Z$ satisfying the formulas $G(x) = a(E^n b(x))$ for $x \in E^n X$ and $G(d'_V(w, t)) = a(E^n B_t(w))$ for $w \in E^n W = V$, where a, b and c are representatives of α, β and γ, respectively, and B_t is a null-homotopy of $b \circ c$. Since $\alpha \circ E^n\beta = 0$, the restriction $G | E^n X$ is null-homotopic, and by the homotopy extension theorem, we have a homotopy $G_s : E^n(X \cup_\gamma CW) \longrightarrow Z$ such that $G_0 = G$ and $G_1(E^n X) = z_0$. Then there exists uniquely a mapping $H_1 : E^{n+1}W \longrightarrow Z$ such that $H_1 \circ E^n p = G_1$. Thus the class λ of H_1 satisfies the condition $(E^n p)^*\lambda = \alpha \circ E^n\beta$. Furthermore, any element satisfying $(E^n p)^*\lambda = \alpha \circ E^n\beta$ is represented by a mapping H_1 given in this way. Denote that $A_t = G_t | E^n X$, then A_t is a null-homotopy of $a \circ E^n b$. Consider a homotopy $H_s : E^{n+1}W \longrightarrow Z$ given by the formula

$$H_s(d_V(w, t)) = \begin{cases} G_s(d'_V(w, (2t-1+s)/(1+s)) & \text{if } (1-s)/2 \leq t \leq 1, \\ A_{1-2t}(E^n c(w)) & \text{if } 0 \leq t \leq (1-s)/2, \end{cases}$$

Then H_s is a homotopy between H_1 and $H_0 = H$ of (1.14). Thus $\lambda = \{H_1\} = \{H\}$ belongs to $\{\alpha, E^n\beta, E^n\gamma\}_n$. In the above discussion the homotopy B_t is fixed but we may use arbitrary null-homotopy A_t of $a \circ E^n b$. Then, as is seen in the proof of Lemma 1.1, the set $\{\lambda\}$ is a coset of $\pi(E^{n+1}X \longrightarrow Z) \circ E^{n+1}\gamma$. Furthermore, by changing B_t, $\{\lambda\}$ runs over the whole of $\{\alpha, E^n\beta, E^n\gamma\}_n$. Consequently Proposition 1.9 is proved.

CHAPTER II

Generalized Hopf Invariant and Secondary Composition.

In the following, we shall devote to the consideration on the homotopy groups of spheres $\pi_p(S^m) = \pi(S^p \to S^m)$.

Denote by $\iota_n \in \pi_n(S^n)$ the homotopy class of the identity mapping of S^n. Obviously, $\iota_n \circ \alpha = \alpha = \alpha \circ \iota_p$ for $\alpha \in \pi_p(S^m)$.

It follows from (1.7)

(2.1) $\alpha \circ (\beta_1 \pm \beta_2) = \alpha \circ \beta_1 \pm \alpha \circ \beta_2$ and $(\alpha_1 \pm \alpha_2) \circ E\beta$
$= \alpha_1 \circ E\beta \pm \alpha_2 \circ E\beta$, in particular, $k(\alpha \circ \beta) = \alpha \circ (k\beta)$
and $k(\alpha \circ E\beta) = (k\,\alpha) \circ E\beta$ for an integer k.

The reduced join $\alpha \# \beta \in \pi_{p+q}(S^{m+n})$ of $\alpha \in \pi_p(S^m)$ and $\beta \in \pi_q(S^n)$ is defined as in the previous chapter by use of the special mappings $\phi_{p,q}$ and $\phi_{m,n}$ of (1.3). In particular, from the definition of E^n it follows that

$$E^n\alpha = \alpha \# \iota_n .$$

Also the secondary composition
$\{\alpha, E^n\beta, E^n\gamma\}_n \in \pi_{i+n+1}(S^m)/(\alpha \circ E^n\pi_{i+1}(S^k) + \pi_{j+n+1}(S^m) \circ E^{n+1}\gamma)$ of $\alpha \in \pi_{k+n}(S^m)$, $\beta \in \pi_j(S^k)$ and $\gamma \in \pi_i(S^j)$ is defined if the condition (1.13) is satisfied, where we use the mapping d_{i+n} of (1.8) in place of d of (1.14).

Of course, all the discussion in the previous chapter may be applied in this particular case.

Now we introduce from [9] the concept of reduced product of S^m and some necessary results. Denote by $(S^m)_\infty$ the reduced product of S^m. $(S^m)_\infty$ is a free semi-group with the set $S^m - e_0$ of generators and the unit e_0. Each point of $(S^m)_\infty$ is represented by a product $x_1 \cdots x_t$ of $x_1, \ldots, x_t \in S^m$. For fixed positive integer t, $(S^m)_t$ denotes the set of all elements $x_1 \cdots x_t$. Then the topology of $(S^m)_t$ is given from the product $S^m \times \cdots \times S^m$ of t-times S^m under the identification: $S^m \times \cdots \times S^m \to (S^m)_t$ of $(x_1, \ldots, x_t) \to x_1 \cdots x_t$. $(S^m)_t - (S^m)_{t-1}$ is an open tm-cell. By giving the weak topology, $(S^m)_\infty$ is a CW-complex and $(S^m)_t$ is its tm-skeleton. We identify $(S^m)_1$ with S^m in the natural way.

Consider a mapping $f : (S^q, e_0) \to (S^m, e_0)$, then a mapping $(f)_t : (S^q)_t \to (S^m)_t$ $(t = 1, 2, \ldots, \infty)$ is given by the formula $(f)_t(y_1 \cdots, y_t) = f(y_1) \cdots f(y_t)$. Obviously $(f)_s = (f)_t | (S^q)_s$ for $s \leq t$.

The canonical injection $i : S^m \to \Omega(S^{m+1})$ is extended to the whole of $(S^m)_\infty$ such a way that $x_1 \cdots x_t$ is mapped to a loop in S^{m+1} which is represented by a suitably weighted sum of the loops $i(x_1), \ldots, i(x_t)$. The resulted mapping is injective, denoted by $i : (S^m)_\infty \to \Omega(S^{m+1})$ and called also by a canonical injection.

The following diagram is homotopically commutative

(2.2)
$$\begin{array}{ccc} (S^q)_\infty & \xrightarrow{i} & \Omega(S^{q+1}) \\ \downarrow & & \downarrow \\ (S^m)_\infty & \xrightarrow{i} & \Omega(S^{m+1}) \end{array} .$$

As one of the main results on reduced product space, the canonical injection induces isomorphisms of homotopy groups: $i_* : \pi_q((S^m)_\infty) \to \pi_q(\Omega(S^{m+1}))$ for all q.

That is to say, $(S^m)_\infty$ and $\Omega(S^{m+1})$ have the same singular homotopy type, and thus we have

Lemma 2.1. $i_* : \pi(K \to (S^m)_\infty) \to \pi(K \to \Omega(S^{m+1}))$ is one-to-one for arbitrary finite cell complex (or more generally CW-complex) K.

Define a one-to-one mapping Ω_1 by

(2.3)
$$\Omega_1 = i_*^{-1} \circ \Omega_0 : \pi(EK \to S^{m+1}) \longleftrightarrow \pi(K \to (S^m)_\infty).$$

In particular, we have isomorphisms $\Omega_1 : \pi_{k+1}(S^{m+1}) \approx \pi_k((S^m)_\infty)$ for all k.

For the injection $i' : S^m \to (S^m)_\infty$, the following diagram is commutative.

(2.4)
$$\begin{array}{ccc} \pi(K \to S^m) & \xrightarrow{E} & \pi(EK \to S^{m+1}) \\ & {\scriptstyle i'_*} \searrow & \downarrow {\scriptstyle \Omega_1} \\ & & \pi(K \to (S^m)_\infty) \end{array} .$$

Consider a mapping $h'_m : ((S^m)_2, S^m) \to (S^{2m}, e_0)$ given by the formula $h'_m(xy) = \phi_{m,m}(x, y)$ for $x, y \in S^m$. Let

(2.5)
$$h_m : ((S^m)_\infty, S^m) \to ((S^{2m})_\infty, e_0)$$

be the combinatorial extension [9] of h'_m, then the following diagram is commutative.

(2.6)
$$\begin{array}{ccc} (S^q)_\infty & \xrightarrow{h_q} & (S^{2q})_\infty \\ \downarrow (f)_\infty & & \downarrow (f\#f)_\infty \\ (S^m)_\infty & \xrightarrow{h_m} & (S^{2m})_\infty . \end{array}$$

In general, the combinatorial extension $h : (S^m)_\infty \to (S^{km})_\infty$ of a mapping $h' : ((S^m)_k, (S^m)_{k-1}) \to (S^{km}, e_0)$ is defined by the formula

$$h(x_1, \ldots, x_t) = \prod_\sigma h'(x_{\sigma(1)}, \ldots, x_{\sigma(k)})$$

where σ is a monotone increasing map of k-letters $\{1, 2, \ldots, k\}$ into t-letters $\{1, 2, \ldots, t\}$ and the order of the product $\prod$ in $(S^{km})_\infty$ is lexicographic from the right (left).

Define a generalized Hopf invariant H by the formula

(2.7) $$H = \Omega_1^{-1} \circ h_{m_*} \circ \Omega_1 : \pi(EK \to S^{m+1}) \to \pi(EK \to S^{2m+1}).$$

In particular, we have homomorphisms $H : \pi_{i+1}(S^{m+1}) \to \pi_{i+1}(S^{2m+1})$.

<u>Proposition</u> 2.2: <u>Let</u> K <u>and</u> L <u>be finite cell complexes</u>. <u>Consider elements</u> $\alpha \in \pi(EK \to S^{m+1})$, $\beta \in \pi(L \to K)$ <u>and</u> $\gamma \in \pi_m(S^r)$, <u>then</u>

$$H(\alpha \circ E\beta) = H(\alpha) \circ E\beta$$

<u>and</u>

$$H(E\gamma \circ \alpha) = E(\gamma \# \gamma) \circ H(\alpha).$$

<u>Proof</u>:
$$\begin{aligned} H(\alpha \circ E\beta) &= (\Omega_0^{-1} \circ i_* \circ h_{m*} \circ i_*^{-1} \circ \Omega_0)(\alpha \circ E\beta) && \text{by (2.7)} \\ &= \Omega_0^{-1}((i_* \circ h_{m_*} \circ i_*^{-1})(\Omega_0\alpha) \circ \beta) && \text{by (1.12)} \\ &= (\Omega_0^{-1} \circ i_* \circ h_{m*} \circ i_*^{-1} \circ \Omega_0)(\alpha) \circ E\beta && \text{by (1.12)} \\ &= H(\alpha) \circ E\beta . \\ H(E\gamma \circ \alpha) &= (\Omega_0^{-1} \circ i_* \circ h_{r*} \circ i_*^{-1} \circ \Omega_0)(E\gamma \circ \alpha) && \text{by (2.7)} \\ &= \Omega_0^{-1}((i_* \circ h_{r*} \circ i_*^{-1})(\Omega E\gamma) \circ \Omega_0\alpha) && \text{by (1.12)} \\ &= \Omega_0^{-1}(\Omega E(\gamma \# \gamma) \circ (i_* \circ h_{r*} \circ i_*^{-1})(\Omega_0\alpha)) \\ & && \text{by (2.2), (2.6)} \\ &= \Omega^{-1}\Omega E(\gamma \# \gamma) \circ (\Omega_1^{-1} \circ h_{m*} \circ \Omega_1)(\alpha) && \text{by (1.12)} \\ &= E(\gamma \# \gamma) \circ H(\alpha) . \end{aligned}$$

<u>Proposition</u>: 2.3. <u>Let</u> L, M <u>and</u> K <u>be finite cell complexes</u>.

Assume that $n \geq 1$ and elements $\alpha \in \pi(E^n K \to S^{m+1})$, $\beta \in \pi(L \to K)$ and $\gamma \in \pi(M \to L)$ satisfy the condition $\alpha \circ E^n\beta = \beta \circ \gamma = 0$. Then

$$H\{\alpha, E^n\beta, E^n\gamma\}_n \subset \{H\alpha, E^n\beta, E^n\gamma\}_n.$$

Proof: By Proposition 1.7, an arbitrary element of $\{\alpha, E^n\beta, E^n\gamma\}_n$ is the composition $(-1)^n \bar{\alpha} \circ E^n\tilde{\gamma}$ of an extension $\bar{\alpha} \in \pi(E^n(K \cup_\beta CL) \to S^{m+1})$ of α and the n-suspension $E^n\tilde{\gamma}$ of a coextension $\tilde{\gamma}$ of γ. By Proposition 2.2, $H((-1)^n(\bar{\alpha} \circ E^n\tilde{\gamma})) = (-1)^n H(\bar{\alpha}) \circ E^n\tilde{\gamma}$. Let $\iota \in \pi(E^n K \to E^n(K \cup_\beta CL))$ be the class of the inclusion, then ι is a suspension element and $\bar{\alpha} \circ \iota = \alpha$. By Proposition 2.2, $H(\alpha) = H(\bar{\alpha} \circ \iota) = H(\bar{\alpha}) \circ \iota$. This shows that $H(\bar{\alpha})$ is an extension of $H(\alpha)$. Then, by use of Proposition 1.7, we have that $(-1)^n H(\bar{\alpha} \circ E^n\gamma) = (-1)^n H(\bar{\alpha}) \circ E^n\tilde{\gamma}$ is contained in $\{H(\alpha), E^n\beta, E^n\gamma\}_n$. Thus the proof of the proposition is established.

We remark that the above discussions on $(S^m)_\infty$ and the results on the generalized Hopf invariant are still true if we replace the sphere S^m by some suitable space, for example finite cell complex.

Theorem 2.4. (James [10], Toda [21]). Let p be a prime and $m > 1$. Let $h : (S^m)_\infty \to (S^{pm})_\infty$ be a continuous mapping which maps $(S^m)_{p-1}$ to e_0 and e^{pm} homeomorphically onto $S^{pm} - e_0$. If m is odd and $p = 2$, then $h_* : \pi_i((S^m)_\infty, S^m) \to \pi_i((S^{2m})_\infty)$ are isomorphisms onto for all i. If m is even, then $h_* : \pi_i((S^m)_\infty, (S^m)_{p-1}) \to \pi_i((S^{pm})_\infty)$ are isomorphisms onto for $i < (p+1)m-1$ and isomorphisms of the p-primary components for all i.

Proof: By use of Theorem 2 of [6], we have that

$$h_* : \pi_i((S^m)_\infty, (S^m)_{p-1}) \to \pi_i((S^{pm})_\infty)$$

are isomorphisms onto for $i < (p+1)m-1$.

Next we recall the cohomological structure of $(S^m)_\infty$, which is computed from the cohomological structure of the k-fold products $S^m \times \dots \times S^m$ of S^m under the identifications to $(S^m)_k$. Then we have that, for generators u_k of $H^{km}((S^m)_\infty)$ dual to the cell e^{km}, the following relations hold. If m is odd, then $u_1 u_{2k} = u_{2k+1}$ and $u_{2k}u_{2h} = \binom{k+h}{k} u_{2(k+h)}$. If m is even, then $u_k u_h = \binom{k+h}{k} u_{k+h}$. Let $\Lambda = Z$ if m is odd and $p = 2$ and $\Lambda = Z_p$ if m is even. Then it is easily verified that the injection homomorphism $i^* : H^i((S^m)_\infty, \Lambda) \to H^i((S^m)_{p-1}, \Lambda)$ is an isomorphism onto for $i < pm$, the induced homomorphism $h^* : H^i((S^{pm})_\infty, \Lambda) \to H^i((S^m)_\infty, \Lambda)$ is an isomorphism into for all i and that we have an isomorphism

$$h^*H^*((S^{pm})_\infty, \Lambda) \otimes G \approx H^*((S^m)_\infty, \Lambda)$$

by means of cup-products, where G is a subgroup of $H^*((S^m)_\infty, \Lambda)$ which is isomorphic $H^*((S^m)_{p-1}, \Lambda)$ under i^*.

Now let X be a set of the pairs (ℓ, x) such that x is a point of $(S^n)_\infty$ and $\ell : I^1 \to (S^{pm})_\infty$ is a path with $\ell(1) = h(x)$. Let $p(\ell, x) = h(x) = \ell(1)$. Then $(X, p, (S^{pm})_\infty)$ is a fibre space, in the sense of Serre [13]. Consider the spectral sequence $\{E_r^{s,t}\}$ associated with this fibre space, then by the main theorem of [13] we have an isomorphism

$$E_2^* \approx B^* \otimes H^*(F, \Lambda),$$

where $B^* = H^*((S^{pm})_\infty, \Lambda)$.

The pairs (ℓ, x) of the constant paths $\ell(I^1) = h(x)$ form an imbedding of $(S^m)_\infty$ into X such that $(S^m)_\infty$ is a deformation retract of X. By the retraction, h is equivalent to p. The subcomplex $(S^m)_{p-1}$ is contained in the fibre $F = p^{-1}(e_0)$.

We shall prove that the injection $i_0 : (S^m)_{p-1} \subset F$ induces isomorphisms $i_0^* : H^*(F, \Lambda) \approx H^*((S^m)_{p-1}, \Lambda)$.

Let $i_1 : F \subset X$ be the injection. Then $i_0^* \circ i_1^*$ is equivalent to i^*. Thus $i_1^* : H^i(X, \Lambda) \to H^i(F, \Lambda)$ is an isomorphism into for $i < pm$. Let G_0 be the subgroup of $H^*(F, \Lambda)$ which corresponds to G by i_1^*. G_0 is in the image of i_1^* and i_0^* maps G_0 isomorphically onto $H^*((S^m)_{p-1}, \Lambda)$. The injection homomorphism i_1^* is equivalent to the composition $: H^*(X, \Lambda) \to E_\infty^{0,*} \to E_2^{0,*} = 1 \otimes H^*(F, \Lambda)$. It follows then that

$$d_r(1 \otimes G_0) = 0 \quad \text{for} \quad r \geq 2$$

and its limit in E_∞^* corresponds to G. Since $d_r(B^* \otimes 1) = 0$ for $r \geq 2$, then we have $d_r(B^* \otimes G_0) = 0$ for $r \geq 2$. Consider the limit of $B^* \otimes G_0$ in E_∞^*, then it is a graded module associated with $H^*(X, \Lambda) \approx p^*H^*((S^{pm})_\infty, \Lambda) \otimes G$. Then it follows that $B^* \otimes G_0$ is isomorphic to E_∞^* and $H^*(X, \Lambda)$ as modules. This means that every d_r-images in $B^* \otimes G_0$ are trivial. Assume that f_0 is an element of minimum dimension in $H^*(F, \Lambda)$ which is not contained in G_0. Then $d_r(1 \otimes f_0)$ is in $B^* \otimes G_0$ and hence $d_r(1 \otimes f_0) = 0$

for $r \geq 2$. Since $1 \otimes f_0$ is not a d_r-image, then $1 \otimes f_0 \neq 0$ in E_∞^*. But this contradicts to the isomorphism $B^* \otimes G_0 \approx E_\infty^*$. Thus such an element f_0 does not exist. Consequently we have proved that $H^*(F, \Lambda) = G_0$ and $i_0^* : H^*(F, \Lambda) \approx H^*((S^m)_{p-1}, \Lambda)$.

Applying generalized J. H. C. Whitehead's theorem in [15], to the injection i_0, we obtain isomorphisms $i_{0*} : \pi_i((S^m)_{p-1}) \to \pi_i(F)$ if m is odd and $p = 2$ and isomorphisms of the p-primary components if m is even.

Compairing the exact sequences of homotopy groups associated with the pairs $((S^m)_\infty, (S^m)_{p-1})$ and (X, F) by the injection (a homotopy equivalence) of $(S^m)_\infty$ into X, it follows from the five lemma that the above statement for i_0^* is true for the injection homomorphism of $\pi_i((S^m)_\infty, (S^m)_{p-1})$ into $\pi_i(X,F)$. Since the diagram

$$\begin{array}{ccc} \pi_i((S^m)_\infty, (S^m)_{p-1}) & \longrightarrow & \pi_i(X, F) \\ & h_* \searrow \quad \swarrow p_* & \\ & \pi_i((S^{pm})_\infty) & \end{array}$$

is commutative and since p_* is an isomorphism onto for all i, then we see that the theorem has been proved. q. e. d.

The mapping $h_m : ((S^m)_\infty, S^m) \to ((S^{2m})_\infty, e_0)$, $m > 1$, satisfies the condition of the above theorem. Thus

$$h_{m*} : \pi_i((S^m)_\infty, S^m) \to \pi_i((S^{2m})_\infty)$$

are isomorphisms onto for all i if m is odd or if $i < 3m-1$ and isomorphisms onto of the 2-primary components for all i if m is even.

Now consider the exact sequence for the pair $((S^m)_\infty, S^m)$:

$$(2.8) \qquad \cdots \to \pi_i(S^m) \to \pi((S^m)_\infty) \to \pi_i((S^m)_\infty, S^m) \overset{\partial}{\to} \pi_{i-1}(S^m) \to \cdots .$$

Define a homomorphism Δ by

$$(2.9) \qquad \Delta = \partial \circ h_{m*}^{-1} \circ \Omega_1 : \pi_{i+1}(S^{2m+1}) \to \pi_{i-1}(S^m)/\partial \operatorname{Ker} h_{m*},$$

where $\partial \operatorname{Ker} h_{m*}$ is trivial if m is odd or if $i < 3m-1$ and it is finite and has no 2-torsion if m is even.

Also define a homomorphism Δ^{-1} by

$$(2.10) \quad \Delta^{-1} = \Omega_1^{-1} \circ h_{m*} \circ \partial^{-1} : \pi_{i-1}(S^m) \to \pi_{i+1}(S^{2m+1})/H\pi_{i+1}(S^{m+1}),$$

where we remark that $H\,\pi_{i+1}(S^{m+1})$ is the kernel of the above Δ.

From the commutativity of (2.4) and the definitions (2.7) and (2.9) of H and Δ, we have an exact sequence

(2.11) $\cdots \to \pi_i(S^m) \xrightarrow{E} \pi_{i+1}(S^{m+1}) \xrightarrow{H} \pi_{i+1}(S^{2m+1}) \xrightarrow{\Delta} \pi_{i-1}(S^m) \xrightarrow{E} \cdots$

for an odd m or for $i < 3m-1$. We may consider that (2.11) is an exact sequence of the 2-primary components in the case that m is even and $i \geq 3m-1$.

Proposition 2.5. $\Delta\alpha \circ \beta \subset \Delta(\alpha \circ E^2\beta)$ *for* $\alpha \in \pi_{i+1}(S^{2m+1})$ *and* $\beta \in \pi_{j-1}(S^{i-1})$. *In particular,* $\Delta(E^2\gamma) \ni \pm [\iota_m, \iota_m] \circ \gamma$ *where* $[\ ,\ ]$ *indicates the product of Whitehead. The symbols* $\subset$ *and* $\ni$ *can be replaced by the equality* $=$ *provided that* m *is odd,* $j < 3m-1$ *or that the elements in the relation are in the 2-primary components.*

Proof: First we recall about relative homotopy groups $\pi_i(X,A,x_0)$, whose element is represented by a mapping $f : (CS^{i-1}, S^{i-1}, e_0) \to (X,A,x_0)$. The boundary homomorphism ∂ is given by taking the restriction of f on S^{i-1}. The relativization : $\pi_i(X,x_0,x_0) = \pi_i(X,x_0)$ is given by associating the composition $f \circ p_i : (S^i, e_0) \to (X,x_0)$ for $f : (CS^{i-1}, S^{i-1}, e_0) \to (X,x_0,x_0)$, where p_i is a shrinking map as in (1.17) given by use of d_{i-1}.

By regarding the composition in relative homotopy groups, we have that $\partial(\alpha' \circ \partial_0^{-1}\beta) = \partial(\alpha') \circ \beta$ and $h_{m*}(\alpha' \circ \partial_0^{-1}\beta) = h_{m*}(\alpha') \circ E\beta$ for $\alpha' \in \pi_i((S^m)_\infty, S^m)$ and the boundary homomorphism $\partial_0 : \pi_j(CS^{i-1}, S^{i-1}) \to \pi_{j-1}(S^{i-1})$ which is an isomorphism onto. Then

$$\begin{aligned}\Delta\alpha \circ \beta &= \partial(h_{m*}^{-1}(\Omega_1\alpha)) \circ \beta = \partial(h_{m*}^{-1}(\Omega_1\alpha) \circ \partial_0^{-1}\beta)\\ &\subset \partial(h_{m*}^{-1}(\Omega_1(\alpha) \circ E\beta)) = \partial(h_{m*}^{-1}(\Omega_1(\alpha \circ E^2\beta)) \qquad \text{by (1.12)}\\ &= \Delta(\alpha \circ E^2\beta).\end{aligned}$$

The element $\Delta(\iota_{2m+1})$ is the class of the attaching map of the 2m-cell $(S^m)_2 - S^m$ onto S^m, which is the image of the attaching map of the cell $(S^m - e_0) \times (S^m - e_0)$ onto $S^m \vee S^m$ under the identification : $(S^m \times S^m, S^m \vee S^m) \to ((S^m)_2, S^m)$. Then it is verified directly from the definition of Whitehead product that $\Delta(\iota_{2m+1}) = \pm [\iota_m, \iota_m]$. Thus $\pm [\iota_m, \iota_m] \circ \gamma = \Delta(\iota_{2m+1}) \circ \gamma \in \Delta(E^2\gamma)$.

Proposition 2.6. *Let* $\alpha \in \pi(K \to S^m)$, $\beta \in \pi_k(K)$ *and* $\gamma \in \pi_i(S^k)$ *satisfy the condition* $E(\alpha \circ \beta) = \beta \circ \gamma = 0$. *Then*

$$H\{E\alpha, E\beta, E\gamma\}_1 = -\Delta^{-1}(\alpha\circ\beta)\circ E^2\gamma.$$

Proof: Let $a : (K, x_0) \to (S^m, e_0)$ and $b : (S^k, e_0) \to (K, x_0)$ be representatives of α and β respectively. Consider the space $K \cup_b CS^k$ and let $\bar{b} : (CS^k, S^k) \to (K \cup_b CS^k, K)$ be the characteristic mapping. Consider the injection i of S^m into $(S^m)_\infty$. By (2.3) and (2.4), $E(\alpha\circ\beta) = 0$ implies that $i_*\alpha \circ \beta = 0$. Thus the composition $a \circ b$ is null-homotopic in $(S^m)_\infty$, and we can extend the mapping a over

$$\bar{a} : (K \cup_b CS^k) \to ((S^m)_\infty, S^m).$$

Then there exists a mapping a' such that the following diagram is commutative:

$$\begin{array}{ccccc}
(CS^k_x, S^k) & \xrightarrow{\ \bar{b}\ } & (K \cup_b CS^k, K) & \xrightarrow{\ \bar{a}\ } & ((S^m)_\infty, S^m) \\
 & \searrow^{p'} & \downarrow p & & \downarrow h_m \\
 & & (S^{k+1}, e_0) & \xrightarrow{\ a'\ } & ((S^{2m})_\infty, e_0),
\end{array}$$

where p and p' are shrinking maps of (1.17) given by use of d_k. Since $a \circ b$ represents $\alpha \circ \beta$, then $\bar{a} \circ \bar{b}$ represents an element of $\partial^{-1}(\alpha\circ\beta)$. Thus $a' \circ p$ and also a' represent an element of $h_{m*}(\partial^{-1}(\alpha\circ\beta))$.

Next consider a representative $\tilde{c} : S^{i+1} \to K \cup_b CS^k$ of a coextension $\tilde{\gamma}$ of γ. By (1.18), the composition $p \circ \tilde{c}$ represents $E\gamma$. By Proposition 1.7, the composition $\bar{a} \circ \tilde{c} : S^{i+1} \to (S^m)_\infty$ represents an element of the secondary composition $\{i_*\alpha, \beta, \gamma\}$. Thus the class $\{h_m \circ \bar{\alpha} \circ \tilde{c}\} = \{a' \circ p \circ \tilde{c}\}$ is a common element of $h_{m*}\{i_*\alpha, \beta, \gamma\}$ and $h_{m*}(\partial^{-1}(\alpha\circ\beta)) \circ E\gamma$. We have

$$\begin{aligned}
H\{E\alpha, E\beta, E\gamma\}_1 &= H\{\Omega_1^{-1}\, i_*\alpha, E\beta, E\gamma\}_1 && \text{by (2.4)} \\
&= -H(\Omega_1^{-1}\{i_*\alpha, \beta, \gamma\}) && \text{by Proposition 1.3} \\
&= -\Omega_1^{-1}\, h_{m*}\{i_*\alpha, \beta, \gamma\} && \text{by (2.7)}
\end{aligned}$$

and

$$\begin{aligned}
\Delta^{-1}(\alpha\circ\beta)\circ E^2\gamma &= \Omega_1^{-1}\, h_{m*}(\partial^{-1}(\alpha\circ\beta))\circ E^2\gamma && \text{by (2.10)} \\
&= \Omega_1^{-1}(h_{m*}(\partial^{-1}(\alpha\circ\beta))\circ E\gamma) && \text{by (1.12)}
\end{aligned}$$

Then it follows that $H\{E\alpha, E\beta, E\gamma\}_1$ and $-\Delta^{-1}(\alpha\circ\beta)\circ E^2\gamma$ have the common element $\Omega_1^{-1}\{a' \circ p \circ \tilde{c}\}$. These two sets are cosets of the same subgroup

$$\begin{aligned}
&H(E\alpha \circ E\pi_{i+1}(K) + \pi_{j+1}(S^{m+1}) \circ E^2\gamma) \\
&= HE(\alpha) \circ E\pi_{i+1}(K) + H\pi_{j+1}(S^{m+1}) \circ E^2\gamma && \text{by Proposition 2.2} \\
&= H\pi_{j+1}(S^{m+1}) \circ E^2\gamma && \text{by (2.11)}.
\end{aligned}$$

Therefore we have the equality $H\{E\alpha, E\beta, E\gamma\}_1 = -\Delta^{-1}(\alpha\circ\beta)\circ E^2\gamma$ and the proposition is proved.

Finally, it is well known that $H[\iota_n, \iota_n] = \pm 2\iota_{2n-1}$ for even n. Then it follows from Proposition 2.5

<u>Proposition</u> 2.7. $H(\Delta(\iota_{2n+1})) = \pm 2\iota_{2n-1}$ <u>for even</u> n. (cf.(11.14)).

CHAPTER III

Reduced Join and Stable Groups

Consider the reduced join $\alpha\#\beta \in \pi_{p+q+k+h}(S^{p+q})$ of $\alpha \in \pi_{p+k}(S^p)$ and $\beta \in \pi_{p+h}(S^q)$, then

Proposition 3.1 (Barratt-Hilton [5].)

$$\alpha\#\beta = (-1)^{(p+k)h}E^q\alpha \circ E^{p+k}\beta = (-1)^{ph}E^p\beta \circ E^{q+h}\alpha.$$

Proof: It was known that $\alpha\#\iota_n = E^n\alpha$. Let $\sigma_{r,s} : S^{r+s} \to S^{r+s}$ be a mapping given by $\sigma_{r,s}(\phi_{r,s}(x,y)) = \phi_{s,r}(y,x)$, then $\sigma_{r,s}$ is a hemeomorphism of degree $(-1)^{rs}$. The relation $\iota_n\#\alpha = \{\sigma_{p,n}\} \circ (\alpha\#\iota_n) \circ \{\sigma_{n,p+k}\} = (-1)^{pn}\iota_{p+n} \circ E^n\alpha \circ (-1)^{(p+k)n}\iota_{p+k+n}$ holds. By (2.1), it follows that $\iota_n\#\alpha = (-1)^{kn}E^n\alpha$. By (1.5),

$$\begin{aligned} \alpha\#\beta &= (\alpha\circ\iota_{p+k}) \# (\iota_q\circ\beta) = (\alpha\#\iota_q) \circ (\iota_{p+k}\#\beta) \\ &= (-1)^{(p+k)h}E^q\alpha \circ E^{p+k}\beta \end{aligned}$$

and

$$\begin{aligned} \alpha\#\beta &= (\iota_p\circ\alpha) \# (\beta\circ\iota_{q+h}) = (\iota_p\#\beta) \circ (\alpha\#\iota_{q+h}) \\ &= (-1)^{ph}E^p\beta\circ E^{q+h}\alpha. \end{aligned}$$

q. e. d.

The difference $E^{q-1}\alpha \circ E^{p+k-1}\beta - (-1)^{kh}E^{p-1}\beta \circ E^{q+h-1}\alpha$ vanishes under the suspension homomorphism E, and thus it is an image of Δ of the sequence (2.11). The following result for this element was proved in [22: Theorem (4.6)].

Proposition 3.2 Let $\alpha \in \pi_{p+k}(S^p)$ and $\beta \in \pi_{q+h}(S^q)$ and assume that $k \leq \text{Min.}(p+q-3.2p-4)$ and $h \leq \text{Min.}(p+q-3,2q-4)$, then $E^{q-1}\alpha \circ E^{p+k-1}\beta - (-1)^{kh}E^{p-1}\beta \circ E^{q+h-1}\alpha = [\iota_{p+q-1},\iota_{p+q-1}] \circ E^{2q-2}H(\alpha) \circ E^{p+k-1}H(\beta)$ and $2([\iota_{p+q-1},\iota_{p+q-1}] \circ E^{2q-2}H(\alpha) \circ E^{p+k-1}H(\beta)) = 0$. (Recently, this formula is established without restrictions on k and h [4].)

By Proposition 2.5, we have $[\iota_{p+q-1},\iota_{p+q-1}] \circ E^{2q-2}H(\alpha) \circ E^{p+k-1} H(\beta) \in \Delta\, (E^{2q}H(\alpha) \circ E^{p+k+1}H(\beta)) \subset \pi_{p+q+k+h-1}(S^{p+q-1})$. From the assumption of Proposition 3.2, $p+q+k+h \leq 3(p+q-1)-3$. Then this is the case that Δ is one valued. Thus we have the following corollary.

Corollary 3.3: Under the same assumption of the previous proposition 3.2, $E^{q-1}\alpha \circ E^{p+k-1}\beta - (-1)^{kh}E^{p-1}\beta \circ E^{q+h-1}\alpha = \Delta(E^{2q}H(\alpha) \circ E^{p+k+1}H(\beta))$.

The reduced join yields us to give some relations in secondary compositions.

Proposition 3.4: (Proposition 4.6 of [25].)

i). Assume that $\alpha\in\pi_{p+h}(S^p)$, $\beta\in\pi_{q+k}(S^q)$ and $\gamma\in\pi_{r+\ell}(S^r)$ satisfy the condition $\alpha\#\beta = \beta\#\gamma = 0$, then the secondary compositions $\{E^{q+r}\alpha, E^{p+h+r}\beta, E^{p+h+q+k}\gamma\}$ and $(-1)^{hk+k\ell+\ell h+1}\{E^{p+q}\gamma, E^{p+r+\ell}\beta, E^{q+k+r+\ell}\alpha\}$ have a common element.

ii). Further assume that $\alpha\#\gamma = 0$, then the sum $(-1)^{h\ell}\{E^{q+r}\alpha, E^{p+h+r}\beta, E^{p+h+q+k}\gamma\} + (-1)^{kh}\{E^{p+r}\beta, E^{p+q+k}\gamma, E^{q+k+r+\ell}\alpha\} + (-1)^{\ell k}\{E^{p+q}\gamma, E^{q+r+\ell}\alpha, E^{p+h+r+\ell}\beta\}$ contains the zero element.

Proof: Let a, b and c be representatives of α, β and γ respectively, and let A_t and B_t be null-homotopies of $a \# b$ and $b \# c$ respectively. Denote by i_n the identity of S^n. Consider mappings $H, H' : S^{s+1} \to S^{p+q+r}$, $s = p+h+q+k+r+\ell$, given by

$$H(d_s(x,t)) = \begin{cases} (a \# i_{q+r})(i_{p+h} \# B_{2t-1})(x) & \text{for } \tfrac{1}{2} \leq t \leq 1 \\ (A_{1-2t} \# i_r)(i_{p+h+q+k} \# c)(x) & \text{for } 0 \leq t \leq \tfrac{1}{2}, \end{cases}$$

$$H'(d_s(x,t)) = \begin{cases} (i_{p+q} \# c)(A_{2t-1} \# i_{r+\ell})(x) & \text{for } \tfrac{1}{2} \leq t \leq 1, \\ (i_p \# B_{1-2t})(a \# i_{q+k+r+\ell})(x) & \text{for } 0 \leq t \leq \tfrac{1}{2}. \end{cases}$$

The mapping H represents an element of

$$\begin{aligned}\{\alpha \# \iota_{q+r}, \iota_{p+h} \# \beta \# \iota_r, \iota_{p+h+q+k} \# \gamma\} &= \\ &= \{E^{q+r}\alpha, (-1)^{(p+h)k} E^{p+h+r}\beta, (-1)^{(p+h+q+k)\ell} E^{p+h+q+k}\gamma\} \\ &= (-1)^{\varepsilon}\{E^{q+r}\alpha, E^{p+h+r}\beta, E^{p+h+q+k}\gamma\},\end{aligned}$$

where $\varepsilon = (p+h)k + (p+h+q+k)\ell = (p+q)\ell + pk + hk + k\ell + \ell h$. The mapping H' represents an element of $\{\iota_{p+q} \# \gamma, \iota_p \# \beta \# \iota_{r+\ell}, \alpha \# \iota_{q+k+r+\ell}\} =$

$$= \{(-1)^{(p+q)\ell} E^{p+q}\gamma, (-1)^{pk} E^{p+r+\ell}\beta, E^{q+k+r+\ell}\alpha\}.$$

It follows from ii) of (1.4) that $H = - H'$. Then the class of $(-1)^{\varepsilon} H$ is the required common element of i).

ii). Let C_t be a null-homotopy of $a \# c$. Then $C_t' = (i_p \# \sigma_{r,q}) \circ (C_t \# i_q) \circ (i_{p+h} \# \sigma_{q,r+\ell})$ and $C_t'' = (i_q \# \sigma_{r,q+k}) \circ (C_t \# i_{q+k}) \circ (i_{p+h} \# \sigma_{q+k,r+\ell})$ are null-homotopies of $a \# i_q \# c = (i_{p+q} \# c) \circ (a \# i_{q+r+\ell})$ and $a \# i_{q+k} \# c = (i_{p+q+k} \# c) \circ (a \# i_{q+k+r+\ell})$ respectively. The relation $C_t' \circ (i_{p+h} \# b \# i_{r+\ell}) = (i_p \# \sigma_{r,q}) \circ (C_t \# b) \circ (i_{p+h} \# \sigma_{q+k,r+\ell}) = (i_p \# b \# i_r) \circ C_t''$ holds. Consider mappings H, H', H'' : $S^{s+1} \to S^{p+q+r}$ given by the formulas :

$$H(d_s(x,t)) = \begin{cases} (a \# i_{q+r})(i_{p+h} \# B_{2t-1})(x) & \text{for } \tfrac12 \leq t \leq 1 \\ (A_{1-2t} \# i_r)(i_{p+h+q+k} \# c)(x) & \text{for } 0 \leq t \leq \tfrac12 \end{cases}$$

$$H'(d_s(x,t)) = \begin{cases} C_{2t-1}'(i_{p+h} \# b \# i_{r+\ell})(x) & \text{for } \tfrac12 \leq t \leq 1 \\ (a \# i_{q+r})(i_{p+h} \# B_{1-2t})(x) & \text{for } 0 \leq t \leq \tfrac12 \end{cases}$$

$$H''(d_s(x,t)) = \begin{cases} (A_{2t-1} \# i_r)(i_{p+q+h+k} \# c)(x) & \text{for } \tfrac12 \leq t \leq 1 \\ (i_p \# b \# i_r)\, C_t''(x) & \text{for } 0 \leq t \leq \tfrac12 \end{cases}$$

Then the mappings H, H', and H'' represent elements of the following secondary compositions respectively : ($\delta = (p+h)k + (p+q+k)\ell$)

$$\{\alpha \# \iota_{q+r}, \iota_{p+h} \# \beta \# \iota_r, \iota_{p+h+q+k} \# \gamma\} = (-1)^{\delta+h\ell}\{E^{q+r}\alpha, E^{p+h+r}\beta, E^{p+h+q+k}\gamma\},$$
$$\{\iota_{p+q} \# \gamma, \alpha \# \iota_{q+h+r+\ell}, \iota_{p+h} \# \beta \# \iota_{r+\ell}\} = (-1)^{\delta+\ell k}\{E^{p+q}\gamma, E^{q+r+\ell}\alpha, E^{p+h+r+\ell}\beta\}$$
$$\{\iota_p \# \beta \# \iota_r, \iota_{p+q+k} \# \gamma, \alpha \# \iota_{q+k+r+\ell}\} = (-1)^{\delta+kh}\{E^{p+r}\beta, E^{p+q+k}\gamma, E^{q+k+r+\ell}\alpha\}.$$

It is verified directly that $H(d_s(x,t)) = H'(d_s(x,1-t))$, $H'(d_s(x,t)) = H''(d_s(x,1-t))$ and $H''(d_s(x,t)) = H(d_s(x,1-t))$ for $\tfrac12 \leq t \leq 1$. Then it is easy to give a null-homotopy of the sum $(H + H') + H''$. Thus the sum of the secondary compositions in ii) contains $(-1)^\delta (\{H\} + \{H'\} + \{H''\}) = 0$.

Now consider the condition

$$(3.1) \qquad (1 - (-1)^k)\, E^{n-h}\alpha \circ E^{n+k-h}\alpha = 0$$

for an element α of $\pi_{n+k}(S^n)$ and an integer $h \geq 0$. By Proposition 3.1, the relation (3.1) holds for $h = 0$. If k is even, (3.1) holds for $0 \leq h \leq n$.

Let $f : S^{n+k} \to S^n$ be a representative of α and construct a cell complex $K = S^n \cup_f CS^{n+k}$ as in Chapter I. Let $d_{n+k}' : S^{n+k} \times I^1 \to CS^{n+k}$ be the identification which defines CS^{n+k}. Then we denote by $d' : S^{n+k} \times I^1 \to K$ the composition of d_{n+k}' and the identification of CS^{n+k} into K. Define a mapping $p : K \to S^{n+k+1}$ by the formulas $p(d'(x,t)) = d_{n+k}(x,t)$ and

$p(S^n) = e_0$. p shrinks the subset S^n of K to a point. Denote by 1_K the identity of K and consider the reduced join

$$1_K \# f : E^{n+k}K = K \# S^{n+k} \to K \# S^n = E^n K.$$

The restriction $1_K \# f \mid S^{2n+k}$ represents $\iota_n \# \alpha = (-1)^{kn} E^n \alpha$. Since $E^n K - S^{2n}$ is a (2n+k+1)-cell attached by $E^n f$ to S^{2n}, then the restriction is homotopic to zero in $E^n K$. By the homotopy extension theorem, we have a homotopy $G_s : E^{n+k}K \to E^n K$ between $G_0 = 1_K \# f$ and a mapping G_1 which maps S^{2n+k} to a point. Then there exists a mapping $F : S^{2n+k+1} \to E^n K$ such that $G_1 = F \circ E^n p$. For the class of such a mapping F, we have the following

<u>Lemma</u> 3.5: <u>Let an element</u> α <u>of</u> $\pi_{n+k}(S^n)$ <u>and an integer</u> $h \geq 0$ <u>satisfy</u> (3.1). <u>Let</u> ι_K <u>be the homotopy class of the identity of</u> K . <u>Assume that</u> $k \leq 2n - 2$, <u>then for any coextension</u> $\gamma \in \pi_{2n+2k-h+1}(E^{n-h}K)$ <u>of</u> $(1 - (-1)^k) E^{n+k-h}\alpha$ <u>there exists an element</u> α^* <u>of</u> $\pi_{2n+2k+1}(S^{2n})$ <u>such that the equality</u> $\iota_K \# \alpha = (E^{n+k}p)^*(i_* \alpha^* + (-1)^{nk} E^h \gamma)$ <u>in</u> $\pi(E^{n+k}K \to E^n K)$ <u>holds, where</u> $i : S^{2n} \to E^n K$ <u>is the injection</u>.

<u>Proof</u>: We shall give precisely a homotopy G_s between $1_K \# f$ and $F \circ E^n p$. Denote by $\phi_m : K \times S^m \to E^m K$ (m = n or n+k) the identification which defines $E^m K = K \# S^m$. By Proposition 3.1, $\iota_n \# \alpha = (-1)^{nk} E^n \alpha = \alpha \# (-1)^{kn} \iota_n$. Then there exists a homotopy $A_s : (S^{2n+k}, e_0) \to (S^{2n+k}, e_0)$ from $A_0 = 1_n \# f$ to $A_1 = f \# \tau$, where 1_n denotes the identity of S^n and $\tau : S^n \to S^n$ is a mapping of degree $(-1)^{kn}$. Now define a homotopy $G_s : E^{n+k}K \to E^n K$ by the following formulas. For a point $\phi_{n+k,n}(x,y)$ of S^{2n+k}, we set

$$G_s(\phi_{n+k,n}(x,y)) = \begin{cases} A_{2s}(\phi_{n+k,n}(x,y)) & \text{if } 0 \leq s \leq \tfrac{1}{2}, \\ \phi_n(d'(x,2s-1), \tau(y)) & \text{if } \tfrac{1}{2} \leq s \leq 1, \end{cases}$$

and then for a point $\phi_{n+k}(d'(u,t),v)$ of $E^{n+k}K$

$$G_s(\phi_{n+k}(d'(u,t),v)) = \begin{cases} \phi_n(d'(u,(2t-s)/(2-s)), f(v)) & \text{if } s/2 \leq t \leq 1, \\ G_{s-2t}(\phi_{n,n+k}(f(u),v)) & \text{if } 0 \leq t \leq s/2. \end{cases}$$

G_s is well defined and a homotopy from $G_0 = 1_K \# f$ to G_1 which maps S^{2n+k} to a point e_0. Since $E^{n+k}p$ is a mapping which shrinks S^{2n+k}, then there exists uniquely a mapping $F : S^{2n+2k+1} \to E^n K$ such that $G_1 = F \circ E^{n+k} p$. Next consider the composition $E^n p \circ F : S^{2n+2k+1} \to S^{2n+k+1}$. It is calculated easily from the above formulas that $E^n p \circ F$ is given by

$$(E^n p \circ F)(\phi_{n+k+1,n+k}(d_{n+k}(u,t),v)) = \begin{cases} \phi_{n+k+1,n}(d_{n+k}(u,2t-1),f(v)) & \text{if } \frac{1}{2} \leq t \leq 1, \\ e_0 & \frac{1}{4} \leq t \leq \frac{1}{2}, \\ \phi_{n+k+1,n+k}(d_{n+k}(x,1-4t),\ \tau(y)) & \text{if } 0 \leq t \leq \frac{1}{4}, \end{cases}$$

where $\phi_{n+k,n}(x,y) = \phi_{n,n+k}(f(u),v)$. We may consider that the above formula indicates the sum $H_1 + (H_0 - H_2)$ of the three mappings H_1, H_0, H_2 with respect to the $(n+k+1)$-th coordinate t, such that by taking homotopy classes $\{E^n p \circ F\} = \{H_1\} + \{H_0\} - \{H_2\}$, where H_0 is the trivial mapping and H_1, H_2 are given by

$$H_1(\phi_{n+k+1,n+k}(d_{n+k}(u,t),v)) = \phi_{n+k+1,n}(d_{n+k}(u,t),f(v)) ,$$
$$H_2(\phi_{n+k+1,n+k}(d_{n+k}(u,t),v)) = \phi_{n+k+1,n}(d_{n+k}(x,t),\tau(y)) .$$

Obviously $\{H_0\} = 0$ and $\{H_1\} = \{1_{n+k+1} \# f\} = \iota_{n+k+1} \# \alpha = (-1)^{(n+k+1)k} E^{n+k+1}\alpha = (-1)^{kn} E^{n+k+1}\alpha$. It is verified directly that $H_2 = (1_{n+k+1} \# \tau) \circ \sigma \circ (Ef \# 1_{n+k})$ for a homeomorphism $\sigma : S^{2n+k+1} \to S^{2n+k+1}$ of degree $(-1)^k$ which is given by $\sigma(\psi_{2n+2k+1}(x_1,\dots,x_{n+k},t,y_1,\dots,y_n)) = \psi_{2n+2k+1}(x_1,\dots,x_n,t,x_{n+1},\dots,x_{n+k},y_1,\dots,y_n)$. It follows that $\{H_2\} = (\iota_{n+k+1} \# (-1)^{kn}\iota_n) \circ (-1)^k \iota_{2n+k+1} \circ (E\alpha \# \iota_{n+k}) = (-1)^{kn+k} E^{n+k+1}\alpha$. Thus $(E^n p)_*\{F\} = (-1)^{kn}(1 - (-1)^k)E^{n+k+1}\alpha$.

On the other hand, it follows from the definition of the coextension, that $(E^{n-h}p)_*(\gamma) = (1 - (-1)^k)E^{n+k-h+1}\alpha$.

By (1.6), $(E^n p)_* E^h\gamma = E^h((E^{n-h}p)_*\gamma) = (1 - (-1)^k)E^{n+k+1}\alpha$, and thus $(E^n p)_*(\{F\} - (-1)^{kn}E^h\gamma) = 0$.

From the assumption $k \leq 2n-2$, we have $2n+2k+1 \leq (2n+k) + (2n-1)$. The pair $(E^n K, S^{2n})$ is $(2n+k)$-connected and S^{2n} is $(2n-1)$-connected. Then it follows from Theorem 2 of [6] that the induced homomorphism

$$(E^n p)_* : \pi_{2n+2k+1}(E^n K, S^{2n}) \to \pi_{2n+k+1}(S^{2n+k+1})$$

is an isomorphism onto. Consider the following commutative diagram of the exact row.

$$\begin{array}{ccccc} \pi_{2n+2k+1}(S^{2n}) & \xrightarrow{i_*} & \pi_{2n+2k+1}(E^nK) & \xrightarrow{j_*} & \pi_{2n+2k+1}(E^nK,S^{2n}) \\ & & & \searrow^{(E^np)_*} & \downarrow (E^np)_* \\ & & & & \pi_{2n+2k+1}(S^{2n+k+1}). \end{array}$$

Then $(E^np)_*(\{F\} - (-1)^{kn}E^h\gamma) = 0$ implies that $\{F\} - (-1)^{kn}E^h\gamma = i_*\alpha^*$ for an element α^* of $\pi_{2n+2k+1}(S^{2n})$. Consequently we have proved that $\iota_K \# \alpha = \{G_0\} = \{G_1\} = (E^{n+k}p)^*\{F\} = (E^{n+k}p)^*(i_* \alpha^* + (-1)^{kn}E^h\gamma)$, and the Lemma is proved.

<u>Theorem</u> 3.6: <u>Let an element</u> α <u>of</u> $\pi_{n+k}(S^n)$ <u>and an integer</u> $h \geq 0$ <u>satisfy</u> (3.1). <u>Assume that</u> $k \leq 2n-2$, <u>then there exists an element</u> $\alpha^* \in \pi_{2n+2k+1}(S^{2n})$ <u>such that, for any element</u> β <u>of</u> $\pi_{n+t}(S^m)$ <u>satisfying</u> $\beta \circ E^t\alpha = 0$, <u>the composition</u> $E^n\beta \circ E^t\alpha^*$ <u>is contained in the sum of cosets</u>

$$(-1)^{km+kt+t}\{E^m\alpha, E^{n+k}\beta, E^{n+k+t}\alpha\}_{n+k} + (-1)^{kn+h+t+1}\{E^n\beta, E^{n+t}\alpha, (1-(-1)^k)E^{n+k+t}\alpha\}_{h+t}.$$

<u>Further if</u> $(1-(-1)^k)E^{n+k+1}\alpha = 0$, <u>in particular if</u> k <u>is even, then</u> $E^n\beta \circ E^t\alpha^* \in (-1)^{km+kt+t}\{E^m\alpha, E^{n+k}\beta, E^{n+k+t}\alpha\}_{n+k}$.

<u>Proof</u>: Since $\beta \circ E^t\alpha = 0$, there exists an extension $\bar\beta \in \pi(E^tK \to S^m)$ of β. By (1.16), we identify E^tK with $S^{n+t} \cup_{\alpha'} CS^{n+k+t}$ where $\alpha' = E^t\alpha$. Let $p' : E^tK = S^{n+t} \cup_{\alpha'} CS^{n+k+t} \to S^{n+k+t+1}$ be a mapping defined as in (1.17), then we have a relation $p' = \sigma \circ E^tp$ for a homeomorphism $\sigma : S^{n+k+t+1} \to S^{n+k+t+1}$ of degree $(-1)^t$ given by $\sigma(\psi_{n+k+t+1}(t_1,\ldots,t_{n+k+t},t)) = \psi_{n+k+t+1}(t_1,\ldots,t_{n+k},t,t_{n+k+1},\ldots,t_{n+k+t})$. Consider the reduced join $\bar\beta \# \alpha \in \pi(E^{n+k+t}K \to S^{m+n})$.

First we have

$$\begin{aligned} \bar\beta \# \alpha &= (\bar\beta \circ E^t\iota_K) \# (\iota_n \circ \alpha) \\ &= (\bar\beta \# \iota_n) \circ ((\iota_K \# \iota_t) \# \alpha) && \text{by (1.5)} \\ &= E^n\bar\beta \circ (\iota_K \# \alpha \# (-1)^{kt}\iota_t) && \text{by Proposition 3.1} \\ &= E^n\bar\beta \circ (-1)^{kt}E^t((E^{n+k}p)^*(i_*\alpha^* + (-1)^{kn}E^h\gamma)) && \text{by Lemma 3.5} \\ &= (-1)^{kt}E^n\bar\beta \circ (E^{n+k+t}p)^*(E^ti_*\alpha^* + (-1)^{kn}E^{h+t}\gamma) && \text{by (1.6)} \\ &= (-1)^{kt}(E^{n+k}p')^*(E^{n+k}\sigma)^*(E^n\beta \circ E^t\alpha^* + (-1)^{kn}E^n\bar\beta \circ E^{h+t}\gamma) \\ &= (-1)^{(k+1)t}(E^{n+k}p')^*(E^n\beta \circ E^t\alpha^* + (-1)^{kn}E^n\bar\beta \circ E^{h+t}\gamma). \end{aligned}$$

By Proposition 1.7,

$$E^n\bar\beta \circ E^{t+h}\gamma \in (-1)^{h+t}\{E^n\beta, E^{n+t}\alpha, (1-(-1)^k)E^{n+t+k}\alpha\}_{h+t}.$$

$$\begin{aligned}\text{Next}\quad \bar\beta \# \alpha &= (\iota_m \circ \bar\beta) \# (\alpha \circ \iota_{n+k}) \\ &= (\iota_m \# \alpha) \circ (\bar\beta \# \iota_{n+k}) && \text{by (1.5)} \\ &= (-1)^{km} E^m\alpha \circ E^{n+k}\bar\beta && \text{by Proposition 3.1} \\ &= (-1)^{km}(E^m\alpha \circ E^{n+k}\bar\beta).\end{aligned}$$

It follows from Proposition 1.9 that any element λ satisfying the condition $(E^{n+k}p')_* \lambda = E^m\alpha \circ E^{n+k}\bar\beta$ belongs to $\{E^m\alpha, E^{n+k}\beta, E^{n+k+t}\alpha\}_{n+k}$. In particular, by the first calculation, $\lambda = (-1)^{km+(k+1)t}(E^n\beta \circ E^t\alpha^* + (-1)^{kn}E^n\bar\beta \circ E^{h+t}\gamma)$ satisfies this condition. Thus $(-1)^{km+(k+1)t}(E^n\beta \circ E^t\alpha^* + (-1)^{kn}E^n\bar\beta \circ E^{h+t}\gamma) \in \{E^m\alpha, E^{n+k}\beta, E^{n+k+t}\alpha\}_{n+k}$.

Combining the above result of $E^n\bar\beta \circ E^{h+t}\gamma$, we have obtained that

$$\begin{aligned}E^n\beta \circ E^t\alpha^* \in\ & (-1)^{km+(k+1)t}\{E^m\alpha, E^{n+k}\beta, E^{n+k+t}\alpha\}_{n+k} \\ & + (-1)^{kn+h+t+1}\{E^n\beta, E^{n+t}\alpha, (1-(-1)^k)E^{n+k+t}\alpha\}_{h+t}.\end{aligned}$$

Consider the case that $(1-(-1)^k)E^{n+k+1}\alpha = 0$. In Lemma 3.5, we may chose a coextension γ of $(1-(-1)^k)E^{n+k+1}\alpha$ as the trivial class 0. Then in the above discussion, the term $E^n\bar\beta \circ E^{h+t}\gamma$ vanishes. Thus

$$E^n\beta \circ E^t\alpha^* \in (-1)^{km+(k+1)t}\{E^m\alpha, E^{n+k}\beta, E^{n+k+t}\alpha\}_{n+k},$$

and Theorem 3.6 has been proved.

<u>Corollary</u> 3.7. <u>Let</u> r <u>be an integer</u>. <u>Assume that</u> $r\beta = 0$ <u>for an element</u> $\beta \in \pi_{m+k}(S^m)$, $k > 0$, <u>then</u>

$$\{r\iota_{m+1}, E\beta, r\iota_{m+k+1}\}_1 \begin{cases} \equiv 0 & \underline{\text{if}}\ \ r \not\equiv 2 \pmod 4, \\ \ni E\beta \circ \eta_{m+k+1} & \underline{\text{if}}\ \ r \equiv 2 \pmod 4, \end{cases}$$

<u>where</u> η_{m+k+1} <u>generates</u> $\pi_{m+k+2}(S^{m+k+1}) \approx Z_2$.

<u>Proof</u>: To prove this, we use the following results. $\pi_{n+1}(S^n) \approx \pi_{n+2}(S^n) \approx Z_2$ for $n \geq 3$ (cf. Chapter V). Let η_n the generator of $\pi_{n+1}(S^n)$, then $E\eta_n = \eta_{n+1}$, $\eta_n \circ \eta_{n+1}$ generates $\pi_{n+2}(S^n)$ and the secondary composition $\{2\iota_n, \eta_n, 2\iota_{n+1}\}$ consists of a single element $\eta_n \circ \eta_{n+1}$. (See [20] and Example 2 of Chapter VIII.)

Now let $\alpha = r\iota_1$ in Theorem 3.6. It follows then

$$E\beta \circ t\eta_{m+k+1} \in \{r\iota_{m+1}, E\beta, r\iota_{m+k+1}\}_1$$

for an integer t which depends on r but does not depend on β. If r is odd, then $E\beta \circ \eta_{m+k+1} = r(E\beta \circ \eta_{m+k+1}) \in r(\pi_{m+k+2}(S^{m+1}))$. Thus $E\beta \circ t\eta_{m+k+1}$ is contained in the same coset as $t.0 = 0$. Then we have obtained the corollary for the case that r is odd. Next let r be even and $r = 2s$. By Proposition 1.2,

$$s^2(\eta_n \circ \eta_{n+1}) \in s\iota_n \circ \{2\iota_n, \eta_n, 2\iota_{n+1}\} \circ s\iota_{n+2} \subset \{r\iota_n, \eta_n, r\iota_{n+1}\} .$$

Also $t(\eta_n \circ \eta_{n+1}) \in \{r\iota_n, \eta_n, r\iota_{n+1}\}$ which consists of a single element. It follows that $s^2(\eta_n \circ \eta_{n+1}) = t(\eta_n \circ \eta_{n+1})$ and thus $s^2 \equiv t \pmod 2$. If s is even, i.e., if $r \equiv 0 \pmod 4$, then $E\beta \circ t\eta_{m+k+1} = E\beta \circ s^2\eta_{m+k+1} = 0$. If s is odd, i.e., $r \equiv 2 \pmod 4$, then $E\beta \circ t\eta_{m+k+1} = E\beta \circ s^2\eta_{m+k+1} = E\beta \circ \eta_{m+k+1}$. Consequently we see that the corollary has proved.

As a consequence of the exactness of the sequence (2.11), we have Freudenthal's suspension theorem:

(3.2) $E : \pi_i(S^m) \to \pi_{i+1}(S^{m+1})$ <u>is an isomorphism onto for</u> $i < 2m - 1$ <u>and a homomorphism onto for</u> $i = 2m - 1$.

Denote by $G_k = \lim_{n \to \infty} \pi_{n+k}(S^n)$ the k-th stable homotopy group of the sphere, and let $E^\infty : \pi_{n+k}(S^n) \to G_k$ be the projection. By (3.2),

(3.3) $E^\infty : \pi_{n+k}(S^n) \to G_k$ <u>is an isomorphism onto if</u> $n > k+1$ <u>and a homomorphism onto if</u> $n = k+1$.

By (1.6), $E^m(\alpha \circ \beta) = E^m\alpha \circ E^m\beta$. Then the composition

$$\circ : G_k \times G_h \to G_{k+h}$$

is defined by the formula $E^\infty(\alpha \circ \beta) = E^\infty\alpha \circ E^\infty\beta$.

(1.7) or (2.1) shows that the composition operator in the stable case is bilinear : $\alpha \circ (\beta_1 \pm \beta_2) = \alpha \circ \beta_1 \pm \alpha \circ \beta_2$ and $(\alpha_1 \pm \alpha_2) \circ \beta = \alpha_1 \circ \beta \pm \alpha_2 \circ \beta$.

Proposition 3.1 shows the anti-commutativity :

(3.4) $\alpha \circ \beta = (-1)^{kh}\beta \circ \alpha$ <u>for</u> $\alpha \in G_k$ <u>and</u> $\beta \in G_h$.

We shall define a secondary composition

$$< \alpha, \beta, \gamma > \in G_{k+h+\ell+1}/(\alpha \circ G_{h+\ell+1} + \gamma \circ G_{k+h+1})$$

of elements $\alpha \in G_k$, $\beta \in G_h$ and $\gamma \in G_k$ which satisfy the condition $\alpha \circ \beta = \beta \circ \gamma = 0$.

For sufficiently large n, $\alpha = E^\infty\alpha'$, $\beta = E^\infty\beta'$, and $\gamma = E^\infty\gamma'$ for some $\alpha' \in \pi_{n+k}(S^n)$, $\beta' \in \pi_{n+k+h}(S^{n+k})$ and $\gamma' \in \pi_{n+k+h+\ell}(S^{n+k+h})$. Then we define that

$$< \alpha, \beta, \gamma > = E^\infty(-1)^{n-1}\{\alpha', \beta', \gamma'\} .$$

In virtue of Proposition 1.3, this definition does not depend on n.

Several properties of the usual secondary composition will be translated to the properties of the stable case.

It follows from Proposition 1.2,

(3.5), 0). $< \alpha,\beta,\gamma > \equiv 0$ if one of α,β and γ is 0.

i). $< \alpha,\beta,\gamma > \circ \delta \subset < \alpha,\beta,\gamma \circ \delta >$

ii). $< \alpha,\beta,\gamma \circ \delta > \subset < \alpha,\beta \circ \gamma,\delta >$

iii). $< \alpha \circ \beta,\gamma,\delta > \subset < \alpha,\beta \circ \gamma,\delta >$

iv). $\alpha \circ < \beta,\gamma,\delta > \subset (-1)^k < \alpha \circ \beta,\gamma,\delta >$, $\alpha \in G_k$.

It follows from Proposition 1.4,

3.6). $$< \alpha,\beta,\gamma > \circ \delta = (-1)^{k+1}\alpha \circ < \beta,\gamma,\delta >, \quad \alpha \in G_k.$$

From Proposition 1.5,

3.7). $$0 \in << \alpha,\beta,\gamma >, \delta,\varepsilon > + (-1)^k < \alpha, < \beta,\gamma,\delta >, \varepsilon > + (-1)^{k+h} < \alpha,\beta, < \gamma,\delta, \varepsilon >>, \quad \alpha \in G_k,\ \beta \in G_h.$$

From Proposition 1.6,

3.8). $$< \alpha,\beta,\gamma > + < \alpha,\beta,\gamma' > \supset < \alpha,\beta,\gamma + \gamma' >,$$
$$< \alpha,\beta,\gamma > + < \alpha,\beta',\gamma > = < \alpha,\beta + \beta',\gamma >,$$
$$< \alpha,\beta,\gamma > + < \alpha',\beta,\gamma > \supset < \alpha + \alpha',\beta,\gamma >.$$

It follows from Proposition 3.4,

3.9), i). $< \alpha,\beta,\gamma > = (-1)^{hk+k\ell+\ell h+1} < \gamma,\beta,\alpha >$,

ii). $(-1)^{k\ell} < \alpha,\beta,\gamma > + (-1)^{hk} < \beta,\gamma,\alpha > + (-1)^{\ell k} < \gamma,\alpha,\beta > \ni 0$,

where $\alpha \in G_k$, $\beta \in G_h$ and $\gamma \in G_\ell$

It follows from Theorem 3.6,

3.10). Let $\alpha \in G_k$, $\beta \in G_h$ and $\alpha \circ \beta = 0$. If k is odd, then $< \alpha,\beta,\alpha >$ and $(-1)^h < \beta,\alpha,2\alpha >$ have a common element. If k is even, then $< \alpha,\beta,\alpha >$ and $\beta \circ G_{2k+1}$ have a common element.

CHAPTER IV

Suspension Sequence mod. 2

It is well known that

(4.1). $\pi_i(S^1) = 0$ *for* $i > 1$, $\pi_i(S^n) = 0$ *for* $i < n$ *and* $\pi_n(S^n)$ *is an infinite cycle group generated by* ι_n.

We recall also

(4.2). (Serre [13]). $\pi_i(S^n)$ *is finite except for the cases that* $i = n$, *and that* n *even and* $i = 2n-1$.

We denote the p-primary component of the group $\pi_i(S^n)$ by $\pi_i(S^n;p)$. (p : prime)

We define a subgroup π_i^n of $\pi_i(S^n)$ as follows.

$$(4.3).\qquad \pi_i^n = \begin{cases} \pi_n(S^n) & \text{if } i = n, \\ E^{-1}(\pi_{2n}(S^{n+1};2)) & \text{if } i = 2n-1, \\ \pi_i(S^n;2) & \text{if } i \neq n,\ 2n-1. \end{cases}$$

The group π_{2n-1}^n is precisely given as follows.

Lemma 4.1: *If* n *is odd, then* $\pi_{2n-1}^n = \pi_{2n-1}(S^n;2)$. *If* n *is even and if* $[\iota_{n-1}, \iota_{n-1}] \neq 0$, *then* π_{2n-1}^n *is the direct sum of* $\pi_{2n-1}(S^n;2)$ *and the infinite cyclic group* $\Delta(\pi_{2n+1}(S^{2n+1}))$. *If* n *is even and if* $[\iota_{n-1}, \iota_{n-1}] = 0$, *then* π_{2n-1}^n *is the direct sum of* $\pi_{2n-1}(S^n;2)$ *and an infinite cyclic group generated by an element* α *such that* $H(\alpha) = \iota_{2n-1}$ *and* $E\alpha \in \pi_{2n}^{n+1} = \pi_{2n}(S^{n+1};2)$. *In all cases,* $[\iota_n, \iota_n]$ *is contained in* π_{2n-1}^n.

Proof: Let n be odd. By the anti-commutativity of Whitehead product, we have that $2\Delta(\iota_{2n+1}) = 2[\iota_n, \iota_n] = 0$. By the exactness of the sequence (2.9), it follows that the kernel of $E : \pi_{2n-1}(S^n) \to \pi_{2n}(S^{n+1})$ is in $\pi_{2n-1}(S^n;2)$. By (4.2), the group $\pi_{2n-1}(S^n)$ is finite. Thus $\pi_{2n-1}^n = E^{-1}(\pi_{2n}(S^{n+1};2))$ coincides with $\pi_{2n-1}(S^n;2)$. Next let n be even and consider the exact sequence $\pi_{2n-2}(S^{n-1}) \xrightarrow{E} \pi_{2n-1}(S^n) \xrightarrow{H} \pi_{2n-1}(S^{2n-1}) \xrightarrow{\Delta} \pi_{2n-3}(S^{n-1})$ of (2.9). Since the image of Δ is generated by $[\iota_{n-1}, \iota_{n-1}]$ and since $2[\iota_{n-1}, \iota_{n-1}] = 0$, it follows that $\pi_{2n-1}(S^n)$ is the direct sum of the finite group $E(\pi_{2n-2}(S^{n-1}))$ and an infinite cyclic group generated by an element β, where $H(\beta) = \iota_{2n-1}$ if $[\iota_{n-1}, \iota_{n-1}] = 0$ and $H(\beta) = 2\iota_{n-1}$ if $[\iota_{n-1}, \iota_{n-1}] \neq 0$. In the case $[\iota_{n-1}, \iota_{n-1}] \neq 0$, we may chose β as $\pm [\iota_n, \iota_n]$ by Proposition 2.7. Since the kernel of $E : \pi_{2n-1}(S^n) \to \pi_{2n}(S^{n+1})$ is generated by $\pm [\iota_n, \iota_n] = \Delta(\iota_{2n+1})$, then $\pi^n_{2n-1} = E^{-1}(\pi_{2n}(S^{n+1};2))$ is the sum of $\Delta(\pi_{2n+1}(S^{2n+1}))$ and the 2-primary component of $E\,\pi_{2n-2}(S^{n-1})$ which coincides with $\pi_{2n-1}(S^n;2)$. Then we have proved the lemma for the case that n is even and $[\iota_{n-1}, \iota_{n-1}] \neq 0$.

Consider the case that n is even and $[\iota_{n-1}, \iota_{n-1}] = 0$. Since $\pi_{2n}(S^{n+1})$ is finite by (4.2), then there exists an integer t such that $(2t+1)E\,\beta \in \pi_{2n}(S^{n+1};2)$. Set $\alpha = \beta - t(\pm [\iota_n, \iota_n])$, where the sign $\pm$ depends on the sign of $H[\iota_n, \iota_n] = \pm 2\iota_{2n-1}$ (Proposition 2.7). It follows that $E(\alpha) \in \pi_{2n}(S^{n+1};2)$ and $H(\alpha) = \iota_{2n-1}$. From the exactness of the above sequence, we have that $\pi_{2n-1}(S^n)$ is the direct sum of $E\,\pi_{2n-2}(S^{n-1})$ and the infinite cyclic group $\{\alpha\}$ generated by α. It is easily seen that $E(\pi_{2n-1}(S^n;2) + \{\alpha\}) = \pi_{2n}(S^{n+1};2)$. Consider the difference $[\iota_n, \iota_n] - 2\alpha$. Since $H([\iota_n, \iota_n] - 2\alpha) = 0$, then $[\iota_n, \iota_n] - 2\alpha \in E\,\pi_{2n-2}(S^{n-1})$. Since $E([\iota_n, \iota_n] - 2\alpha) = -2E\,\alpha \in \pi_{2n}(S^{n+1};2)$, and since $E : \pi_{2n-1}(S^n) \to \pi_{2n}(S^{n+1})$ is an isomorphism of $E\,\pi_{2n-2}(S^{n-1})$ into $\pi_{2n}(S^{n+1})$, then we have that $[\iota_n, \iota_n] \in E\,\pi_{2n-2}(S^n;2) + \{\alpha\} = \pi_{2n-1}(S^n;2) + \{\alpha\}$. If follows then $\pi^n_{2n-1} = E^{-1}\pi_{2n}(S^{n+1};2) = \pi_{2n-1}(S^n;2) + \{\alpha\}$. Consequently the lemma 4.1 has proved.

The following exact sequence is the main tool for calculating the 2-primary components of homotopy groups $\pi_{n+k}(S^n)$.

<u>Proposition</u> 4.2: <u>The following sequence is exact</u>.

(4.4) $$\cdots \to \pi^n_i \xrightarrow{E} \pi^{n+1}_{i+1} \xrightarrow{H} \pi^{2n+1}_{i+1} \xrightarrow{\Delta} \pi^n_{i-1} \xrightarrow{E} \pi^{n+1}_i \xrightarrow{H} \cdots .$$

<u>Proof</u>: By Theorem 2.4 and by the exactness of the sequence (2.9), a subsequence of the sequence (2.11) is an exact sequence of the 2-primary components if the groups in the subsequence are finite. Then we have the

exactness of the subsequences of (4.4) in which the groups π_n^n, π_{n+1}^{n+1}, π_{2n+1}^{2n+1}, π_{2n+1}^{n+1} and π_{2n-1}^{n} do not appear. By (4.1) and by the fact that $E : \pi_n(S^n) \to \pi_{n+1}(S^{n+1})$ are isomorphisms for all n, we have the exactness of the sequences $\to \pi_n^n \xrightarrow{E} \pi_{n+1}^{n+1} \to$. Then the only problem is the exactness of the sequence

$$\pi_{2n}^n \xrightarrow{E} \pi_{2n+1}^{n+1} \xrightarrow{H} \pi_{2n+1}^{2n+1} \xrightarrow{\Delta} \pi_{2n-1}^{n} \xrightarrow{E} \pi_{2n}^{n+1} .$$

But, the proof of the exactness of this sequence was done essentially in the proof of the previous lemma 4.1. Then we have the exactness of (4.4).

Let p be a prime, then denote by $(G_k;p)$ the p-primary component of the stable group G_k. By the exactness of the sequence (4.4) or by (3.2), we have

(4.5). $E^{m-n} : \pi_{n+k}^n \to \pi_{m+k}^m$, $m \geq n$, and $E^\infty : \pi_{n+k}^n \to (G_k;2)$ are isomorphisms for $n \geq k+2$.

Lemma 4.3. Let $n \geq 1$. If α and γ are in the p-primary components, then so are $\alpha \circ E^n\beta$ and $\beta \circ \gamma$. Further, if $\alpha \circ E^n\beta = \beta \circ \gamma = 0$, then each element of $\{\alpha, E^n\beta, E^n\gamma\}_n$ is in the p-primary components.

This lemma states for elements of the homotopy groups of spheres, but the statement is true for general case as follows.

(4.6). Let $n \geq 1$. If $p^r\alpha = 0$ and $p^s\gamma = 0$ for positive integers r,s and for $\alpha \in \pi(E^nY \to Z)$, and $\gamma \in \pi(EW \to X)$, then $p^r(\alpha \circ E^n\beta) = 0$ and $p^s(\beta \circ \gamma) = 0$ for $\beta \in \pi(X \to Y)$. Further, if $\alpha \circ E^n\beta = 0$ and $\beta \circ \gamma = 0$, then $p^{r+s}\ \{\alpha, E^n\beta, E^n\gamma\}_n = 0$.

Proof: By (1.7), $p^r(\alpha \circ E^n\beta) = p^r\alpha \circ E^n\beta = 0$ and $p^s(\beta \circ \gamma) = \beta \circ p^s\gamma = 0$. Let ι be the class of the identity of EW. Then it follows from Proposition 1.4 and (1.7) that

$$\begin{aligned} p^{r+s}\{\alpha, E^n\beta, E^n\gamma\}_n &= p^r(\{\alpha, E^n\beta, E^n\gamma\}_n \circ E^{n+1}p^s\iota) \\ &= p^r(\alpha \circ \pm E^n\{\beta, \gamma, p^s\iota\}) \\ &= p^r\alpha \circ \pm E^n\{\beta, \gamma, p^s\iota\} = 0 \end{aligned}$$

Thus, (4.6) is proved. Lemma 4.3 is, of course, a particular case of (4.6).

For the case $p = 2$, we have the following

(4.7). Let $n \geq 1$. Assume that $\alpha \in \pi_{k+n}(S^m;2)$ and $\gamma \in \pi_i(S^j;2)$, then $\alpha \circ E^n\pi_j(S^k) = \alpha \circ E^n\pi_j^k \subset \pi_{j+n}^m$ and $\pi_j(S^k) \circ \gamma = \pi_j^k \circ \gamma \subset \pi_i^k$. Further, assume that $\alpha \circ E^n\beta = \beta \circ \gamma = 0$, then the secondary composition $\{\alpha, E^n\beta, E^n\gamma\}_n$ is a coset of $\alpha \circ E^n\pi^k_{i+1} + \pi^m_{j+n+1} \circ E^{n+1}\gamma$ and it is a subset of π^m_{i+n+1}.

It is well known that for $n = 2$, 4 or 8, there is a Hopf fibering $h : S^{2n-1} \to S^n$ of a fibre S^{n-1} and that

$$E + h_* : \pi_{i-1}(S^{n-1}) \oplus \pi_i(S^{2n-1}) \to \pi_i(S^n)$$

is an isomorphism onto for all i [17]. This is true if we replace h by a mapping whose Hopf invariant is ± 1. Because of the above isomorphisms are induced by a mapping $i \cdot \Omega h : S^{n-1} \times \Omega(S^{2n-1}) \to \Omega(S^n)$, where "$\cdot$" denotes a loop-multiplication in $\Omega(S^n)$. This mapping induces the isomorphisms of cohomology groups and thus those of homology and homotopy. The details of the proof are left to readers. In our case, we have

Proposition 4.4. *Let α be an element of π^n_{2n-1} such that $H(\alpha) = \iota_{2n-1}$ or $-\iota_{2n-1}$, then the correspondence $(\beta, \gamma) \to E\beta + \alpha \circ \gamma$ gives isomorphisms of $\pi^{n-1}_{i-1} \oplus \pi^{2n-1}_i$ onto π^n_i for all i. In particular $E : \pi^{n-1}_{i-1} \to \pi^n_i$ is an isomorphism into for all i.*

We shall apply following

Lemma 4.5. *Let $n = 2$, 4 or 8. $E : \pi_{i-1}(S^{n-1}) \to \pi_i(S^n)$ is an isomorphism into for all i. $r\,\iota_{n-1} \circ \beta = r\beta$ for an arbitrary integer r and for $\beta \in \pi_{i-1}(S^{n-1})$.*

Proof: The first part is obvious by the Hopf decomposition of $\pi_i(S^n)$. Since $E(r\iota_{n-1} \circ \beta) = E(r\,\beta)$, by (2.1), it follows then the second part.

It follows easily from the proof of Lemma 4.1,

(4.8). *Let n be even. If $\Delta(\iota_{2n-1}) = [\iota_{n-1}, \iota_{n-1}] \neq 0$, then $\pi^n_{2n-1} = E\,\pi^{n-1}_{2n-2} \oplus \Delta\,\pi^{2n+1}_{2n+1} \approx E\,\pi^{n-1}_{2n-2} \oplus Z$ and $\pi^{n+1}_{2n} = E^2\pi^{n-1}_{2n-2} \approx E\,\pi^{n-1}_{2n-2}$. If $\Delta(\iota_{2n-1}) = [\iota_{n-1}, \iota_{n-1}] = 0$, then $E^2 : \pi^{n-1}_{2n-2} \to \pi^{n+1}_{2n}$ is an isomorphism into and $\pi^{n+1}_{2n}/E^2\pi^{n-1}_{2n-2} \approx Z_2$. An element α of π^n_{2n-1} is mapped into $E^2\pi^{n-1}_{2n-2}$ by E if and only if $H(\alpha) \in 2\pi^{2n-1}_{2n-1}$.*

We have

(4.9). $\pi^n_m \circ \pi^m_k \subset \pi^n_k$.

Proof: If $k \neq m$ and $k \neq 2m-1$, then $\pi^m_k = \pi_k(S^m;2)$ and (4.9)

is true by (4.7). If $k = m$, then $\pi^n_m \circ \pi^m_m = \pi^n_m$ and (4.9) is true. If $k = 2m-1$ and m is odd, then $\pi^m_k = \pi_k(S^m;2)$ by Lemma 4.1 and (4.9) is a consequence of (4.7). Now assume that $k = 2m-1$ and m is even. Consider an element α of π^m_{2m-1}. Since $H[\iota_m, \iota_m] = \pm 2\iota_{2m-1}$, there exists an integer r such that $H(2\alpha) = H(r[\iota_m, \iota_m])$. By the exactness of the sequence (4.4), we have that $2\alpha = r[\iota_m, \iota_m] + E\gamma$ for some $\gamma \in \pi^{m-1}_{2m-2}$. Now let β be an arbitrary element of π^n_m and let 2^t be the order of β. We shall show that $2^{2t+1}(\beta \circ \alpha) = 0$, then $\beta \circ \alpha \in \pi_{2m-1}(S^n;2) \subset \pi^n_{2m-1}$ and (4.9) is proved for this case.

By (2.1) and by the bilinearity and naturality of the Whitehead product [28],

$$\begin{aligned} 2^{2t+1}(\beta \circ \alpha) &= \beta \circ r2^{2t}[\iota_m, \iota_m] + \beta \circ 2^{2t}E\gamma \\ &= \beta \circ r[2^t\iota_m, 2^t\iota_m] + 2^{2t}\beta \circ E\gamma \\ &= \beta \circ 2^t\iota_m \circ r[\iota_m, \iota_m] + 0 \\ &= 2^t\beta \circ r[\iota_m, \iota_m] = 0 . \end{aligned}$$

CHAPTER V

Auxiliary Calculation of $\pi_{n+k}(S^n;2)$ for $1 \leq k \leq 7$.

In this chapter, we shall recall the known results [7][8][14][16][20] on the 2-primary components of $\pi_{n+k}(S^n)$ for $1 \leq k \leq 7$, in the connection with the sequence (4.4).

We shall denote by Z the group of integers and by Z_r the group of the integers modulo r. We shall also denote by $\{\alpha\} \subset \pi_i(S^n)$ the cyclic subgroup of $\pi_i(S^n)$ which is generated by an element α of $\pi_i(S^n)$.

We say again from (4.1)

(5.1). $\pi_i^1 = 0$ for $i > 1$. $\pi_i^n = 0$ for $i < n$. $G_k = (G_k;2) = 0$ for $k < 0$. $\pi_n^n = \{\iota_n\} \approx Z$ and $G_0 = \{\iota\} \approx Z$, where $\iota = E^\infty \iota_n$.

i). The Groups π_{n+1}^n. By (5.1), $\pi_2^1 = 0$. Then it follows from Proposition 4.4 that $H : \pi_3^2 \rightarrow \pi_3^3$ is an isomorphism onto. Let $\eta_2 \in \pi_3^2$ be the class such that $H(\eta_2) = \iota_3$, then η_2 generates $\pi_3^2 \approx Z$. Denote that $\eta_n = E^{n-2}\eta_2$ for $n \geq 2$ and $\eta = E^\infty \eta_2$.

By Proposition 2.7, $H(\Delta(\iota_5)) = \pm H[\iota_2, \iota_2] = \pm 2\iota_3$. Thus $\Delta(\iota_5) = \pm 2\eta_3$, since $H : \pi_3^2 \rightarrow \pi_3^3$ is an isomorphism.

Consider the exact sequence $\pi_5^5 \xrightarrow{\Delta} \pi_3^2 \xrightarrow{E} \pi_4^3 \rightarrow \pi_5^4 = 0$ of (4.4). Then E is onto and its kernel is generated by $2\eta_2$. Thus $\pi_4^3 = \{\eta_3\} \approx Z_2$, and also by (4.5), $\pi_{n+1}^n = \{\eta_n\} \approx Z_2$ for $n \geq 3$ and $(G_1;2) = \{\eta\} \approx Z_2$. Consequently we have obtained

Proposition 5.1. $\pi_3^2 = \{\eta_2\} \approx Z$, $\pi_{n+1}^n = \{\eta_n\} \approx Z_2$ for $n \geq 3$ and $(G_1, 2) = \{\eta\} \approx Z_2$. We have relations $H(\eta_2) = \iota_3$ and $\Delta(\iota_5) = \pm 2\eta_3$.

By (5.1), $\pi_{i-1}^1 = 0$ for $i \geq 3$. It follows then by Proposition 4.4

(5.2). The composition : $\alpha \rightarrow \eta_2 \circ \alpha$ defines an isomorphism $\pi_i^3 \approx \pi_i^2$ for $i \geq 3$.

The following lemma will be useful in the following.

<u>Lemma</u> 5.2. <u>Assume that</u> $2\alpha = 0$ <u>for an element</u> α <u>of</u> $\pi_i(S^3)$, <u>then for an arbitrary element</u> β <u>of</u> $\{\eta_3, 2\iota_4, E\alpha\}_1$, <u>the relations</u> $H(\beta) = E^2\alpha$ <u>and</u> $2\beta = \eta_3 \circ E\alpha \circ \eta_{i+1}$ <u>hold</u>. <u>Such a</u> β <u>belongs to</u> π^3_{i+2}. $\Delta(E^2\alpha) = 0$.

<u>Proof</u>: By Lemma 4.5, $2\iota_3 \circ \alpha = 2\alpha = 0$. Apply Proposition 2.6 tó the second composition $\{\eta_3, 2E\iota_3, E\alpha\}_1$, then

$$H(\beta) \in \Delta^{-1}(\eta_2 \circ 2\iota_3) \circ E^2\alpha = \Delta^{-1}(2\eta_2) \circ E^2\alpha$$
$$= \pm\, \iota_5 \circ E^2\alpha = E^2\alpha.$$

Next $2\beta \in 2\{\eta_3, 2\iota_4, E\alpha\}_1 = \{\eta_3, 2\iota_4, E\alpha\}_1 \circ 2\iota_{i+2}$

$= \eta_3 \circ E\{2\iota_3, \alpha, 2\iota_i\}$ by Proposition 1.4

$\subseteq \eta_3 \circ -\{2\iota_4, E\alpha, 2\iota_{i+1}\}_1$ by Proposition 1.3

$\ni \eta_3 \circ -(E\alpha \circ \eta_{i+1}) = \eta_3 \circ E\alpha \circ \eta_{i+1}$

by Corollary 3.7.

The composition $\eta_3 \circ -\{2\ _4, E\alpha, 2\iota_{i+1}\}_1$ is a coset of the subgroup $\eta_3 \circ \pi_{i+2}(S^4) \circ 2\iota_{i+2} + \eta_3 \circ 2\iota_4 \circ E\pi_{i+2}(S^4) = \eta_3 \circ \pi_{i+2}(S^4) \circ 2\iota_{i+2}$. For an element γ of $\pi_{i+2}(S^4)$, $E(\eta_3 \circ \gamma \circ 2\iota_{i+2}) = \eta_4 \circ 2E\gamma = 2\eta_4 \circ E\gamma = 0$ by (2.1). By Lemma 4.5, it follows that $\eta_3 \circ \gamma \circ 2\iota_{i+2} = 0$, and thus the subgroup $\eta_3 \circ \pi_{i+2}(S^4) \circ 2\iota_{i+2} = 0$. This shows that the composition $\eta_3 \circ -\{2\iota_4, E\alpha, 2\iota_{i+1}\}_1$ consists of a single element. Therefore we have that $2\beta = \eta_3 \circ E\alpha \circ \eta_{i+2}$. Obviously $4\beta = 0$ and $\beta \in \pi^3_{i+2}$. $\Delta(E^2\alpha) = \Delta H \beta = 0$.

q. e. d.

The element η_3 of $\pi_4(S^3)$ satisfies the condition $2\iota_3 \circ \eta_3 = 0$ of the above lemma. Consider an element ν' of $\{\eta_3, 2\iota_4, \eta_4\}_1$, then we have

(5.3). $\nu' \in \pi^3_6$, $H(\nu') = \eta_5$ <u>and</u> $2\nu' = \eta_3 \circ \eta_4 \circ \eta_5$.

ii). <u>The Group</u> π^n_{n+2}.

Denote that $\eta^2_n = \eta_n \circ \eta_{n+1}$ and $\eta^2 = \eta \circ \eta$, then $E\eta^2_n = \eta^2_{n+1}$ and $E^\infty \eta^2_n = \eta^2$.

<u>Proposition</u> 5.3. $\pi^n_{n+2} = \{\eta^2_n\} \approx Z_2$ <u>for</u> $n \geq 2$. $(G_2;2) = \{\eta^2\} \approx Z_2$.

<u>Proof</u>: By (5.2), $\pi^2_4 = \{\eta^2_2\} \approx Z_2$. Consider the exact sequence $\pi^3_6 \xrightarrow{H} \pi^5_6 \xrightarrow{\Delta} \pi^2_4 \xrightarrow{E} \pi^3_5 \xrightarrow{H} \pi^5_5 \xrightarrow{\Delta} \pi^2_3$ of (4.4). By Proposition 5.1, $\Delta : \pi^5_5 \to \pi^2_3$ is an isomorphism into. Then $H(\pi^3_5) = 0$ and E is onto. By (5.3) and Proposition 5.1, $H : \pi^3_6 \to \pi^5_6$ is onto. Then $\Delta(\pi^5_6) = 0$ and E is an

isomorphism into. Therefore, we have proved that $E : \pi_4^2 \to \pi_5^3$ is an isomorphism onto and that the proposition is true for $n = 3$. Next consider the exact sequence $\pi_5^3 \xrightarrow{E} \pi_6^4 \to \pi_6^7$. $\pi_6^7 = 0$ by (5.1), then E is onto. By Proposition 4.4, E is an isomorphism into. Thus $E : \pi_5^3 \to \pi_6^4$ is an isomorphism onto, and the proposition is proved for $n = 4$. For $n > 4$ and for $(G_2;2)$, the proposition is proved by (4.5).

iii). <u>The Group</u> π_{n+3}^n.

<u>Lemma</u> 5.4. <u>There exists an element</u> ν_4 <u>of</u> π_7^4 <u>such that</u>

$$H(\nu_4) = \iota_7 \quad \underline{\text{and}} \quad 2E\nu_4 = E^2\nu'.$$

<u>Proof</u>: Apply Theorme 3.6 for $\alpha = \eta_2$. η_2 satisfies the condition $(1-(-1)^1)E^4\eta_2 = 2\eta_6 = 0$. Then there exists an element α^* of $\pi_7(S^4)$ such that $E^2\beta \circ E^t\alpha^* \in (-1)^m \{E^m\eta_2, E^3\beta, E^{t+3}\eta_2\}_3$ for any element $\beta \in \pi_{t+2}(S^m)$ such that $\beta \circ E^t\eta_2 = 0$.

Consider the case that $\beta = 2\iota_3$ and $t = 1$. By (2.1), $2\iota_3 \circ E\eta_2 = 2E\eta_2 = 2\eta_3 = 0$. Thus $E^2(2\iota_3) \circ E\alpha^* = 2E\alpha^*$ is contained in $-\{\eta_5, 2\iota_6, \eta_6\}_3$. Now we show that

(5.4) <u>for</u> $n \geq 3$ <u>and</u> $t \leq n-2$ <u>the secondary construction</u> $\{\eta_n, 2\iota_{n+1}, \eta_{n+1}\}_t$ <u>consists of two elements</u> $E^{n-3}\nu'$ <u>and</u> $-E^{n-3}\nu'$.

First consider the case $t \geq 1$. By (4.7), Proposition 5.3 and by (5.3), we have that the above secondary composition is a coset of $\eta_n \circ \pi_{n+3}^{n+1} + \pi_{n+2}^2 \circ \eta_{n+2}$ which is generated by $\eta_n \circ \eta_{n+1} \circ \eta_{n+2} = 2E^{n-3}\nu'$. By the definition of ν', Proposition 1.3 and by (1.15), we have $E^{n-3}\nu' \in \{E^{n-3}\eta_3, 2\iota_4, \eta_4\}_1 \subset (-1)^{n-3}\{\eta_n, 2\iota_{n+1}, \eta_{n+1}\}_{n-2} \subset (-1)^{n-3}\{\eta_n, 2\iota_{n+1}, \eta_{n+1}\}_t$. Then (5.4) is verified easily for the case $t \geq 1$.

By (3.2), $\pi_{n+3}(S^{n+1}) = E\,\pi_{n+2}(S^n)$. Then it follows that $\{\eta_n, 2\iota_{n+1}, \eta_{n+1}\}_1$ and $\{\eta_n, 2\iota_{n+1}, \eta_{n+1}\}$ are cosets of the same subgroup $\eta_n \circ E\,\pi_{n+2}(S^n) + \pi_{n+2}(S^n) \circ \eta_{n+2}$. By (1.15), we have that these secondary compositions coincide. Then (5.4) is proved for the case $t = 0$.

Now the element $2E\alpha^*$ is $E^2\nu'$ or $-E^2\nu'$. Since $H : \pi_6^3 \to \pi_6^5$ maps ν' to the generator η_5 of π_6^5, then ν' is not divisible by 2. As $2E\alpha^* = \pm E^2\nu'$, $E^2\nu'$ is divisible by 2. Then it follows from (4.8) that $H(\alpha^*) = (2s+1)\iota_7$ for an integer s. As $H[\iota_4, \iota_4] = (-1)^u 2\iota_7$, we set

$$\nu_4 = \alpha^* - (-1)^u s[\iota_4, \iota_4] \qquad \text{if} \quad 2E\alpha^* = E^2\nu' ,$$
$$\nu_4 = -\alpha^* + (-1)^u(s+1)[\iota_4, \iota_4] \qquad \text{if} \quad 2E\alpha^* = -E^2\nu' .$$

Then it is verified easily that the element ν_4 satisfies Lemma 5.4.

Lemma 5.5. *For the element* ν_4 *of Lemma* 5.4, *we have that the secondary composition* $\{\eta_{m+2}, E^3\beta, \eta_{t+5}\}_3$ *contains* $E^2\beta \circ E^t\nu_4$ *or* $-E^2\beta \circ E^t\nu_4$, *where* β *is an element of* $\pi_{t+2}(S^m)$ *such that* $\beta \circ \eta_{t+2} = 0$ *and* $t > 0$.

Proof: In the proof of the previous lemma, we see that $E^2\beta \circ E^t\alpha^*$ or $-E^2\beta \circ E^t\alpha^*$ is in the secondary composition of this lemma. Since $E[\iota_4, \iota_4] = 0$, then $E^t\nu_4 = E^t\alpha^*$ or $E^t\nu_4 = -E^t\alpha^*$ by the definition of ν_4. Then the lemma is proved.

We denote that $\nu_n = E^{n-4}\nu_4$ for $n \geq 4$ and $\nu = E^\infty\nu_4$.

We also denote that

$$\eta_n^3 = \eta_n \circ \eta_{n+1} \circ \eta_{n+2} \quad \text{for} \quad n \geq 2 \quad \text{and} \quad \eta^3 = \eta \circ \eta \circ \eta .$$

By Lemma 5.4 and by (5.3), we have

(5.5). *For* $n \geq 5$, $2\nu_n = E^{n-3}\nu'$ *and* $4\nu_n = \eta_n^3$. $4\nu = \eta^3$.

It follows from Proposition 4.4 and Lemma 5.4

(5.6). *The correspondence* $(\alpha, \beta) \to E\alpha + \nu_4 \circ \beta$ *gives an isomorphism of* $\pi_{i-1}^3 \oplus \pi_i^7$ *onto* π_i^4.

Proposition 5.6.
$$\pi_5^2 = \{\eta_2^3\} \approx Z_2,$$
$$\pi_6^3 = \{\nu'\} \approx Z_4,$$
$$\pi_7^4 = \{\nu_4\} \oplus \{E\nu'\} \approx Z \oplus Z_4,$$
$$\pi_{n+3}^n = \{\nu_n\} \approx Z_8 \quad \textit{for} \quad n \geq 5,$$
and
$$(G_3;2) = \{\nu\} \approx Z_8 .$$

Proof: By (5.2) and by Proposition 5.3, we have $\pi_5^2 = \{\eta_2 \circ \eta_3^2\} = \{\eta_2^3\} \approx Z_2$. Consider the exact sequence $\pi_7^3 \xrightarrow{H} \pi_7^5 \xrightarrow{\Delta} \pi_5^2 \xrightarrow{E} \pi_6^3 \xrightarrow{H} \pi_6^5$ of (4.4). By Proposition 2.2 and by (5.3), $H(\nu' \circ \eta_6) = H(\nu') \circ \eta_6 = \eta_5 \circ \eta_6 = \eta_5^2$. Then it follows from Proposition 5.3, that $H : \pi_7^3 \to \pi_7^5$ is onto and thus E is an isomorphism into. Then $\eta_3^3 = E\eta_2^3$ is of order 2. It follows from (5.3) that ν' is of order 4. Since $\pi_6^5 = \{\eta_5\} \approx Z_2$, ν' is a generator of π_6^3. Thus $\pi_6^3 = \{\nu'\} \approx Z_4$. By (5.6), $\pi_7^4 = \{\nu_4\} \oplus \{E\nu'\} \approx Z \oplus Z_4$. By (4.8), $\pi_8^5 / E^2\pi_6^3 \approx Z_2$ and $E^2 : \pi_6^3 \to \pi_8^5$ is an isomorphism into. It follows from (5.5) that ν_5 is of order 8 and it generates π_8^5. By (4.5), $\pi_{n+3}^n = \{\nu_n\} \approx Z_8$ for $n \geq 5$ and $(G_3;2) = \{\nu\} \approx Z_8$. q. e. d.

In this proof we see that

(5.7). $$H(\nu' \circ \eta_6) = \eta_5^2$$

It is verified easily that the kernel of $E : \pi_7^4 \to \pi_8^5$ is generated by $2\nu_4 - E\nu'$. It follows from the exactness of the sequence (4.4) that

(5.8) $$\Delta(\iota_9) = \pm (2\nu_4 - E\nu'). \quad (= \pm [\iota_4, \iota_4]).$$

iv). <u>The Group</u> π_{n+4}^n.

By (5.2) and Proposition 5.6, we have $\pi_6^2 = \{\eta_2 \circ \nu'\} \approx Z_4$.

<u>Lemma</u> 5.7. <u>If</u> $E^2\alpha \in 2\iota_5 \circ \pi_{i+2}(S^5)$ <u>for an element</u> $\alpha \in \pi_i(S^3)$, <u>then</u> $E(\eta_2 \circ \alpha) = 0$. <u>In particular</u> $E(\eta_2 \circ \nu') = 0$ <u>and</u> $\Delta(\nu_5) = \pm (\eta_2 \circ \nu')$.

<u>Proof</u>: We have $E^2(\eta_2 \circ \alpha) = \eta_4 \circ E^2\alpha \in \eta_4 \circ 2\iota_5 \circ \pi_{i+2}(S^5) = 0$. Then it follows from Lemma 4.5 that $E(\eta_2 \circ \alpha) = 0$. By Lemma 5.4, $E^2\nu' = 2\nu_5 = 2\iota_5 \circ E\nu_4$, then it follows that $E(\eta_2 \circ \nu') = 0$. By the exactness of the sequence (4.4), we have that $\pi_6^2 = \Delta(\pi_8^5) = \{\Delta(\nu_5)\}$. Thus $\Delta(\nu_5) = \pm (\eta_2 \circ \nu')$. q. e. d.

Since $E\pi_6^2 = \{E(\eta_2 \circ \nu')\} = 0$, then we have an exact sequence $0 \to \pi_7^3 \xrightarrow{H} \pi_7^5$ from (4.4). By (5.7) and Proposition 5.3, we have that H is an isomorphism onto and $\pi_7^3 = \{\nu' \circ \eta_6\} \approx Z_2$. Now the first two parts of the following proposition is proved.

<u>Proposition</u> 5.8.
$$\pi_6^2 = \{\eta_2 \circ \nu'\} \approx Z_4,$$
$$\pi_7^3 = \{\nu' \circ \eta_6\} \approx Z_2,$$
$$\pi_8^4 = \{\nu_4 \circ \eta_7\} \oplus \{E\nu' \circ \eta_7\} \approx Z_2 \oplus Z_2,$$
$$\pi_9^5 = \{\nu_5 \circ \eta_8\} \approx Z_2,$$
$$\pi_{n+4}^n = 0 \text{ \underline{for} } n \geq 6 \text{ \underline{and} } (G_4;2) = 0.$$

<u>Proof</u>: The result for π_8^4 follows from (5.6) and Proposition 5.1. Consider the exact sequence $\pi_{10}^9 \xrightarrow{\Delta} \pi_8^4 \xrightarrow{E} \pi_9^5 \xrightarrow{H} \pi_9^9 \xrightarrow{\Delta} \pi_7^4$ of (4.4). By (5.8) and Proposition 5.6, we have that $\Delta : \pi_9^9 \to \pi_7^4$ is an isomorphism into. It follows by the exactness of the sequence that E is onto. The kernel of E is generated by
$$\begin{aligned}\Delta(\eta_9) &= \Delta(\iota_9) \circ \eta_7 && \text{(by Proposition 2.5)}\\ &= (2\nu_4 - E\nu') \circ \eta_7 && \text{(by (5.8))}\\ &= \nu_4 \circ (2\eta_7) - E\nu' \circ \eta_7 = E\nu' \circ \eta_7 .\end{aligned}$$

Then we have that $\pi_9^5 = E\pi_8^4 = \{\nu_5 \circ \eta_8\} \approx Z_2$. Next consider the exact sequence $\pi_{11}^{11} \xrightarrow{\Delta} \pi_9^5 \xrightarrow{E} \pi_{10}^6 \xrightarrow{H} \pi_{10}^{11} = 0$ of (4.4). Then E is onto and π_{10}^6 is generated by $\nu_6 \circ \eta_9$. Now we shall prove the formulas

(5.9). $\eta_3 \circ \nu_4 = \nu' \circ \eta_6$, $\eta_n \circ \nu_{n+1} = 0$ for $n \geq 5$ and $\nu_n \circ \eta_{n+3} = 0$ for $n \geq 6$.

First consider the element $\eta_3 \circ \nu_4$. Since $2E(\eta_3 \circ \nu_4) = 2(\eta_4 \circ \nu_5) = 2\eta_4 \circ \nu_5 = 0$, then it follows from Lemma 4.5 that $2(\eta_3 \circ \nu_4) = 0$. Thus $\eta_3 \circ \nu_4 \in \pi_7^3$. Since $\nu' \circ \eta_6$ generates π_7^3, then $\eta_3 \circ \nu_4 = x\, \nu' \circ \eta_6$ for $x = 0$ or 1. Applying H, we have

$$\begin{aligned} H(\eta_3 \circ \nu_4) &= E(\eta_2 \# \eta_2) \circ H(\nu_4) && \text{by Proposition 2.2} \\ &= \pm (\eta_4 \circ \eta_5) \circ \iota_7 && \text{by Proposition 3.1 and Lemma 5.4} \\ &= \eta_4^2 \ (\neq 0) && \text{by Proposition 5.3,} \\ \text{and} \quad &= H(x\, \nu' \circ \eta_6) = x\eta_4^2 && \text{by (5.7).} \end{aligned}$$

Then this implies that $x = 1$ and $\eta_3 \circ \nu_4 = \nu' \circ \eta_6$.

Next $\eta_5 \circ \nu_6 = E^2\nu' \circ \eta_8 = 2\nu_5 \circ \eta_8 = 0$ by (5.5). Thus $\eta_n \circ \nu_{n+1} = 0$ for $n \geq 5$. By Proposition 3.1, $\eta_2 \# \nu_4 = \eta_6 \circ \nu_7 = \nu_6 \circ \eta_9$. Then $\nu_6 \circ \eta_9 = \eta_6 \circ \nu_7 = 0$ and $\nu_n \circ \eta_{n+3} = 0$ for $n \geq 6$. Consequently all the formulas of (5.9) are established.

By the last result, $\pi_{10}^6 = \{\nu_6 \circ \eta_9\} = 0$.

By (4.5), $\pi_{n+4}^n = (G_4;2) = 0$ for $n \geq 6$. Then the proof of the Proposition 5.8 is finished.

In this proof, we see that $\Delta : \pi_{11}^{11} \to \pi_9^5$ is onto, since $E : \pi_9^5 \to \pi_{10}^6$ is trivial. Since π_9^5 is generated by $\nu_5 \circ \eta_8$, it follows then

(5.10). $$\Delta(\iota_{11}) = \nu_5 \circ \eta_8 .$$

We have also in the proof

(5.11). $$\Delta(\eta_9) = E\nu' \circ \eta_7 .$$

v). The Group π_{n+5}^n.

Proposition 5.9.
$$\begin{aligned} \pi_7^2 &= \{\eta_2 \circ \nu' \circ \eta_6\} \approx Z_2 , \\ \pi_8^3 &= \{\nu' \circ \eta_6^2\} \approx Z_2 , \\ \pi_9^4 &= \{\nu_4 \circ \eta_7^2\} \oplus \{E\nu' \circ \eta_7^2\} \approx Z_2 \oplus Z_2 , \\ \pi_{10}^5 &= \{\nu_5 \circ \eta_8^2\} \approx Z_2 , \\ \pi_{11}^6 &= \{\Delta(\iota_{13})\} \approx Z , \\ \pi_{n+5}^n &= (G_5;2) = 0 \quad \text{for } n \geq 7 . \end{aligned}$$

Proof: The first statement follows from (5.2) and Proposition 5.8. By Lemma 5.7, $E\pi_7^2 = \{E(\eta_2 \circ \nu') \circ \eta_7\} = 0$. It follows from the exactness of (4.4) that the sequence $0 \to \pi_8^3 \xrightarrow{H} \pi_8^5 \xrightarrow{\Delta} \pi_6^2$ is exact. By Lemma 5.7, Proposition 5.6 and by (5.5), we have that the kernel of Δ is generated by $4\nu_5 = \eta_5^3$. By Proposition 2.2 and by (5.3),

$$H(\nu' \circ \eta_6^2) = H(\nu') \circ \eta_6^2 = \eta_5 \circ \eta_6^2 = \eta_5^3 .$$

It follows from the exactness of the above sequence the second statement of the proposition. By (5.6), we have the statement of π_9^4.

Next consider the exact sequence

$$\pi_{11}^9 \xrightarrow{\Delta} \pi_9^4 \xrightarrow{E} \pi_{10}^5 \xrightarrow{H} \pi_{10}^9 \xrightarrow{\Delta} \pi_8^4 \xrightarrow{E} \pi_9^5$$

of (4.4). By Proposition 5.8 and (5.11), $E : \pi_8^4 \to \pi_9^5$ has the kernel isomorphic to Z_2. Then $\pi_{10}^9 \approx Z_2$ has to be mapped by Δ isomorphically into π_8^4. It follows that $E : \pi_9^4 \to \pi_{10}^5$ is onto. π_{11}^9 is generated by η_9^2 and

$$\Delta(\eta_9^2) = \Delta(\eta_9) \circ \eta_8 = E\nu' \circ \eta_7^2$$

by Proposition 2.5 and (5.11). Then we have that

$$\pi_{10}^5 = E\pi_9^4 = \{\nu_5 \circ \eta_8^2\} \approx Z_2$$

By (5.9), $E(\nu_5 \circ \eta_8^2) = \nu_6 \circ \eta_9 \circ \eta_{10} = 0$ and thus $E\pi_{10}^5 = 0$. In fact $\Delta(\eta_{11}) = \nu_5 \circ \eta_8^2$ by Proposition 2.5 and (5.10). Then we have an exact sequence $0 \to \pi_{11}^6 \xrightarrow{H} \pi_{11}^{11} \xrightarrow{\Delta} \pi_9^5$ from (4.4). By Proposition 5.8 and by (5.10), Δ is onto and its kernel is generated by $2\iota_{11}$. By Proposition 2.7, $H(\Delta(\iota_{13})) = \pm 2\iota_{11}$. It follows that $\pi_{11}^6 = \{\Delta(\iota_{13})\} \approx Z$.

From the exactness of the sequence $\pi_{13}^{13} \xrightarrow{\Delta} \pi_{11}^6 \xrightarrow{E} \pi_{12}^7 \longrightarrow \pi_{12}^{13} = 0$ we have that $\pi_{12}^7 = 0$. By (4.5), we have the last assertion of the proposition. q. e. d.

In this proof, we see that

(5.12). $\Delta : \pi_{n+7}^{2n+1} \to \pi_{n+5}^n$ <u>are isomorphisms into for</u> $n = 4$, 5 <u>and</u> 6 .

Next we shall prove

Lemma 5.10. $\Delta(\iota_{13}) \in \{\nu_6, \eta_9, 2\iota_{10}\} \in \pi_{11}(S^6) / 2\pi_{11}(S^6)$.

Proof: It is easy to see that the secondary composition is defined and it is a coset of $\nu_6 \circ \pi_{11}(S^9) + \pi_{11}(S^6) \circ 2\iota_{11} = \nu_6 \circ \pi_{11}^9 + 2\pi_{11}(S^6) = 2\pi_{11}(S^6)$. (cf. (4.7),(5.9)). Applying the relation (5.10) to Proposition 2.6, we have that $H\{\nu_6, \eta_9, 2\iota_{10}\}$ contains $\iota_{11} \circ E(2\iota_{10}) = 2\iota_{11}$. We have also $H(\Delta(\iota_{13})) = \pm 2\iota_{11}$. It follows from the exactness of the sequence $\pi_{10}(S^5) \xrightarrow{E} \pi_{11}(S^6) \xrightarrow{H} \pi_{11}(S^{11})$ of (2.8) that $\Delta(\iota_{13}) \in \{\nu_6, \eta_9, 2\iota_{10}\} + E\pi_{10}(S^5)$. Since $E\pi_{10}(S^5)$ is finite and it has the vanishing 2-primary component $E\pi^5_{10} = 0$, then $E\pi_{10}(S^5) \subset 2\pi_{11}(S^6)$. Therefore $\Delta(\iota_{13})$ and $-\Delta(\iota_{13})$ are contained in $\{\nu_6, \eta_9, 2\iota_{10}\}$.

vi). The Groups π^n_{n+6}.

Denote that $\nu^2_n = \nu_n \circ \nu_{n+3}$ for $n \geq 4$ and $\nu^2 = \nu \circ \nu$.

Proposition 5.11.
$$\pi^2_8 = \{\eta_2 \circ \nu' \circ \eta^2_6\} \approx Z_2,$$
$$\pi^3_9 = 0,$$
$$\pi^4_{10} = \{\nu^2_4\} \approx Z_8,$$
$$\pi^n_{n+6} = \{\nu^2_n\} \approx Z_2, \quad \underline{\text{for}} \quad n \geq 5,$$
$$(G_6;2) = \{\nu^2\} \approx Z_2.$$

Proof: The first assertion is proved by Proposition 5.9 and (5.2). By Lemma 5.7, $E\pi^2_8 = \{E(\eta_2 \circ \nu') \circ \eta^2_6\} = 0$. We have also $E\pi^2_7 = 0$ in Proposition 5.9. It follows from the exactness of (4.4) that the sequence $0 \to \pi^3_9 \xrightarrow{H} \pi^5_9 \xrightarrow{\Delta} \pi^2_7 \to 0$ is exact. By Proposition 5.8 and Proposition 5.9, we have that Δ is an isomorphism onto. Then it follows that $\pi^3_9 = 0$. By Proposition 5.6 and by (5.6) we have that $\pi^4_{10} = \{\nu^2_4\} \approx Z_8$. Consider the exact sequences $\pi^{2n-1}_{n+8} \xrightarrow{\Delta} \pi^n_{n+6} \xrightarrow{E} \pi^{n+1}_{n+7} \xrightarrow{H} \pi^{2n-1}_{n+7} \xrightarrow{\Delta} \pi^n_{n+5}$ for $n = 4$, 5 and 6. It follows from (5.12) that H are trivial and E are homomorphisms onto. Next we shall prove

(5.13). $\Delta(\nu_9) = \pm 2(\nu^2_4)$, $\Delta(\eta^2_{11}) = 0$ and $\Delta(\eta_{13}) = 0$.

By Proposition 2.5 and (5.8),
$$\Delta(\nu_9) = \Delta(\iota_9) \circ \nu_7 = \pm (2\nu_4 - E\nu') \circ \nu_7$$
$$= \pm (2\nu^2_4 - E(\nu' \circ \nu_6)).$$

Since $\nu' \circ \nu_6 \in \pi^3_9 = 0$, then $\Delta(\nu_9) = \pm 2(\nu^2_4)$. We have also
$$\Delta(\eta^2_{11}) = \Delta(\iota_{11}) \circ \eta^2_9 \quad \text{by Proposition 2.5}$$
$$= \nu_5 \circ \eta_8 \circ \eta^2_9 = \nu_5 \circ \eta^3_8 \quad \text{by (5.10)}$$
$$= 4(\nu_5 \circ \nu_8) = 4E(\nu^2_4) \quad \text{by (5.5)}$$
$$= E\Delta(\pm 2\nu_9) = 0.$$
$$\Delta(\eta_{13}) = \Delta(\iota_{13}) \circ \eta_{11} \quad \text{by Proposition 2.5}$$
$$\in \{\nu_6, \eta_9, 2\iota_{10}\} \circ \eta_{11} \quad \text{by Lemma 5.10}$$

$= \nu_6 \circ \{\eta_9, 2\iota_{10}, \eta_{10}\}$ by Proposition 1.4.

By (5.4), we have that $\nu_6 \circ \{\eta_9, 2\iota_{10}, \eta_{10}\}$ consists of $\nu_6 \circ E^6\nu'$ and $-(\nu_6 \circ E^6\nu')$. By (5.5), $\nu_6 \circ E^6\nu' = 2\nu_6^2 = E^2\Delta(\nu_9) = 0$. Thus $\Delta(\eta_{13}) = 0$, and (5.13) is proved.

Now apply (5.13) to the above exact sequences, then it follows that the proposition is true for $n = 5$, 6 and 7.

By Proposition 4.4, π_{14}^8 is isomorphic to $\pi_{13}^7 + \pi_{14}^{15} = \pi_{13}^7$ by E. Thus the proposition is proved for $n = 8$. For $n \geq 9$ and for the stable group, the proposition is proved by (4.5). Consequently the Proposition 5.11 has esteblished. q. e. d.

Lemma 5.12. $\{\eta_n, \nu_{n+1}, \eta_{n+4}\}$ consists of a single element ν_n^2 for $n \geq 6$.

Proof: By (5.9), this secondary composition is defined and it is a coset of $\eta_n \circ \pi_{n+6}(S^{n+1}) + \pi_{n+5}(S^n) \circ \eta_{n+5}$. By (3.2), $\pi_{n+6}(S^{n+1}) = E\pi_{n+5}(S^n)$. Then by (4.7), and by Proposition 5.9, $\eta_n \circ \pi_{n+6}(S^{n+1}) = \eta_n \circ \pi_{n+6}^{n+1} = 0$. Similarly $\pi_{n+5}(S^n) \circ \eta_{n+5} = \pi_{n+5}^n \circ \eta_{n+5}$, and this group is trivial if $n > 6$ and generated by $\Delta(\iota_{13}) \circ \eta_{11}$ if $n = 6$. As is seen in the proof of (5.13), $\Delta(\iota_{13}) \circ \eta_{11} = \Delta(\eta_{13}) = 0$. In any case, the secondary composition consists of a single element, which is $x\nu_n^2$ for an integer $x = 0$ or 1. By (1.15) and Proposition 1.3, $E\{\eta_n, \nu_{n+1}, \eta_{n+4}\} \subset -\{\eta_{n+1}, \nu_{n+2}, \eta_{n+5}\}$. This shows that the integer x does not depend on n. Now apply Lemma 5.5 for the case $\beta = \nu_m$ and $m \geq 6$. Then it follows that $E^2\nu_m \circ E^{m+1}\nu_4 = \nu_{m+2}^2$ is contained in $\{\eta_{m+2}, E^3\nu_m, \eta_{m+6}\}_3 \subset \{\eta_{m+2}, \nu_{m+3}, \eta_{m+6}\}$. Thus $\nu_{m+2}^2 = x\nu_{m+2}^2$ and $x = 1$. Consequently the lemma is proved.

Remark. The Lemma 5.12 is true for $n = 5$.

vii). The Groups π_{n+7}^n.

By (5.2) and Proposition 5.11, $\pi_9^2 = 0$.

It follows from (4.4) that the sequence

$$0 = \pi_9^2 \rightarrow \pi_{10}^3 \xrightarrow{H} \pi_{10}^5 \xrightarrow{\Delta} \pi_8^2 \xrightarrow{E} \pi_9^3 = 0$$

is exact. By Proposition 5.9 and Proposition 5.11, we see that Δ is an isomorphism onto. It follows then $\pi_{10}^3 = 0$.

By (5.6) and by Proposition 5.8, $\pi_{11}^4 \approx \pi_{10}^3 + \pi_{11}^7 = 0$.

Next consider the exact sequence $0 = \pi_{11}^4 \xrightarrow{E} \pi_{12}^5 \xrightarrow{H} \pi_{12}^9 \xrightarrow{\Delta} \pi_{10}^4$ of (4.4). By (5.13), Proposition 5.11 and Proposition 5.6, we have that Δ is a homomorphism of degree 2 and its kernel is generated by $4\nu_9 = \eta_9^3$. Thus we have an isomorphism $H : \pi_{12}^5 \to 4\pi_{12}^9 \approx Z_2$.

Lemma 5.13. The secondary composition $\{\nu_5, 8\iota_8, \nu_8\}_t$ consists of a single element, denoted by σ''', such that $H(\sigma''') = 4\nu_9 = \eta_9^3$. The element σ''' generates $\pi_{12}^5 \approx Z_2$. $(0 \leq t \leq 3)$

Proof: The secondary composition is a coset of $\nu_5 \circ \pi_{12}(S^8) + \pi_9(S^5) \circ \nu_9$. By (3.2), (4.7) and Proposition 5.8, $\nu_5 \circ \pi_{12}(S^8) = \nu_5 \circ E\pi_{11}(S^7) = \nu_5 \circ \pi_{11}^7 = 0$. By (3.2), Proposition 5.8 and by (5.9), $\pi_9(S^5) \circ \nu_9 = \pi_9^5 \circ \nu_9 = \{\nu_5 \circ \eta_8 \circ \nu_9\} = 0$. It follows then that the secondary composition consists of a single element. By (1.15),

$$\{\nu_5, 8\iota_8, \nu_8\}_t \subset \{\nu_5, 8\iota_8, \nu_8\}.$$

Then $\{\nu_5, 8\iota_8, \nu_8\}_t$ is also consists of the same single element σ'''.

By (4.7), the element σ''' is in π_{12}^5.

By Proposition 2.6, $H(\sigma''') = H\{\nu_5, 8\iota_8, \nu_8\}_1 = \Delta^{-1}(\nu_4 \circ 8\iota_7) \circ \nu_9$.

By (5.8), $\Delta(4\iota_9) = \pm 4(2\nu_4 - E\nu') = \pm (8\nu_4 - E4\nu') = \pm 8\nu_4 = \pm (\nu_4 \circ 8\iota_7)$. Then $\Delta^{-1}(\nu \circ 8\iota_7) \circ \nu_9$ contains $\pm (4\iota_9) \circ \nu_9 = \pm 4\nu_9 = \eta_9^3$. Thus $H(\sigma''') = 4\nu_9$ and the lemma is proved by the isomorphism $H : \pi_{12}^5 \approx 4\pi_{12}^9 \approx Z_2$. q. e. d.

Next we prove

(5.14). The sequences $0 \to \pi_{12}^5 \xrightarrow{E} \pi_{13}^6 \xrightarrow{H} \pi_{13}^{11} \to 0$

and $0 \to \pi_{13}^6 \xrightarrow{E} \pi_{14}^7 \xrightarrow{H} \pi_{14}^{13} \to 0$

are exact.

By the exactness of (4.4), it is sufficient to prove that $\Delta(\pi_{14}^{11}) = \Delta(\pi_{13}^{11}) = \Delta(\pi_{15}^{13}) = \Delta(\pi_{14}^{13}) = 0$. The last two formulas of (5.13) show that $\Delta(\pi_{13}^{11}) = \Delta(\pi_{14}^{13}) = 0$. By Proposition 2.5 and (5.13), $\Delta(\eta_{13}^2) = \Delta(\eta_{13}) \circ \eta_{12} = 0$ and this shows $\Delta(\pi_{15}^{13}) = 0$. By Proposition 2.5, (5.10) and by (5.9),

$$\Delta(\nu_{11}) = \Delta(\iota_{11}) \circ \nu_9 = \nu_5 \circ \eta_8 \circ \nu_9 = 0.$$

This proves with Proposition 5.6 that $\Delta(\pi_{14}^{11}) = 0$. Consequently the exactness of (5.14) is established.

Lemma 5.14. There exist elements $\sigma_8 \in \pi_{15}^8$, $\sigma' \in \pi_{14}^7$ and

$\sigma'' \in \pi^6_{13}$ <u>such that</u> $H(\sigma_8) = \iota_{15}$, $2E\sigma_8 = E^2\sigma''$,
$H(\sigma') = \eta_{13}$, $2\sigma' = E\sigma''$,
<u>and</u> $H(\sigma'') = \eta^2_{11}$, $2\sigma'' = E\sigma'''$.

<u>Proof</u>: Apply Theorem 3.6 for $\alpha = \nu_4$. By Proposition 5.11, the element ν_4 satisfies the condition $2E^{4-3}\nu_4 \circ E^{7-3}\nu_4 = 2\nu_5^2 = 0$ of (3.1) for $h = 3$. Then there exists an element α^* of $\pi_{15}(S^8)$ such that

$$E^4\beta \circ E^t\alpha^* \in (-1)^m\{\nu_{m+4}, E^7\beta, \nu_{t+11}\}_7 + (-1)^t\{E^4\beta, \nu_{t+8}, 2\nu_{t+11}\}_{t+3}$$

for any element $\beta \in \pi_{t+4}(S^m)$ satisfying $\beta \circ \nu_{t+4} = 0$.

Now let $\beta = 8\iota_5$. Then $8\iota_5 \circ \nu_5 = 8\nu_5 = 0$ by (2.1) and $8E\alpha^* = E^4 8\iota_5 \circ E\alpha^* \in -\{\nu_9, 8\iota_{12}, \nu_{12}\}_7 - \{8\iota_9, \nu_9, 2\nu_{12}\}_4$.

The secondary composition $\{\nu_9, 8\iota_{12}, \nu_{12}\}_7$ is a coset of $\nu_9 \circ E^7\pi_9(S^5) + \pi_{12}(S^9) \circ \nu_{13} = \nu_9 \circ E^7\pi^5_{13} \circ \nu_{13} = 0$ by (4.7) and by Proposition 5.8, and thus it consists of a single element. By Proposition 1.3 and Lemma 5.13, $E^4\sigma''' \in E^4\{\nu_5, 8\iota_8, \nu_8\}_3 \subset \{\nu_9, 8\iota_{12}, \nu_{12}\}_7$. It follows then $-\{\nu_9, 8\iota_{12}, \nu_{12}\}_7$ consists of $-E^4\sigma''' = E^4\sigma'''$.

By Proposition 1.2, $\{8\iota_9, \nu_9, 2\nu_{12}\}_4 \subset \{2\iota_9, 4\iota_9 \circ \nu_9 \circ 2\iota_{12}, \nu_{12}\}_4$ $= \{2\iota_9, 0, \nu_{12}\}_4 = 2\iota_9 \circ E^4\pi_{12}(S^5) + \pi_{13}(S^9) \circ \nu_{13} = 2E^4\pi_{12}(S^5) + \pi^9_{13} \circ \nu_{13} =$ $= 2E^4\pi_{12}(S^5)$. Thus we have that $8\,E\alpha^* - E^4\sigma''' \in 2\,E^4\pi_{12}(S^5)$.

Since $\pi^5_{12} = \pi_{12}(S^5;2) \approx Z_2$, then $2E^4\pi_{12}(S^5)$ is a finite group of odd order. Thus there exists an odd integer x such that

$$8x\ E\alpha^* = x\ E^4\sigma''' = E^4\sigma'''.$$

Now, from (5.14) and Lemma 5.13, it follows that the groups π^6_{13} and π^7_{14} have 4 and 8 elements respectively and that $E^2 : \pi^5_{12} \to \pi^7_{14}$ is an isomorphism into. Also it follows from (4.8) that π^9_{16} has 16 elements and $E^4 : \pi^5_{12} \to \pi^9_{16}$ is an isomorphism into.

Then $8x\ E\alpha^* = E^4\sigma''' \neq 0$ and $16x\ E\alpha^* = 0$. This shows that $xE\alpha^*$ is of order 16 and it generates π^9_{16} which is isomorphic to Z_{16}. There exist uniquely elements $\sigma' \in \pi^7_{14}$ and $\sigma'' \in \pi^6_{13}$ such that $E^2\sigma' = 2xE\alpha^*$ and $E^3\sigma'' = 4xE\alpha^*$. Then it follows that σ' and σ'' generate respectively the groups $\pi^7_{14} \approx Z_8$ and $\pi^6_{13} \approx Z_4$. Obviously $2\sigma' = E\sigma''$ and $2\sigma'' = E\sigma'''$. From the exactness of the sequences of (5.14), it follows that $H(\sigma') = \eta_{13}$ and $H(\sigma'') = \eta^2_{12}$.

Next, the element $xE\alpha^*$ is not in the image of E^2. Then, by (4.8), $x\alpha^*$ has an odd Hopf invariant. Since $H(\Delta(\iota_{17})) = \pm 2\iota_{15}$, there exists an integer y such that $H(x\alpha^*) = H(y\Delta(\iota_{17})) + \iota_{15}$. Then we define σ_8 by setting $\sigma_8 = x\alpha^* - y\Delta(\iota_{17})$. Then we have that $H(\sigma_8) = \iota_{15}$ and $E\sigma_8 = xE\alpha^*$.

Consequently we see that the Lemma 5.14 has established. q. e. d.

We denote that $\sigma_n = E^{n-8}\sigma_8$ for $n \geq 8$ and $\sigma = E^{\infty}\sigma_8$.

By (4.5), we have that π^n_{n+7} for $n \geq 8$ and $(G_7;2)$ are isomorphic to $\pi^9_{16} \approx Z_{16}$ and they are generated by σ_n and σ respectively. Now the groups π^n_{n+7} are all computed except π^8_{15}. The group π^8_{15} is computed by the following (5.15), which follows directly from Proposition 4.4.

(5.15) <u>The correspondence</u> $(\alpha, \beta) \to E\alpha + \sigma_8 \circ \beta$ <u>gives isomorphisms of</u> $\pi^7_{i-1} \oplus \pi^{15}_i$ <u>onto</u> π^8_i.

The results on π^n_{n+7} are listed as follows.

<u>Proposition</u> 5.15.

$$\pi^2_9 = \pi^3_{10} = \pi^4_{11} = 0,$$
$$\pi^5_{12} = \{\sigma'''\} \approx Z_2,$$
$$\pi^6_{13} = \{\sigma''\} \approx Z_4,$$
$$\pi^7_{14} = \{\sigma'\} \approx Z_8,$$
$$\pi^8_{15} = \{\sigma_8\} \oplus \{E\sigma'\} \approx Z \oplus Z_8,$$
$$\pi^n_{n+7} = \{\sigma_n\} \approx Z_{16} \quad \underline{\text{for}}\ n \geq 9,$$
$$(G_7;2) = \{\sigma\} \approx Z_{16}.$$

Consider the exact sequence $\pi^{17}_{17} \xrightarrow{\Delta} \pi^8_{15} \xrightarrow{E} \pi^9_{16}$. The kernel of E is generated by the element $2\sigma_8 - E\sigma'$, since the above result. Then it follows from the above exact sequence

(5.16). $$\Delta(\iota_{17}) = \pm (2\sigma_8 - E\sigma').$$

It follows from the beginning of the proof of Lemma 5.14 and from the definition of σ_8,

<u>Lemma</u> 5.16. <u>Let</u> $t > 0$ <u>and let</u> β <u>be an element of</u> $\pi_{t+4}(S^m)$ <u>such that</u> $\beta \circ \nu_{t+4} = 0$, <u>then</u>

$$E^4\beta \circ \sigma_{t+8} \in (-1)^m x\{\nu_{m+4}, E^n\beta, \nu_{t+11}\}_7 + (-1)^t x\{E^4\beta, \nu_{t+8}, 2\nu_{t+11}\}_{t+3},$$

<u>where</u> x <u>is an odd integer</u>.

CHAPTER VI.

Some Elements Given By Secondary Compositions.

i). The Elements ε_n.

Consider a secondary composition $\{\eta_3, E\nu', \nu_7\}_1$.
The condition $\eta_3 \circ E\nu' = \nu' \circ \nu_6 = 0$ of (1.13) is verified by Lemma 5.7 and Proposition 5.11. Then the above secondary composition is defined and it is a coset of the subgroup

$$\eta_3 \circ E\pi_{10}(S^3) + \pi_8(S^3) \circ \nu_8 = \eta_3 \circ E\pi^3_{10} + \pi^3_8 \circ \nu_8 \quad \text{by (4.7)}$$

$$= \{\nu' \circ \eta^2_6 \circ \nu_8\} = 0 \quad \text{by Prop. 5.15, 5.9 and by (5.9).}$$

Then the secondary composition consists of a single element which is denoted by $\varepsilon_3 = \{\eta_3, E\nu', \nu_7\}_1 \in \pi_{11}(S^3)$.

We denote also that

$$\varepsilon_n = E^{n-3}\,\varepsilon_3 \quad \text{for } n \geq 3 \quad \text{and} \quad \varepsilon = E^\infty \varepsilon_3 .$$

<u>Lemma</u> 6.1: $H(\varepsilon_3) = \nu^2_5$ <u>and</u> $2\varepsilon_n = 2\varepsilon = 0, \quad n \geq 3$.

<u>Proof</u>: By Lemma 5.7, $\eta_2 \circ \nu' = \pm\Delta(\nu_5)$. Then it follows from Proposition 2.6 that

$$\nu^2_5 = \pm\,\nu_5 \circ \nu_8 \in \Delta^{-1}(\eta_2 \circ \nu') \circ \nu_8 = H\{\eta_3, E\nu', \nu_7\}_1 .$$

Since $\{\eta_3, E\nu', \nu_7\}_1$ consists of ε_3, then $H(\varepsilon_3) = \nu^2_5$. Next

$$\begin{aligned} 2\varepsilon_3 &= 2\iota_3 \circ \varepsilon_3 \in 2\iota_3 \circ \{\eta_3, E\nu', \nu_7\} && \text{by Lemma 4.5} \\ &= \{2\iota_3, \eta_3, E\nu'\} \circ \nu_8 \\ &\subset \pi_8(S^3) \circ \nu_8 = \pi^3_8 \circ \nu_8 && \text{by (4.7)} \\ &= \{\nu' \circ \eta^2_6 \circ \nu_8\} && \text{by Proposition 5.9} \\ &= 0 && \text{by (5.9).} \end{aligned}$$

Then, obviously, $2\varepsilon_n = 2\varepsilon = 0$. q. e. d.

We shall show

(6.1). <u>The following secondary compositions contain</u> ε_n: $\{\eta_n, 2\nu_{n+1}, \nu_{n+4}\}_t$

<u>for</u> $n \geq 4$ <u>and</u> $0 \leq t \leq n-2$; $\{\eta_n, \nu_{n+1}, 2\nu_{n+4}\}_t$, $\{\eta_n, 2\iota_{n+1}, \nu_{n+1}^2\}_t$ <u>and</u> $\{\eta_n, \nu_{n+1}^2, 2\iota_{n+7}\}_t$ <u>for</u> $n \geq 5$ <u>and</u> $0 \leq t \leq n-5$.

<u>Proof</u>: By Proposition 1.3 and (1.15); $\varepsilon_n = (-1)^{n-3} E_n^{n-3} \varepsilon_3 \in (-1)^{n-3} E^{n-3} \{\eta_3, E\nu', \nu_7\}_1 \subset \{\eta_n, 2\nu_{n+1}, \nu_{n+4}\}_{n-2} \subset \{\eta_n, 2\nu_{n+1}, \nu_{n+4}\}_t$ for $n \geq 4$ and $0 \leq t \leq n-2$. Next consider the secondary composition $\{\eta_5, 2\nu_6, \nu_9\}_1$ which is a coset of $\eta_5 \circ E\pi_{12}^5 + \pi_{10}^5 \circ \nu_{10}$ by (4.7). By Proposition 5.15 and Lemma 5.14, $\eta_5 \circ E\pi_{12}^5$ is generated by $\eta_5 \circ E\sigma''$. By (5.14), $E : \pi_{13}^6 \to \pi_{14}^7$ is an isomorphism into. It follows that $E(2\iota_6 \circ \sigma'' - E\sigma''') = 2E\sigma'' - E2\sigma'' = 0$ implies $2\iota_6 \circ \sigma'' = E\sigma'''$. Then $\eta_5 \circ E\sigma''' = \eta_5 \circ 2\iota_6 \circ \sigma'' = 0$ and $\eta_5 \circ E\pi_{12}^5 = 0$. By Proposition 5.9, and (5.9) $\pi_{10}^5 \circ \nu_{10} = \{\nu_5 \circ \eta_8^2 \circ \nu_{10}\} = 0$. Therefore the secondary composition $\{\eta_5, 2\nu_6, \nu_9\}_1$ is a coset of the trivial subgroup 0, and it consists of a single element which has to be ε_5. By Proposition 1.2,

$$\{\eta_5, \nu_6, 2\nu_9\}_1 \subset \{\eta_5, 2\nu_6, \nu_9\}_1.$$

It follows that $\{\eta_5, \nu_6, 2\nu_9\}_1$ consists of ε_5. By Proposition 1.2,

$$\{\eta_5, 2\iota_6, \nu_6^2\}_1 \subset \{\eta_5, 2\nu_6, \nu_9\}_1 .$$

It follows that $\{\eta_5, 2\iota_6, \nu_6^2\}_1$ consists of ε_5. Similarly,

$$\varepsilon_5 \in \{\eta_5, \nu_6, 2\nu_9\}_1 \subset \{\eta_5, \nu_6^2, 2\iota_{12}\}_1 .$$

Now we have that (6.1) is true for $n = 5$. Then the proof for the case $n \geq 6$ is similar to the first part of this proof. q. e. d.

Next we have

(6.2) $$\sigma''' \circ \nu_{12} \equiv \eta_5^2 \circ \varepsilon_7 \mod 4(\nu_5 \circ E\sigma') .$$

<u>Proof</u>: Consider the secondary composition $\{\nu_5, 8\iota_8, \nu_8^2\}_3$ which is a coset of $\nu_5 \circ E^3\pi_{12}^5 + \pi_9^5 \circ \nu_9^2$ by (4.7). By Proposition 5.15 and Lemma 5.14, $\nu_5 \circ E^3\pi_{12}^5 = \{\nu_5 \circ E^3\sigma'''\} = \{4(\nu_5 \circ E\sigma')\}$. By Proposition 5.8 and (5.9), $\pi_9^5 \circ \nu_9^2 = \{\nu_5 \circ \eta_8 \circ \nu_9^2\} = 0$. Then the secondary composition is a coset of $\{4(\nu_5 \circ E\sigma')\}$. By Proposition 1.2 and by Lemma 5.13,

$$\sigma''' \circ \nu_{12} \in \{\nu_5, 8\iota_8, \nu_8\}_3 \circ \nu_{12} \subset \{\nu_5, 8\iota_8, \nu_8^2\}_3.$$

By (6.1) and Proposition 1.2,

$$\begin{aligned}\eta_5^2 \circ \varepsilon_7 &\in \eta_5^2 \circ \{\eta_7, 2\iota_8, \nu_8^2\}_3 \\ &\subset \{\eta_5^3, 2\iota_8, \nu_8^2\}_3 = \{4\nu_5, 2\iota_8, \nu_8^2\}_3 \\ &\subset \{\nu_5, 8\iota_8, \nu_8^2\}_3 .\end{aligned}$$

Therefore the elements $\sigma''' \circ \nu_{12}$ and $\eta_5^2 \circ \varepsilon_7$ are in the same coset of $4(\nu_5 \circ E\sigma')$. Then (6.2) is verified. q. e. d.

ii). <u>The elements</u> $\bar{\nu}_n$.

Consider a secondary composition $\{\nu_6, \eta_9, \nu_{10}\}_t$, $0 \leq t \leq 4$, which is defined by the relations of (5.9) and is a coset of $\pi_{11}^6 \circ \nu_{11} + \nu_6 \circ E^t\pi_{14-t}(S^{9-t})$ by (4.7). Since $\pi_{14}(S^9)$ is stable : $\pi_{14}(S^9) = E\pi_{13}(S^8)$, then $\nu_6 \circ E^t\pi_{14-t}(S^{9-t}) \subset \nu_6 \circ \pi_{14}(S^9) = \nu_6 \circ \pi_{14}^9 = 0$ by Proposition 5.9 and (4.7). We have also $\pi_{11}^6 \circ \nu_{11} = \{\Delta(\iota_{13}) \circ \nu_{11}\} = \{\Delta(\nu_{13})\}$ by Proposition 5.9.

Denote an element of $\{\nu_6, \eta_9, \nu_{10}\}_t$ by

$$\bar{\nu}_6 \in \{\nu_6, \eta_9, \nu_{10}\}_t \in \pi_{14}^6/\{\Delta(\nu_{13})\} ,$$

and denote that

$$\bar{\nu}_n = E^{n-6}\,\bar{\nu}_6 \quad \text{for } n \geq 6 \text{ and } \bar{\nu} = E^{\infty}\,\bar{\nu}_6 .$$

<u>Lemma</u> 6.2: $H(\bar{\nu}_6) \equiv \nu_{11} \mod 2\nu_{11}$, $\Delta(\nu_{13}) = \pm 2\bar{\nu}_6$, $8\bar{\nu}_6 = 0$ <u>and</u> $2\bar{\nu}_n = 2\bar{\nu} = 0$ <u>for</u> $n \geq 7$. <u>The secondary composition</u> $\{\nu_n, \eta_{n+3}, \nu_{n+4}\}_t$, $0 \leq t \leq n-2$, <u>consists of a single element</u> $\bar{\nu}_n$ <u>if</u> $n \geq 7$.

<u>Proof</u>: By Proposition 2.6 and by (5.10),

$$\nu_{11} = \iota_{11} \circ \nu_{11} \in \Delta^{-1}(\nu_5 \circ \eta_8) \circ \nu_{11} = H\{\nu_6, \eta_9, \nu_{10}\}_1 .$$

By Proposition 2.7, $H(\pm\ \Delta\ (\nu_{13})) = H(\pm\ \Delta(\iota_{13}) \circ \nu_{11}) = 2\iota_{11} \circ \nu_{11} = 2\nu_{11}$. It follows that $H(\bar{\nu}_6) \equiv \nu_{11} \mod 2\nu_{11}$.

By Proposition 1.4, (4.7) and by Proposition 5.9,

$$8\bar{\nu}_6 \in \{\nu_6, \eta_9, \nu_{10}\}_1 \circ 8\iota_{14} = \nu_6 \circ E\{\eta_8, \nu_9, 8\iota_{12}\}$$
$$\subset \nu_6 \circ E\pi_{13}(S^8) = \nu_6 \circ \pi_{14}^9 = 0.$$

It is easy to see that the secondary composition $\{\nu_6, \eta_9, 2\nu_{10}\}$ is a coset of $\{2\Delta(\nu_{13})\}$. By Proposition 1.2 and Lemma 5.10,

$$2\bar{\nu}_6 = \bar{\nu}_6 \circ 2\iota_{14} \in \{\nu_6, \eta_9, \nu_{10}\} \circ 2\iota_{14} \subset \{\nu_6, \eta_9, 2\nu_{10}\}$$

and

$$\Delta(\nu_{13}) = \Delta(\iota_{13}) \circ \nu_{11} \in \{\nu_6, \eta_9, 2\iota_{10}\} \circ \nu_{11} \subset \{\nu_6, \eta_9, 2\nu_{10}\} .$$

If follows that $\Delta(\nu_{13}) \equiv 2\bar{\nu}_6 \mod 2\Delta(\nu_{13})$. Then it is verified easily that $4\bar{\nu}_6 = 2\Delta(\nu_{13})$ and $2\bar{\nu}_6 = \Delta(\nu_{13})$ or $= -\Delta(\nu_{13})$, since $8\bar{\nu}_6 = 4\Delta(\nu_{13}) = \Delta(\eta_{13}^3) = \Delta(\eta_{13}) \circ \eta_{12}^2 = \Delta H(\sigma') \circ \eta_{12}^2 = 0$.

Since $E\,\Delta = 0$, then $2\bar{\nu}_7 = E(\pm\ \Delta(\nu_{13})) = 0$ and $2\bar{\nu}_n = 2\bar{\nu} = 0$ for $n \geq 7$. By Proposition 1.3, and (1.15),

$$\bar{\nu}_n = (-1)^n E^{n-6}\,\bar{\nu}_6 \in (-1)^n E^{n-6}\{\nu_6, \eta_9, \nu_{10}\}_4$$
$$\subset \{\nu_n, \eta_{n+1}, \nu_{n+4}\}_{n-2} \subset \{\nu_n, \eta_{n+1}, \nu_{n+4}\}_t$$

for $n \geq 7$ and $0 \leq t \leq n-2$. By (4.7) and by Proposition 5.9, the last secondary composition is a coset of $\pi^n_{n+5} \circ \nu_{n+5} + \nu_n \circ \pi^{n+3}_{n+8} = 0$. It follows that the secondary composition consists of a single element which has to be $\bar{\nu}_n$. q. e. d.

<u>Lemma</u> 6.3: <u>For</u> $n \geq 6$, <u>we have</u> $\bar{\nu}_n \circ \eta_{n+8} = \eta_n \circ \bar{\nu}_{n+1} = \nu^3_n$, <u>where</u> $\nu^3_n = \nu_n \circ \nu_{n+3} \circ \nu_{n+6}$.

<u>Proof</u>: By Proposition 1.4,

$$\bar{\nu}_n \circ \eta_{n+8} \in \{\nu_n, \eta_{n+3}, \nu_{n+4}\} \circ \eta_{n+8} = \nu_n \circ \{\eta_{n+3}, \nu_{n+4}, \eta_{n+7}\}$$

and $\eta_n \circ \bar{\nu}_{n+1} \in \eta_n \circ \{\nu_{n+1}, \eta_{n+4}, \nu_{n+5}\} = \{\eta_n, \nu_{n+1}, \eta_{n+4}\} \circ \nu_{n+6}$.
It follows from Lemma 5.12 that these secondary compositions consist of a single element ν^3_n. Then we have the equalities of the lemma.

<u>Lemma</u> 6.4: $\eta_9 \circ \sigma_{10} = \bar{\nu}_9 + \varepsilon_9$ <u>and for</u> $n \geq 10$, $\eta_n \circ \sigma_{n+1} = \sigma_n \circ \eta_{n+7} = \bar{\nu}_n + \varepsilon_n$.

<u>Proof</u>: By (5.9), $\eta_5 \circ \nu_6 = 0$. Then it follows from Lemma 5.16 that $\eta_9 \circ \sigma_{10} \in -x\{\nu_9, \eta_{12}, \nu_{13}\}_7 + x\{\eta_9, \nu_{10}, 2\nu_{13}\}_5$, where x is an odd integer. By Lemma 6.2, the first secondary composition consists of $\bar{\nu}_9$. The second secondary composition is a coset of $\eta_9 \circ E^5\pi^5_{12} + \pi^9_{14} \circ 2\nu_{14} = \{\eta_9 \circ E^5\sigma'''\} + 0 = \{8(\eta_9 \circ \sigma_{10})\} = 0$, by (4.7), Proposition 5.15, Proposition 5.9 and Lemma 5.14. Then it consists of a single element which is ε_9 by (6.1). Therefore we have that $\eta_9 \circ \sigma_{10} = -x\bar{\nu}_9 + x\bar{\varepsilon}_9 = \bar{\nu}_9 + \bar{\varepsilon}_9$. By Proposition 3.1, $\eta_2 \# \sigma_8 = \eta_{10} \circ \sigma_{11} = \sigma_{10} \circ \eta_{17}$. Then $\sigma_n \circ \eta_{n+7} = \eta_n \circ \sigma_{n+1} = \bar{\nu}_n + \bar{\varepsilon}_n$ for $n \geq 10$. q. e. d.

iii). <u>The elements</u> μ_n.

We shall give an element μ_3 of π^3_{12} such that $H(\mu_3) = \sigma'''$ by means of secondary compositions.

First consider a secondary composition

$$\{\eta_3, E\nu', 4\iota_7\}_1 \subset \pi_8(S^3)$$

which is a coset of $\eta_3 \circ E\pi^3_7 + \pi_8(S^3) \circ 4\iota_8$, by (4.7). Since the group $\pi_8(S^3)$ is finite and its 2-primary component π^3_8 is isomorphic to Z_2, by Proposition 5.8, then $\pi_8(S^3) \circ 4\iota_8 = 4\pi_8(S^3)$ coincides with the odd component of $\pi_8(S^3)$. Thus the above secondary composition contains an element of π_8^3. Since $2\pi_8^3 = 0$, it follows that

$$0 \in 2\{\eta_3, E\nu', 4\iota_7\}_1 = \{\eta_3, E\nu, 4\iota_7\}_1 \circ 2\iota_8$$
$$\subset \{\eta_3, E\nu', 8\iota_7\}_1, \qquad \text{by Proposition 1.2.}$$

If follows from Proposition 1.9 that there exists an extension $\beta \in \pi(EK \to S^3)$ of ν' such that $\eta_3 \circ E\beta = Ep_0{*}0 = 0$, where $K = S^5 \cup_\lambda CS^5$ for $\lambda = 8\iota_5$ and $p_0 : K \to S^6$ is a shrinking map given as in (1.17).

Next consider a cell complex $M = S^3 \cup_\beta CEK$ and its subcomplex $L = S^3 \cup_{\nu'} CS^6$. Let $i : L \to M$ and $i' : S^5 \to K$ be the injections and let $p : M \to E^2K$ and $p' : L \to S^7$ be shrinking maps given as in (1.17).

Since $\eta_3 \circ E\beta = 0$, then there exists extension $\alpha \in \pi(EM \to S^3)$ of η_3. Obviously, $(Ei){*}\alpha \in \pi(EL \to S^3)$ is an extension of η_3. We shall prove

(6.3) $H(\alpha) = Ep{*}\alpha'$ <u>for some</u> $\alpha' \in \pi(E^2K \to S^5)$, $H(Ei{*}\alpha) = Ei{*}H(\alpha) = (Ep'){*}((Ei')^*\alpha')$ <u>and</u> $(Ei')^*\alpha' = x\nu_5$ <u>for an odd integer</u> x.

<u>Proof</u>: $H(\alpha)$ is represented by a mapping of EM into S^5. Since any mapping of S^4 into S^5 is inessential, then $H(\alpha)$ is represented by a mapping which maps S^4 to e_0. It follows then $H(\alpha) = Ep{*}\alpha'$ for some α', since Ep is a mapping which shrinks $S^4 = ES^3$ of EM to a point e_0.

By Proposition 2.2,

$$H(Ei{*}\alpha) = H(\alpha \circ \{Ei\}) = H(\alpha) \circ \{Ei\} = Ei{*}H(\alpha)$$
$$= Ei{*}\, Ep{*}\alpha' = E(p \circ i)^* \alpha' = E(i' \circ p')^* \alpha$$
$$= (Ep')^*((Ei')^* \alpha') .$$

$(Ei')^* \alpha'$ is an element of $\pi_8(S^5)$ and α' is its extension. The existnece of an extension of $(Ei')^* \alpha'$ implies that $(Ei')^* \alpha' \circ 8\iota_8 = 8(Ei')^* \alpha' = 0$. Thus the element $(Ei')^* \alpha'$ is an element of $\pi_8^5 = \{\nu_5\}$. Let x be an integer such that $x\nu_5 = (Ei')^* \alpha'$.

In order to estimate the integer x, consider the secondary composition $\{\eta_3, E\nu', \nu_7\}_1$ which consists of the elements ε_3 only. It follows from Proposition 1.7 that $\varepsilon_3 = (Ei)^* \alpha \circ E\gamma'$ for the extension $(Ei)^* \alpha$ of η_3 and a coextension $\gamma' \in \pi_{10}(L)$ of ν_6.

Applying the homomorphism H to the last equation,

$$\nu_5^2 = H(\varepsilon_3) = H((Ei)^* \alpha \circ E\gamma') \qquad \text{by Lemma 6.1,}$$
$$= H(Ei^* \alpha) \circ E\gamma' = (Ep')^* x\nu_5 \circ E\gamma' \qquad \text{by Proposition 2.2,}$$

$$= x\nu_5 \circ Ep'_*E\gamma' = x\nu_5 \circ E(p'_* \gamma')$$

$$= x\nu_5 \circ E^2\nu_6 = x\nu_5^2 \qquad \text{by (1.18).}$$

Since ν_5^2 is an element of order 2, then we have that x is odd. Consequently (6.3) is proved. q. e. d.

Next consider a coextension $\gamma \in \pi_9(K)$ of ν_5, then it follows from Proposition 1.7 that $\beta \circ E\gamma$ is an element of $\{\nu', 8\iota_6, \nu_6\}_1$, which is a subset of $\pi_{10}^3 = 0$ by (4.7) and Proposition 5.15. Thus $\beta \circ E\gamma = 0$ and the secondary composition

$$\{\eta_3, E\beta, E^2\gamma\}_1 \in \pi_{12}(S^3)/(\eta_3 \circ E\pi_{11}(S^3) + \pi(E^3K \to S^3) \circ E^3\gamma)$$

is defined. By (4.7), $\eta_3 \circ E\pi_{11}(S^3) = \eta_3 \circ E\pi_{11}^3$. We shall prove

(6.4) $\pi(E^3K \to S^3) \circ E^3\gamma = 0. \quad \pi(E^4K \to S^4) \circ E^4\gamma = \{\nu_4^3\}$.

<u>Proof</u>: Consider an element λ of $\pi(E^3K \to S^3)$. λ is an extension of an element λ' of $\pi_8(S^3)$. It follows from Proposition 1.7, that $-(\lambda \circ E^3\gamma) \in \{\lambda', 8\iota_8, \nu_8\}_3$. Since $\lambda' \circ 8\iota_8 = 8\lambda' = 0$, then λ' is an element of $\pi_8^3 = \{\nu' \circ \eta_6^2\}$ (Proposition 5.9). Thus $\lambda' = y\nu' \circ \eta_6^2$ for $y = 0$ or 1. By Proposition 1.7, (4.7) and by Propositions 5.8 and 5.11, we have

$$\{y\nu' \circ \eta_6^2, 8\iota_8, \nu_8\}_3 \subset \{y\nu', \eta_6^2 \circ 8\iota_8, \nu_8\}_3$$
$$= \{y\nu', 0, \nu_8\}_3 = y\nu' \circ E^3\pi_9^3 + \pi_9^3 \circ \nu_9$$
$$= 0.$$

Therefore we have that $\lambda \circ E^3\gamma = 0$ and the first statement of (6.4) is established. The second statement is proved similarly, by using that $\pi_{10}^4 \circ \nu_{10} = \{\nu_4^2\} \circ \nu_{10} = \{\nu_4^3\}$. q. e. d.

Now we choose an element of this secondary composition and denote it by $\mu_3 \in \{\eta_3, E\beta, E^2\gamma\}_1 \in \pi_{12}(S^3)/(\eta_3 \circ E\pi_{11}^3)$.

We denote also that $\mu_n = E^{n-3}\mu_3$ for $n \geq 3$ and $\mu = E^\infty \mu_3$.

These elements μ_n have the following properties.

<u>Lemma</u> 6.5: $H(\mu_3) = \sigma'''$, $2\mu_n = 2\mu = 0$ <u>for</u> $n \geq 3$ <u>and</u>

$\mu_n \in \{\eta_n, 2\iota_{n+1}, E^{n-4}\sigma'''\}_{n-4} + \{\nu_n^3\}$ <u>for</u> $n \geq 4$.

<u>Proof</u>: It follows from Proposition 1.7 that $\mu_3 = \alpha \circ E\tilde{\gamma}$ for an extension α of η_3 satisfying (6.3) and for a coextension $\tilde{\gamma} \in \pi_{11}(M)$ of $E\gamma$. Then

$$H(\mu_3) = H(\alpha \circ E\tilde{\gamma}) = H(\alpha) \circ E\tilde{\gamma} \qquad \text{by Proposition 2.2,}$$
$$= Ep^*\alpha' \circ E\tilde{\gamma} = \alpha' \circ Ep_*E\tilde{\gamma} \qquad \text{by (6.3),}$$

$$= \alpha' \circ E(p_* \tilde{\gamma}) = \alpha' \circ E^3\gamma \qquad \text{by (1.18).}$$

By (6.3), $(Ei')^* \alpha' = x\nu_5$. This means that α' is an extension of $x\nu_5$. Since γ is a coextension of ν_5, it follows from Proposition 1.7, $\alpha' \circ E^3\gamma \in \{x\nu_5, 8\iota_8, \nu_8\}_3$.

By Lemma 5.13 and Proposition 1.2,

$$x\iota_5 \circ \sigma''' \in x\iota_5 \circ \{\nu_5, 8\iota_8, \nu_8\}_3 \subset \{x\nu_5, 8\iota_8, \nu_8\}_3 .$$

As is seen in the proof of Lemma 5.13, the last secondary composition consists of a single element. Thus $H(\mu_3) = x\iota_5 \circ \sigma'''$. Since $E : \pi^5_{12} \to \pi^6_{13}$ is an isomorphism into by (5.14) and since $E(x\iota_5 \circ \sigma''') = E(x\sigma''') = E\sigma'''$ by (2.1), (6.3) and by $2\sigma''' = 0$, then we have that $x\iota_5 \circ \sigma''' = \sigma'''$ and

$$H(\mu_3) = \sigma''' .$$

We have

$$\begin{aligned} 2\mu_3 = 2\iota_3 \circ \mu_3 &\in 2\iota_3 \circ \{\eta_3, E\beta, E^2\gamma\} && \text{by Lemma 4.5} \\ &= \{2\iota_3, \eta_3, E\beta\} \circ (-E^3\gamma) && \text{by Proposition 1.4} \\ &\subset \pi(E^3K \to S^3) \circ E^3\gamma = 0 && \text{by (6.4).} \end{aligned}$$

Thus $2\mu_n = 2\mu = 0$ for $n \geq 3$.

Next consider the element $\mu_4 = -\mu_4 \in -E\{\eta_3, E\beta, E^2\gamma\}_1 \subset \{\eta_4, E^2\beta, E^3\gamma\}$ (cf. Proposition 1.3 and (1.15)).

The element $E^2\beta \in \pi(E^3K \to S^5)$ is an extension of $E^3\nu' = 2\nu_5$, by Lemma 5.4. Since $\nu_5 \circ 8\iota_8 = 8\nu_5 = 0$, there exists an extension $\beta_1 \in \pi(E^3K \to S^5)$ of ν_5. Then the composition $2\iota_5 \circ \beta_1$ is an extension of $2\iota_5 \circ \nu_5 = 2\nu_5$, by (2.1). Then the difference $E^2\beta - 2\iota_5 \circ \beta_1$ is an extension of $2\nu_5 - 2\nu_5 = 0$. Thus there exists an element β_2 of $\pi_9(S^5)$ such that

$$E^2\beta - 2\iota_5 \circ \beta_1 = p_0^* \beta_2$$

for a shrinking map $p_0 : E^3K \to S^9$ of (1.17). We have

$$\text{(6.5)} \quad \{\eta_4, E^2\beta, E^3\gamma\} = \{\eta_4, 2\iota_5 \circ \beta_1, E^3\gamma\} + \{\eta_4, p_0^* \beta_2, E^3\gamma\} \qquad \text{by Proposition 1.6}$$

$$\supset \{\eta_4, 2\iota_5, \beta_1 \circ E^3\gamma\} + \{\eta_4, \beta_2, p_0 E^3\gamma\} \qquad \text{by Proposition 1.2.}$$

The element $\beta_1 \circ E^3\gamma$ is contained in $\{\nu_5, 8\iota_8, \nu_8\}_3$ by Proposition 1.7. It follows from Lemma 5.13 that

$$\beta_1 \circ E^3\gamma = \sigma''' .$$

By (1.18), $p_{0*} E^3\gamma = \nu_9$. Since $\pi_9(S^5)$ is finite, there exists an integer r such that $r \equiv 1 \pmod 8$ and $r\beta_2 \in \pi_9^5$. By Proposition 5.8, Proposition 5.9, and by (5.9), for the generator $\nu_5 \circ \eta_8$ of $\pi_9^5 \approx Z_2$, the composition $\eta_4 \circ \nu_5 \circ \eta_8 = E\nu' \circ \eta_7^2$ is not zero. Since $\eta_4 \circ r\beta_2 = 0$, it follows that $r\beta_2 = 0$. By Proposition 1.2,

$$\{\eta_4, \beta_2, r\nu_9\} \subset \{\eta_4, r\beta_2, \nu_9\} = \{\eta_4, 0, \nu_9\} \equiv 0 .$$

By (4.7) and Proposition 5.11, the secondary compositions $\{\eta_4, \beta_2, p_{0*}E^3\gamma\} = \{\eta_4, \beta_2, \nu_9\} = \{\eta_4, \beta_2, r\nu_9\}$ and $\{\eta_4, 0, \nu_9\}$ are cosets of the same subgroup $\eta_4 \circ \pi_{13}(S^5) + \pi_{10}^4 \circ \nu_{10} = \eta_4 \circ \pi_{13}(S^5) + \{\nu_4^3\}$. Thus $\{\eta_4, \beta_2, p_{0*}E^3\gamma\} = \eta_4 \circ \pi_{13}(S^5) + \{\nu_4^3\}$. Also it follows from (4.7) and (6.4) that $\{\eta_4, E^2\beta, E^3\gamma\}$ is a coset of $\eta_4 \circ \pi_{13}(S^5) + \{\nu_4^3\}$. Then the inclusion of (6.5) has to be the equality and

$$\mu_4 \in \{\eta_4, E^2\beta, E^3\gamma\} = \{\eta_4, 2\iota_5, \sigma''\} + \{\nu_4^3\} .$$

By use of Proposition 1.3,

$$\mu_n = (-1)^{n-4} \mu_4 \in (-1)^{n-4} E^{n-4} (\{\eta_4, 2\iota_5, \sigma''\} + \{\nu_4^3\})$$
$$\subset \{\eta_n, 2\iota_{n+1}, E^{n-4} \sigma''\}_{n-4} + \{\nu_n^3\} .$$

Then the Lemma 6.5 is proved. q. e. d.

iv). <u>The element</u> ε'.

By (4.7) and Proposition 5.11, the compositions $\nu' \circ \nu_6$ and $\nu' \circ 2\nu_6$, are in $\pi_9^3 = 0$. Thus $\nu' \circ 2\nu_6 = \nu' \circ \nu_6 = 0$. Since $E^3\nu' = 2\nu_6$, the secondary composition $\{\nu', 2\nu_6, \nu_9\}_3$ is defined and it is a coset of $\nu' \circ E^3\pi_{10}^3 + \pi_{10}^3 \circ \nu_{10} = 0$, by (4.7) and Proposition 5.15. Then this secondary composition consists of a single element which is denoted by

$$\varepsilon' = \{\nu', 2\nu_6, \nu_9\}_3 \in \pi_{13}(S^3).$$

<u>Lemma</u> 6.6: $H(\varepsilon') = \varepsilon_5$, $2\varepsilon' = \eta_3^2 \circ \varepsilon_5$ <u>and</u> $E^2\varepsilon'$ <u>is divisible by</u> 2.

<u>Proof</u>: By Proposition 2.3 and by (5.3),

$$H(\varepsilon') \in H\{\nu', 2\nu_6, \nu_9\}_3 \subset \{H(\nu'), 2\nu_6, \nu_9\}_3 = \{\eta_5, 2\nu_6, \nu_9\}_3 .$$

By (6.1), $\varepsilon_5 \in \{\eta_5, 2\nu_6, \nu_9\}_3$.

By (4.7), Proposition 5.15, Proposition 5.9 and by (5.9) we have $\{\eta_5, 2\nu_6, \nu_9\}_3$ is a coset of

$$\eta_5 \circ E^3\pi_{10}^3 + \pi_{10}^5 \circ \nu_{10} = 0 + \{\nu_5 \circ \eta_8^2 \circ \nu_{10}\} = 0 .$$

Thus the coset consists of a single element which is $H(\varepsilon') = \varepsilon_5$.

Similarly, we have that $\{\eta_3^3, 2\nu_6, \nu_9\}_3$ consists of a single element. Then the relation $2\varepsilon' = \eta_3^2 \circ \varepsilon_5$ follows from the following two relations.

$$\begin{aligned} 2\varepsilon' &= 2\iota_3 \circ \varepsilon' \in 2\iota_3 \circ \{\nu', 2\nu_6, \nu_9\}_3 && \text{by Lemma 4.5,} \\ &\subset \{2\iota_3 \circ \nu', 2\nu_6, \nu_9\}_3 && \text{by Proposition 1.2,} \\ &= \{2\nu', 2\nu_6, \nu_9\}_3 = \{\eta_3^3, 2\nu_6, \nu_9\}_3 && \text{by Lemma 4.5 and (5.3),} \end{aligned}$$

and

$$\begin{aligned} \eta_3^2 \circ \varepsilon_5 &\in \eta_3^2 \circ \{\eta_5, 2\nu_6, \nu_9\}_3 && \text{by (6.1),} \\ &\subset \{\eta_3^3, 2\nu_6, \nu_9\}_3 && \text{by Proposition 1.2.} \end{aligned}$$

Next consider the secondary compositions $\{\nu_5, 2\nu_8, \nu_{11}\}_5$ and $\{\nu_5, 4\nu_8, \nu_{11}\}_5$, which are cosets of $\nu_5 \circ E^5\pi_{10}^3 + \pi_{12}^5 \circ \nu_{12} = \{\sigma''' \circ \nu_{12}\}$, by (4.7) and Proposition 5.15. By use of Proposition 1.2, Proposition 1.3 and Lemma 5.4,

$$\begin{aligned} E^2\varepsilon' &\in E^2\{\nu', 2\nu_6, \nu_9\}_3 \subset \{E^2\nu', 2\nu_8, \nu_{11}\}_5 \\ &= \{2\nu_5, 2\nu_8, \nu_{11}\}_5 \subset \{\nu_5, 4\nu_8, \nu_{11}\}_5 \end{aligned}$$

and

$$\begin{aligned} 2\{\nu_5, 2\iota_8, \nu_{11}\}_5 &= \{\nu_5, 8\iota_8, \nu_{11}\}_5 \circ 2\iota_{15} \\ &\subset \{\nu_5, 2\nu_8, 2\nu_{11}\}_5 \subset \{\nu_5, 2\nu_8, \nu_{11}\}_5. \end{aligned}$$

It follows that $2\alpha = E^2\varepsilon' + x\sigma''' \circ \nu_{12}$ for an element α of $\{\nu_5, 2\nu_8, \nu_{11}\}_5$ and an integer x. By (6.2), $\sigma''' \circ \nu_{12} = \eta_5^2 \circ \varepsilon_7 + 4y\,\nu_5 \circ E\sigma' = 2(E^2\varepsilon' + 2y\,\nu_5 \circ E\sigma')$ for an integer y. Thus we have that for any α of $\{\nu_5, 2\nu_8, \nu_{11}\}_5$,

(6.6) $$E^2\varepsilon' = 2(\alpha + x\,E^2\varepsilon' + 2xy\,\nu_5 \circ E\sigma') .$$

Then Lemma 6.6 is proved. q. e. d.

v). <u>The elements</u> ζ_n.

Consider the secondary composition $\{\nu_5, 8\iota_8, E\sigma'\}_1 \subset \pi_{16}^5$. $\nu_5 \circ 8\iota_8 = 8\nu_5 = 0$. $E(8\iota_8 \circ \sigma') = 8E\sigma' = 0$ by (2.1). It follows from Lemma 4.5 that $8\iota_7 \circ \sigma' = 0$. Then the above secondary composition is defined and it is a coset of $\nu_5 \circ E\pi_{15}^7 + \pi_9^5 \circ E^2\sigma'$, by (4.7). By Proposition 5.8 and Lemma 5.14, $\pi_9^5 \circ E^2\sigma' = \{\nu_5 \circ \eta_8 \circ 2\sigma_9\} = \{\eta_5 \circ 2\eta_8 \circ \sigma_9\} = 0$. Choose an element of this secondary composition and denote it by

$$\zeta_5 \in \{\nu_5, 8\iota_8, E\sigma'\}_1 \quad \in \pi_{16}^5/(\nu_5 \circ E\pi_{15}^7).$$

We denote also that

$$\zeta_n = E^{n-5}\,\zeta_5 \quad \text{for } n \geq 5 \quad \text{and} \quad \zeta = E^\infty\zeta_5 .$$

<u>Lemma</u> 6.7: $H(\zeta_5) = 8\sigma_9$ <u>and</u> $4\zeta_5 \equiv \eta_5^2 \circ \mu_7 \bmod \nu_5 \circ 2E\pi_{15}^7$.

$8\zeta_n = 8\zeta = 0$ for $n \geq 5$ provided if $4E\pi_{15}^7 = 0$.

Proof: By Proposition 2.4,

$$H(\zeta_5) \in H\{\nu_5, 8\iota_8, E\sigma'\}_1 = \Delta^{-1}(\nu_4 \circ 8\iota_7) \circ E^2\sigma'.$$

By (5.8), $\pm\Delta(4\iota_9) = 4(2\nu_4 - \nu') = 8\nu_4 = \nu_4 \circ 8\iota_7$. Thus $H\{\nu_5, 8\iota_8, E\sigma'\}_1$ contains $\pm 4\iota_9 \circ E^2\sigma' = \pm 4(2\sigma_9) = 8\sigma_9$, by Lemma 5.13. Since $H\{\nu_5, 8\iota_8, E\sigma'\}_1$ is a coset of $H(\nu_5 \circ E\pi_{15}^7) = HE(\nu_4 \circ \pi_{15}^7) = 0$, then we have proved that $H(\zeta_5) = 8\sigma_9$.

Next, we see that the secondary composition $\{\nu_5, 8\iota_8, E^2\sigma''\}_1$ is a coset of $\nu_5 \circ E\pi_{15}^7$ and the secondary composition $\{2\nu_5, 4\iota_8, E^3\sigma'''\}_1$ is a coset of $2\nu_5 \circ E\pi_{15}^7 = \nu_5 \circ 2E\pi_{15}^7$, by a similar way for $\{\nu_5, 8\iota_8, E\sigma'\}_1$. We have by Proposition 1.2,

$$2\zeta_5 \in \{\nu_5, 8\iota_8, E\sigma'\}_1 \circ 2\iota_{16} \subset \{\nu_5, 8\iota_8, 2E\sigma'\}_1$$
$$= \{\nu_5, 8\iota_8, E^2\sigma''\}_1$$

and

$$\{2\nu_5, 4\iota_8, E^2\sigma''\}_1 \subset \{\nu_5, 8\iota_8, E^2\sigma''\}_1 .$$

It follows that

$$2\zeta_5 \in \{2\nu_5, 4\iota_8, E^2\sigma''\}_1 + \nu_5 \circ E\pi_{15}^7$$

and

$$4\zeta_5 \in \{2\nu_5, 4\iota_8, E^2\sigma''\}_1 \circ 2\iota_{16} + 2(\nu_5 \circ E\pi_{15}^7)$$
$$\subset \{2\nu_5, 4\iota_8, 2E^2\sigma''\}_1 + 2\nu_5 \circ E\pi_{15}^7$$
$$= \{2\nu_5, 4\iota_8, E^3\sigma'''\}_1.$$

By Lemma 6.5, Proposition 1.2 and by (5.9),

$$\eta_5^2 \circ \mu_7 \in \eta_5^2 \circ \{\eta_7, 2\iota_8, E^3\sigma'''\}_3 + \{\eta_5^2 \circ \nu_7^3\}$$
$$\subset \{\eta_5^3, 2\iota_8, E^3\sigma'''\}_1 + \{0\}$$
$$= \{2\nu_5 \circ 2\iota_8, 2\iota_8, E^3\sigma'''\}_1$$
$$\subset \{2\nu_5, 4\iota_8, E^3\sigma'''\}_1.$$

It follows that

$$4\zeta_5 \equiv \eta_5^2 \circ \mu_7 \mod \nu_5 \circ 2E\pi_{15}^7.$$

If $4E\pi_{15}^7 = 0$, then $8\zeta_5 = 2\eta_5^2 \circ \mu_7 = 0$, and thus $8\zeta_n = 8\zeta = 0$ for $n \geq 5$. q. e. d.

CHAPTER VII

2-Primary Components of $\pi_{n+k}(S^n)$ for $8 \leq k \leq 13$.

i). The groups π^n_{n+8} and π^n_{n+9}.

The results for π^n_{n+8} and π^n_{n+9} are stated as follows.

Theorem 7.1: $\pi^2_{10} = 0$,

$\pi^n_{n+8} = \{\varepsilon_n\} \approx Z_2$ for $n = 3, 4, 5$,

$\pi^6_{14} = \{\bar{\nu}_6\} \oplus \{\varepsilon_6\} \approx Z_8 \oplus Z_2$,

$\pi^7_{15} = \{\sigma' \circ \eta_{14}\} \oplus \{\bar{\nu}_7\} + \{\varepsilon_7\} \approx Z_2 \oplus Z_2 \oplus Z_2$,

$\pi^8_{16} = \{\sigma_8 \circ \eta_{15}\} \oplus \{E\sigma' \circ \eta_{15}\} \oplus \{\bar{\nu}_8\} \oplus \{\varepsilon_8\} \approx Z_2 \oplus Z_2 \oplus Z_2 \oplus Z_2$,

$\pi^9_{17} = \{\sigma_9 \circ \eta_{16}\} \oplus \{\bar{\nu}_9\} \oplus \{\varepsilon_9\} \approx Z_2 \oplus Z_2 \oplus Z_2$,

$\pi^n_{n+8} = \{\bar{\nu}_n\} \oplus \{\varepsilon_n\} \approx Z_2 \oplus Z_2$, for $n \geq 10$,

$(G_8;2) = \{\bar{\nu}\} \oplus \{\varepsilon\} \approx Z_2 \oplus Z_2$.

Theorem 7.2: $\pi^2_{11} = \{\eta_2 \circ \varepsilon_3\} \approx Z_2$,

$\pi^3_{12} = \{\mu_3\} \oplus \{\eta_3 \circ \varepsilon_4\} \approx Z_2 \oplus Z_2$,

$\pi^n_{n+9} = \{\nu^3_n\} \oplus \{\mu_n\} \oplus \{\eta_n \circ \varepsilon_{n+1}\} \approx Z_2 \oplus Z_2 \oplus Z_2$ for $n = 4,5,6$,

$\pi^7_{16} = \{\sigma' \circ \eta^2_{14}\} \oplus \{\nu^3_7\} \oplus \{\mu_7\} \oplus \{\eta_7 \circ \varepsilon_8\} \approx Z_2 \oplus Z_2 \oplus Z_2 \oplus Z_2$,

$\pi^8_{17} = \{\sigma_8 \circ \eta^2_{15}\} \oplus \{E\sigma' \circ \eta^2_{15}\} \oplus \{\nu^3_8\} \oplus \{\mu_8\} \oplus \{\eta_8 \circ \varepsilon_9\}$

$\approx Z_2 \oplus Z_2 \oplus Z_2 \oplus Z_2 \oplus Z_2$,

$\pi^9_{18} = \{\sigma_9 \circ \eta^2_{16}\} \oplus \{\nu^3_9\} \oplus \{\mu_9\} \oplus \{\eta_9 \circ \varepsilon_{10}\} \approx Z_2 \oplus Z_2 \oplus Z_2 \oplus Z_2$,

$\pi^{10}_{19} = \{\Delta(\iota_{21})\} \oplus \{\nu^3_{10}\} \oplus \{\mu_{10}\} \oplus \{\eta_{10} \circ \varepsilon_{11}\} \approx Z \oplus Z_2 \oplus Z_2 \oplus Z_2$,

$\pi^n_{n+9} = \{\nu^3_n\} \oplus \{\mu_n\} \oplus \{\eta_n \circ \varepsilon_{n+1}\} \approx Z_2 \oplus Z_2 \oplus Z_2$, for $n \geq 11$

$(G_9;2) = \{\nu^3\} \oplus \{\mu\} \oplus \{\eta \circ \varepsilon\} \approx Z_2 \oplus Z_2 \oplus Z_2$.

First we have $\pi^2_{10} = 0$ by (5.2) and Proposition 5.15. In the exact sequence $0 = \pi^2_{10} \longrightarrow \pi^3_{11} \xrightarrow{H} \pi^5_{11}$ of (4.4), $H(\varepsilon_3) = \nu^2_5$ by Lemma 6.1 and this image generates π^5_{11} by Proposition 5.11. It follows then

$$\pi^3_{11} = \{\varepsilon_3\} \approx Z_2, \text{ and thus } \pi^2_{11} = \{\eta_2 \circ \varepsilon_3\} \approx Z_2 \text{ by (5.2).}$$

By (5.6) and Proposition 5.9, $\pi_{12}^4 = \{\varepsilon_4\} \approx Z_2$.

By Propositions 5.8 and 5.9, we have an exact sequence

$$0 = \pi_{14}^9 \longrightarrow \pi_{12}^4 \xrightarrow{E} \pi_{13}^5 \longrightarrow \pi_{13}^9 = 0$$

from (4.4). Then E is an isomorphism and $\pi_{13}^5 = \{\varepsilon_5\} \approx Z_2$.

Next consider the exact sequence

$$\pi_{13}^5 \xrightarrow{\Delta} \pi_{11}^2 \xrightarrow{E} \pi_{12}^3 \xrightarrow{H} \pi_{12}^5 \longrightarrow \pi_{10}^2 = 0$$

of (4.4). By Lemma 5.2, $\Delta\pi_{13}^5 = \{\Delta(\varepsilon_5)\} = \{\Delta(E^2\varepsilon_3)\} = 0$ since $2\varepsilon_3 = 0$. It follows that E is an isomorphism into. By Proposition 5.15 and by Lemma 6.5, we have that H is onto and that

$$\pi_{12}^3 = \{\mu_3\} \oplus \{\eta_3 \circ \varepsilon_4\} \approx Z_2 \oplus Z_2 .$$

By (5.6) and Proposition 5.11,

$$\pi_{13}^4 = \{\nu_4^3\} \oplus \{\mu_4\} \oplus \{\eta_4 \circ \varepsilon_5\} \approx Z_2 \oplus Z_2 \oplus Z_2 .$$

By Proposition 2.2, Proposition 3.1 and by Lemma 5.14,

$$H(\nu_5 \circ \sigma_8) = E(\nu_4 \# \nu_4) \circ H(\sigma_8) = E(\nu_8 \circ \nu_{11}) \circ \iota_{15} = \nu_9^2 .$$

Then in the exact sequence

$$\pi_{15}^5 \xrightarrow{H} \pi_{15}^9 \xrightarrow{\Delta} \pi_{13}^4 \xrightarrow{E} \pi_{14}^5 \longrightarrow \pi_{14}^9$$

H is onto by Proposition 5.11. It follows that E is an isomorphism into. By Proposition 5.9, $\pi_{14}^9 = 0$. Then E is onto and

$$\pi_{14}^5 = \{\nu_5^3\} \oplus \{\mu_5\} \oplus \{\eta_5 \circ \varepsilon_6\} \approx Z_2 \oplus Z_2 \oplus Z_2$$

By Propositions 5.8 and 5.9, it follows from (4.4) the following two exact sequences:

$$0 \longrightarrow \pi_{14}^5 \xrightarrow{E} \pi_{15}^6 \longrightarrow 0 ,$$

$$0 \longrightarrow \pi_{13}^5 \xrightarrow{E} \pi_{14}^6 \xrightarrow{H} \pi_{14}^{11} .$$

From the first sequence, we have the result for π_{15}^6 in Theorem 7.2. By Lemma 6.2 and Proposition 5.6, we have that H is onto and π_{14}^6 has 16 elements and it is generated by $\bar{\nu}_6$ and ε_6. Since $8\bar{\nu}_6 = 0$, we have

$$\pi_{14}^6 = \{\bar{\nu}_6\} \oplus \{\varepsilon_6\} \approx Z_8 \oplus Z_2 .$$

By Proposition 5.8, $\pi_{17}^{13} = 0$. Then we have the exact sequence

$$0 = \pi_{17}^{13} \longrightarrow \pi_{15}^6 \xrightarrow{E} \pi_{16}^7 \xrightarrow{H} \pi_{16}^{13} \xrightarrow{\Delta} \pi_{14}^6 \xrightarrow{E} \pi_{15}^7 \xrightarrow{H} \pi_{15}^{13}$$

of (4.4) By Lemma 5.14, and Proposition 2.2,

$$H(\sigma' \circ \eta_{14}) = H(\sigma') \circ \eta_{14} = \eta_{13}^2$$

and

$$H(\sigma' \circ \eta_{14}^2) = H(\sigma') \circ \eta_{14}^2 = \eta_{13}^3 = 4\nu_{13} .$$

Since η_{13}^2 generates $\pi_{15}^{13} \approx Z_2$ and since $2(\sigma' \circ \eta_{14}) = \sigma' \circ 2\eta_{14} = 0$, then we have that

$$\pi_{15}^7 = \{\sigma' \circ \eta_{14}\} \oplus E\pi_{14}^6 \approx Z_2 \oplus \pi_{14}^6/\Delta\pi_{16}^{13} .$$

By Lemma 6.2, and Proposition 5.6, $\Delta\pi_{16}^{13}$ is generated by $2\bar{\nu}_6$. It follows that $E\pi_{14}^6 = \{\bar{\nu}_7\} \oplus \{\varepsilon_7\} \approx Z_2 \oplus Z_2$. Then the statement for π_{15}^7 in Theorem 7.1 is proved. The kernel of $\Delta : \pi_{16}^{13} \to \pi_{14}^6$ is generated by $4\nu_{13} = H(\sigma' \circ \eta_{14}^2)$. Then it follows from the exactness of the above sequence and from $2(\sigma' \circ \eta_{14}^2) = 0$ that

$$\pi_{16}^7 = \{\sigma' \circ \eta_{14}^2\} \oplus E\pi_{15}^6 ,$$

where E is an isomorphism into. Thus we have the statement for π_{16}^7 in Theorem 7.2.

The results for π_{16}^8 and π_{17}^8 are verified directly from the results for π_{15}^7 and π_{16}^7 and from Propositions 5.1 and 5.3, by use of (5.15).

Next consider the exact sequence

$$\pi_{19}^{17} \xrightarrow{\Delta} \pi_{17}^8 \xrightarrow{E} \pi_{18}^9 \xrightarrow{H} \pi_{18}^{17} \xrightarrow{\Delta} \pi_{16}^8 \xrightarrow{E} \pi_{17}^9 \xrightarrow{H} \pi_{17}^{17} \xrightarrow{\Delta} \pi_{15}^8$$

of (4.4). By Proposition 5.15 and (5.16), we have that $\Delta : \pi_{17}^{17} \to \pi_{15}^8$ is an isomorphism into. Then it follows that $E : \pi_{16}^8 \to \pi_{17}^9$ is onto. It is verified easily from (5.16) and Proposition 2.5 that

$$\Delta(\eta_{17}) = E\sigma' \circ \eta_{15} \quad \text{and} \quad \Delta(\eta_{17}^2) = E\sigma' \circ \eta_{15}^2 .$$

Thus the first two Δ of the above sequence are isomorphisms into. It follows from the exactness of the sequence that

$$\pi_{17}^9 = E\pi_{16}^8 \approx \pi_{16}^8/\{E\sigma' \circ \eta_{14}\}$$

and

$$\pi_{18}^9 = E\pi_{17}^8 \approx \pi_{17}^8/\{E\sigma' \circ \eta_{14}^2\} .$$

Then we have the results for π_{17}^9 and π_{18}^9.

Consider the exact sequence

$$\pi_{20}^{19} \xrightarrow{\Delta} \pi_{18}^9 \xrightarrow{E} \pi_{19}^{10} \xrightarrow{H} \pi_{19}^{19} \xrightarrow{\Delta} \pi_{17}^9 \xrightarrow{E} \pi_{18}^{10} \to \pi_{18}^{19} = 0$$

of (4.4). We have obtained that $\sigma_9 \circ \eta_{16}$, $\bar{\nu}_9$ and ε_9 are independent generators of π_{17}^9. Thus $\sigma_9 \circ \eta_{16} + \bar{\nu}_9 + \varepsilon_9 \neq 0$. By Lemma 6.4, $E(\sigma_9 \circ \eta_{16} + \bar{\nu}_9 + \varepsilon_9) = 0$. Since $\Delta\pi_{19}^{19}$ is a cyclic subgroup and since any non-trivial subgroup of π_{17}^9 is of order 2, then it follows from the exactness of the sequence

(7.1). $$\Delta(\iota_{19}) = \sigma_9 \circ \eta_{16} + \bar{\nu}_9 + \varepsilon_9 .$$

Thus $\pi_{18}^{10} = E\pi_{17}^{9} = \{\bar{\nu}_{10}\} \oplus \{\varepsilon_{10}\} \approx Z_2 \oplus Z_2$.

By Proposition 2.5, (7.1), Proposition 3.1 and by Lemma 6.3,

$$\Delta(\eta_{19}) = \Delta(\iota_{19}) \circ \eta_{17} = \sigma_9 \circ \eta_{16}^2 + \bar{\nu}_9 \circ \eta_{17} + \varepsilon_9 \circ \eta_{17}$$
$$= \sigma_9 \circ \eta_{16}^2 + \nu_9^3 + \eta_9 \circ \varepsilon_{10} .$$

Thus $E\pi_{18}^{9} = \{\nu_{10}^3\} \oplus \{\mu_{10}\} \oplus \{\eta_{10} \circ \varepsilon_{11}\} \approx Z_2 \oplus Z_2 \oplus Z_2$. By (7.1), the kernel of $\Delta : \pi_{19}^{19} \to \pi_{17}^{9}$ is generated by $2\iota_{19}$ which is $H(\pm \Delta(\iota_{21}))$ by Proposition 2.7. Then we have the results for π_{19}^{10}. We have also the result for π_{20}^{11} from the exactness of the sequence

$$\pi_{21}^{21} \xrightarrow{\Delta} \pi_{19}^{10} \xrightarrow{E} \pi_{20}^{11} \to \pi_{20}^{21} = 0 .$$

Finally the last two resutls of Theorem 7.1 and those of Theorem 7.2 are the consequences from (4.5) and the results of π_{18}^{10} and π_{20}^{11} . Then the proof of Theorem 7.1 and Theorem 7.2 is established.

In the above discussion we see that $\Delta : \pi_{n+11}^{2n+1} \to \pi_{n+9}^{n}$ are isomorphsims into for $n = 8$, 9 and $n = 10$. It follows from the exactness of the sequence (4.4) that

(7.2). $E : \pi_{n+10}^{n} \to \pi_{n+11}^{n+1}$ <u>are homomorphisms onto for</u> $n = 8, 9, 10$.

Next we shall verify the following relation in compositions.

(7.3). $$\eta_5 \circ \bar{\nu}_6 = \nu_5^3 \quad \text{(cf. Lemma 6.3).}$$

(7.4). $$\eta_7 \circ \sigma_8 = \sigma' \circ \eta_{14} + \bar{\nu}_7 + \varepsilon_7 ,$$

$$\eta_6 \circ \sigma' = \sigma'' \circ \eta_{13} = 4\bar{\nu}_6 ,$$

<u>and</u> $$\eta_5 \circ \sigma'' = \sigma''' \circ \eta_{12} = \eta_4 \circ \sigma''' = 0 .$$

(7.5). $$\varepsilon_n \circ \eta_{n+8} = \eta_n \circ \varepsilon_{n+1} \quad \underline{\text{for}} \quad n \geq 3 .$$

First remark that these compositions are in the 2-primary components, by (4.9).

The first relation (7.3) is proved by the relation $\eta_6 \circ \bar{\nu}_7 = \nu_6^3$ of Lemma 6.3 and the fact that $E : \pi_{14}^{5} \to \pi_{15}^{6}$ is an isomorphism onto.

The last relation (7.5) is true for $n \geq 5$, in virtue of Proposition 3.1. Then the relation (7.5) for $n = 3$ or 4 are proved by the isomorphisms $E : \pi_{11}^{3} \to \pi_{12}^{4}$ and $E : \pi_{12}^{4} \to \pi_{13}^{5}$.

By Lemma 6.4, $E^2(\eta_7 \circ \sigma_8) = E^2(\bar{\nu}_7 + \varepsilon_7)$. Since the kernel of $E^2 : \pi_{15}^{7} \to \pi_{17}^{9}$ is generated by $\sigma' \circ \eta_{14}$, then

$$\eta_7 \circ \sigma_8 = x\sigma' \circ \eta_{14} + \bar{\nu}_7 + \varepsilon_7$$

for $x = 0$ or 1. Apply the homomorphism H to the both sides of this equation, then $H(\eta_7 \circ \sigma_8) = E(\eta_6 \# \eta_6) \circ H(\sigma_8) = E\eta_{12}^2 \circ \iota_{15} = \eta_{13}^2$ and

$H(x\sigma' \circ \eta_{14} + \bar{\nu}_7 + \varepsilon_7) = xH(\sigma') \circ \eta_{14} + HE(\bar{\nu}_6 + \varepsilon_6) = x\eta_{13} \circ \eta_{14} = x\eta_{13}^2$, by Proposition 2.2, Proposition 3.1 and Lemma 5.14. Since $\eta_{13}^2 \neq 0$, then $x = 1$. Thus the first relation of (7.4) is proved.

By use of Lemma 5.14, it is verified that the compositions $\eta_6 \circ \sigma'$ and $\sigma'' \circ \eta_{13}$ are in the kernel of $E^2 : \pi_{14}^6 \to \pi_{16}^8$ which is generated by $2\bar{\nu}_6$. By Lemma 5.14 and Proposition 2.2, we have that $H(\eta_6 \circ \sigma') = H(\sigma'' \circ \eta_{13}) = \eta_{11}^3 = 4\nu_{11} = H(4\bar{\nu}_6)$. Then the second relation of (7.4) is proved. The last relation of (7.4) is proved by use of the isomorphisms $E : \pi_{12}^4 \to \pi_{13}^5$ and $E : \pi_{13}^5 \to \pi_{14}^6$ and by checking the relation $E(\eta_5 \circ \sigma'') = E(\sigma''' \circ \eta_{12}) = E(\eta_4 \circ \sigma''') = 0$ from Lemma 5.14.

Next we shall prove

(7.6). <u>The secondary compositions</u> $\{\nu', \nu_6, \eta_9\}$ <u>and</u> $\{E\nu', \nu_7, \eta_{10}\}_1$ <u>consists of</u> ε_3 <u>and</u> ε_4 <u>respectively. The following secondary compositions contains</u> ε_n : $\{2\nu_n, \nu_{n+3}, \eta_{n+6}\}_t$, $\{\nu_n, 2\nu_{n+3}, \eta_{n+6}\}_t$, $\{\nu_n^2, 2\iota_{n+6}, \eta_{n+6}\}_t$, $\{2\iota_n, \nu_n^2, \eta_{n+6}\}_t$, <u>where</u> $n \geq 5$ <u>and</u> t <u>are appropriate integers such that the secondary compositions are defined</u>.

<u>Proof</u>: First consider $\{E\nu', \nu_7, \eta_{10}\}_1$, which is a coset of $E\nu' \circ E\pi_{11}^6 + \pi_{11}^4 \circ \eta_{11} = 0$ by (4.7), Proposition 5.9 and Proposition 5.15. Thus this secondary composition consists of a single element of $\pi_{12}^4 = \{\varepsilon_4\} \approx Z_2$, which is $x\varepsilon_4$ for $x = 0$ or 1. Consider $E^\infty : \pi_{12}^4 \to (G_8;2)$, then $E^\infty x\varepsilon_4 = x\varepsilon$ is an element of $\langle 2\nu, \nu, \eta \rangle$ which is a coset of $G_7 \circ \eta = \{\sigma \circ \eta\} = \{\bar{\nu} + \varepsilon\}$. By i) of (3.9), $x\varepsilon$ belongs to $\langle \eta, \nu, 2\nu \rangle = E^\infty\{\eta_n, \nu_{n+1}, 2\nu_{n+4}\}$ which contains $E^\infty\varepsilon_n = \varepsilon$ by (6.1). Thus

$$x\varepsilon \equiv \varepsilon \mod \bar{\nu} + \varepsilon .$$

This implies that $x = 1$ and that $\{E\nu', \nu_7, \eta_{10}\}_1$ consists of a single element ε_4. By Proposition 1.3,

$$E\{\nu', \nu_6, \eta_9\} \subset \{E\nu', \nu_7, \eta_{10}\}_1 .$$

Since $E : \pi_{11}(S^3) \to \pi_{12}(S^4)$ is an isomorphism into by Lemma 4.5, then it follows that $\{\nu', \nu_6, \eta_9\}$ consists of a single element $E^{-1}\varepsilon_4 = \varepsilon_3$. The remaining part of (7.6) is proved by similar methods to (6.1).

ii). <u>The groups</u> π_{n+10}^n <u>and</u> π_{n+11}^n.

Apply Lemma 5.2 for the element $\mu_3 \in \pi_{12}^3$, then we have an element $\mu' \in \{\eta_3, 2\iota_4, \mu_4\}_1$ such that $H(\mu') = E^2\mu_3 = \mu_5$ and

$2\mu' = \eta_3 \circ \mu_4 \circ \eta_{13}$. By Proposition 3.1, $\eta_2 \# \mu_3 = \eta_5 \circ \mu_6 = \mu_5 \circ \eta_{14}$ and $E(2\mu') = \eta_4 \circ \mu_5 \circ \eta_{14} = \eta_4^2 \circ \mu_6 = E(\eta_3^2 \circ \mu_5)$. It follows from Lemma 4.5 that $2\mu' = \eta_3^2 \circ \mu_5$. We have obtained

(7.7). $$H(\mu') = \mu_5 \quad \underline{\text{and}} \quad 2\mu' = \eta_3^2 \circ \mu_5 .$$

Then the results for the groups π_{n+10}^n and π_{n+11}^n are stated as follows.

<u>Theorem</u> 7.3: $\pi_{12}^2 = \{\eta_2^2 \circ \varepsilon_4\} \oplus \{\eta_2 \circ \mu_3\} \approx Z_2 \oplus Z_2$,

$\pi_{13}^3 = \{\varepsilon'\} \oplus \{\eta_3 \circ \mu_4\} \approx Z_4 \oplus Z_2$,

$\pi_{14}^4 = \{\nu_4 \circ \sigma'\} \oplus \{E\varepsilon'\} \oplus \{\eta_4 \circ \mu_5\} \approx Z_8 \oplus Z_4 \oplus Z_2$,

$\pi_{n+10}^n = \{\nu_n \circ \sigma_{n+3}\} \oplus \{\eta_n \circ \mu_{n+1}\} \approx Z_8 \oplus Z_2$ <u>for</u> $n = 5,6$ <u>and</u> 7

$\pi_{18}^8 = \{\sigma_8 \circ \nu_{15}\} \oplus \{\nu_8 \circ \sigma_{11}\} \oplus \{\eta_8 \circ \mu_9\} \approx Z_8 \oplus Z_8 \oplus Z_2$,

$\pi_{19}^9 = \{\sigma_9 \circ \nu_{16}\} \oplus \{\eta_9 \circ \mu_{10}\} \approx Z_8 \oplus Z_2$,

$\pi_{20}^{10} = \{\sigma_{10} \circ \nu_{17}\} \oplus \{\eta_{10} \circ \mu_{11}\} \approx Z_4 \oplus Z_2$,

$\pi_{21}^{11} = \{\sigma_{11} \circ \nu_{18}\} \oplus \{\eta_{11} \circ \mu_{12}\} \approx Z_2 \oplus Z_2$,

$\pi_{n+10}^n = \{\eta_n \circ \mu_{n+1}\} \approx Z_2$, <u>for</u> $n \geq 12$,

$(G_{10};2) = \{\eta \circ \mu\} \approx Z_2$.

<u>Theorem</u> 7.4: $\pi_{13}^2 = \{\eta_2 \circ \varepsilon'\} \oplus \{\eta_2^2 \circ \mu_4\} \approx Z_4 \oplus Z_2$,

$\pi_{14}^3 = \{\mu'\} \oplus \{\varepsilon_3 \circ \nu_{11}\} \oplus \{\nu' \circ \varepsilon_6\} \approx Z_4 \oplus Z_2 \oplus Z_2$,

$\pi_{15}^4 = \{\nu_4 \circ \sigma' \circ \eta_{14}\} \oplus \{\nu_4 \circ \bar{\nu}_7\} \oplus \{\nu_4 \circ \varepsilon_7\} \oplus \{E\mu'\} \oplus \{\varepsilon_4 \circ \nu_{12}\}$
$\oplus \{E\nu' \circ \varepsilon_7\} \approx Z_2 \oplus Z_2 \oplus Z_2 \oplus Z_4 \oplus Z_2 \oplus Z_2$,

$\pi_{16}^5 = \{\zeta_5\} \oplus \{\nu_5 \circ \bar{\nu}_8\} \oplus \{\nu_5 \circ \varepsilon_8\} \approx Z_8 \oplus Z_2 \oplus Z_2$,

$\pi_{17}^6 = \{\zeta_6\} \oplus \{\bar{\nu}_6 \circ \nu_{14}\} \approx Z_8 \oplus Z_4$,

$\pi_{n+11}^n = \{\zeta_n\} \oplus \{\bar{\nu}_n \circ \nu_{n+8}\} \approx Z_8 \oplus Z_2$, <u>for</u> $n = 7,8$ <u>and</u> 9 ,

$\pi_{n+11}^n = \{\zeta_n\} \approx Z_8$ <u>for</u> $n = 10,11$ <u>and for</u> $n \geq 13$,

$\pi_{23}^{12} = \{\Delta(\iota_{25})\} \oplus \{\zeta_{12}\} \approx Z \oplus Z_8$,

$(G_{11};2) = \{\zeta\} \approx Z_8$.

First we have $\pi_{12}^2 = \{\eta_2 \circ \mu_3\} \oplus \{\eta_2^2 \circ \varepsilon_4\} \approx Z_2 \oplus Z_2$ by (5.2) and by Theorem 7.2. Next we have

(7.8) $$H(\nu' \circ \varepsilon_6) = \eta_5 \circ \varepsilon_6 \quad \underline{\text{and}} \quad H(\varepsilon_3 \circ \nu_{11}) = \nu_5^3 .$$

For, $H(\nu' \circ \varepsilon_6) = H(\nu') \circ \varepsilon_6 = \eta_5 \circ \varepsilon_6$ and $H(\varepsilon_3 \circ \nu_{11}) = H(\varepsilon_3) \circ \nu_{11} = \nu_5^2 \circ \nu_{11} = \nu_5^3$, by Proposition 2.2, (5.3) and Lemma 6.1.

The images of H in (7.7) and (7.8) generate π_{14}^5 , by Theorem 7.2. Thus in the exact sequence

$$\pi^3_{14} \xrightarrow{H} \pi^5_{14} \xrightarrow{\Delta} \pi^2_{12} \xrightarrow{E} \pi^3_{13} \xrightarrow{H} \pi^5_{13}$$

of (4.4), the homomorphism $H : \pi^3_{14} \longrightarrow \pi^5_{14}$ is onto. Then we have that E is an isomorphsim into. By Lemma 6.6, and Theorem 7.1 we have that $H : \pi^3_{13} \longrightarrow \pi^5_{13}$ is onto and that the group π^3_{13} is of order 8 and generated by ε', $\eta^2_3 \circ \varepsilon_5 = 2\varepsilon'$ and $\eta_3 \circ \mu_4$. Thus $\pi^3_{13} = \{\varepsilon'\} \oplus \{\eta_3 \circ \mu_4\} \approx Z_4 \oplus Z_2$.

By (5.6) and Proposition 5.15,

$$\pi^4_{14} = \{\nu_4 \circ \sigma'\} \oplus \{E\varepsilon'\} \oplus \{\eta_4 \circ \mu_5\} \approx Z_8 \oplus Z_4 \oplus Z_2 \ .$$

Consider the composition of homomorphisms $H \circ \Delta : \pi^9_{16} \longrightarrow \pi^7_{14}$, where $\pi^9_{16} = \{\sigma_9\} \approx Z_{16}$ and $\pi^7_{14} = \{\sigma'\} \approx Z_8$ by Proposition 5.15. By Lemma 5.14, Proposition 2.5 and by Proposition 2.7,

$$\begin{aligned}(H \circ \Delta)(4\sigma_9) &= H(\Delta(E^3\sigma'')) = H(\Delta(\iota_9) \circ E\sigma'') \\ &= H(\Delta(\iota_9)) \circ E\sigma'' = \pm 2\iota_7 \circ E\sigma'' \\ &= \pm 2E\sigma'' = 4\sigma' \ .\end{aligned}$$

Since $H \circ \Delta$ is a homomorphism, this relation implies that the homomorphism $H \circ \Delta$ is onto and that the image of $\Delta : \pi^9_{16} \to \pi^4_{14}$ is a cyclic group of order 8 or 16. Then it follows from the result of the group π^4_{14} that

$$\Delta(\sigma_9) = x(\nu_4 \circ \sigma') + yE\varepsilon' + z\eta_4 \circ \mu_5$$

for an odd integer x and some integers y and z. It is verified that

$$\pi^4_{14} = E\pi^3_{13} \oplus \Delta\pi^9_{16} \ .$$

Then it follows from the exactness of the sequence

$$\pi^9_{16} \xrightarrow{\Delta} \pi^4_{14} \xrightarrow{E} \pi^5_{15} \xrightarrow{H} \pi^9_{15}$$

that the sequence

$$0 \longrightarrow \pi^3_{13} \xrightarrow{E^2} \pi^5_{15} \xrightarrow{H} \pi^9_{15} \quad (= \{\nu^2_9\} \approx Z_2)$$

is exact. By Lemma 6.6, $E^2\varepsilon'$ is divisible by 2. Let α be an element of π^5_{15} such that $2\alpha = E^2\varepsilon'$. Then it follows form the exactness of the last sequence that $H(\alpha) = \nu^2_9$ and

$$\pi^5_{15} = \{\alpha\} \oplus \{\eta_5 \circ \mu_6\} \approx Z_8 \oplus Z_2 \ .$$

Furthermore, any element α of $H(\alpha) = \nu^2_9$ satisfies the above decomposition of π^5_{15} and $2\alpha = \pm E^2\varepsilon'$. In particular, we may take $\nu_5 \circ \sigma_8$ as α, since

(7.9). $$H(\nu_5 \circ \sigma_8) = \nu^2_9$$

which has been proved already in the proof of Theorem 7.2. Thus

$$\pi^5_{15} = \{\nu_5 \circ \sigma_8\} \oplus \{\eta_5 \circ \mu_6\} \approx Z_8 \oplus Z_2$$

and we have

(7.10) $\quad 2(\nu_5 \circ \sigma_8) = \pm E^2\varepsilon'$ <u>and</u> $4(\nu_n \circ \sigma_{n+3}) = \eta^2_n \circ \varepsilon_{n+2}$ <u>for</u> $n \geq 5$.

We have by (5.2),

$$\pi^2_{13} = \{\eta_2 \circ \varepsilon'\} \oplus \{\eta^2_2 \circ \mu_4\} \approx Z_4 \oplus Z_2 .$$

Consider the exact sequence

$$\pi^5_{15} \xrightarrow{\Delta} \pi^2_{13} \xrightarrow{E} \pi^3_{14} \xrightarrow{H} \pi^5_{14}$$

of (4.4), in which we know that H is onto and $\pi^3_{14}/E\pi^2_{13}$ is generated by the classes of μ', $\nu' \circ \varepsilon_6$ and $\varepsilon_3 \circ \nu_{11}$. By Lemma 5.2, $\Delta(\eta_5 \circ \mu_6) = 0$. Thus the kernel of E is generated by $\Delta(\nu_5 \circ \sigma_8)$. By Proposition 2.5 and Lemma 5.1, $2\Delta(\nu_5 \circ \sigma_8) = \Delta(\pm E^2\varepsilon') = \Delta(\iota_5) \circ (\pm \varepsilon') = 2(\eta_2 \circ \varepsilon')$. It follows that

(7.11) $$\Delta(\nu_5 \circ \sigma_8) \equiv \pm (\eta_2 \circ \varepsilon') \mod \eta^2_2 \circ \mu_4$$

and $$E\pi^2_{12} = \{\eta^2_3 \circ \mu_5\} \approx Z_2 .$$

It is easy to see that $2(\nu' \circ \varepsilon_6) = 2(\varepsilon_3 \circ \nu_{11}) = 0$. Then it follows from the relation $2\mu' = \eta^2_3 \circ \mu_5$ of (7.7) that

$$\pi^3_{14} = \{\mu'\} \oplus \{\nu' \circ \varepsilon_6\} \oplus \{\varepsilon_3 \circ \nu_{11}\} \approx Z_4 \oplus Z_2 \oplus Z_2.$$

The result for π^4_{15} in Theorem 7.4 follows from (5.6) and Theorem 7.1.

Next we prove

(7.12). $$\varepsilon_3 \circ \nu_{11} = \nu' \circ \bar{\nu}_6 \quad \underline{\text{and}} \quad \varepsilon' \circ \eta_{13} = \nu' \circ \varepsilon_6 .$$

By (7.6) and Proposition 1.4,

$$E(\varepsilon_3 \circ \nu_{11}) \in \{E\nu', \nu_7, \eta_{10}\} \circ \nu_{12} = E\nu' \circ -\{\nu_7, \eta_{10}, \nu_{11}\} .$$

Since $\{\nu_7, \eta_{10}, \nu_{11}\}$ consists of a single element $\bar{\nu}_7$, by Lemma 6.2, then we have $E(\varepsilon_3 \circ \nu_{11}) = E\nu' \circ \bar{\nu}_7 = E(\nu' \circ \bar{\nu}_6)$. It follows from Lemma 4.5 that $\varepsilon_3 \circ \nu_{11} = \nu' \circ \bar{\nu}_6$. Similarly,

$$\nu' \circ \varepsilon_6 \in \nu' \circ E^3\{\nu', \nu_6, \eta_9\} = \{\nu', 2\nu_6, \nu_9\}_3 \circ \eta_{13} .$$

As is seen in the definition of ε', $\{\nu', 2\nu_6, \nu_9\}_3$ consists of a single element ε'. Thus $\nu' \circ \varepsilon_6 = \varepsilon' \circ \eta_{13}$ and (7.12) is proved.

Consider the exact sequence

$$\pi^9_{17} \xrightarrow{\Delta} \pi^4_{15} \xrightarrow{E} \pi^5_{16}$$

of (4.4), where $\pi^9_{17} = \{\sigma_9 \circ \eta_{16}\} \oplus \{\bar{\nu}_9\} \oplus \{\varepsilon_9\} \approx Z_2 \oplus Z_2 \oplus Z_2$ by Theorem 7.1. By use of Proposition 2.5, (5.8), Proposition 3.1 and (7.12), we have the following relations.

$$\Delta(\sigma_9 \circ \eta_{16}) = \Delta(\sigma_9) \circ \eta_{14} = x(\nu_4 \circ \sigma') \circ \eta_{14} + yE\varepsilon' \circ \eta_{14} + z\eta_4 \circ \mu_5 \circ \eta_{14}$$

$$= \nu_4 \circ \sigma' \circ \eta_{14} + yE\nu' \circ \varepsilon_7 + z\eta^2_4 \circ \mu_6 (x : \text{odd}),$$

$$\Delta(\bar{\nu}_9) = \Delta(\iota_9) \circ \bar{\nu}_7 = \pm (2\nu_4 - E\nu') \circ \bar{\nu}_7$$

$$= E\nu' \circ \bar{\nu}_7 = \varepsilon_4 \circ \nu_{12},$$

$$\Delta(\varepsilon_9) = \Delta(\iota_9) \circ \bar{\nu}_7 = \pm (2\nu_4 - E\nu') \circ \varepsilon_7 = E\nu' \circ \varepsilon_7 .$$

These images are independent. It follows that

(7.13) $\Delta : \pi^9_{17} \to \pi^4_{15}$ <u>is an isomorphism into</u>, $\Delta(\bar{\nu}_9) = \varepsilon_4 \circ \nu_{12}$ <u>and</u> $\Delta(\varepsilon_9) = E\nu' \circ \varepsilon_7$ <u>and that</u>

$$E\pi^4_{15} = \{\nu_5 \circ \bar{\nu}_8\} \oplus \{\nu_5 \circ \varepsilon_8\} \oplus \{E^2\mu'\} \approx Z_2 \oplus Z_2 \oplus Z_4 .$$

Now we have an exact sequence

$$0 \longrightarrow E\pi^4_{15} \longrightarrow \pi^5_{16} \xrightarrow{H} \pi^9_{16} \xrightarrow{\Delta} \pi^4_{14}$$

We have seen that Δ maps $\pi^9_{16} = \{\sigma_9\} \approx Z_{16}$ onto a cyclic subgroup of order 8. Then the kernel of Δ is generated by $8\sigma_9$, which is $H(\zeta_5)$ by Lemma 6.7. $2E\pi^7_{15} = 0$ by Theorem 7.1. It follows from Lemma 6.7 that $4\zeta_5 = \eta_5^2 \circ \mu_7$ and that ζ_5 is of order 8. Then, by the exactness of the above sequence, we have

$$\pi^5_{16} = \{\zeta_5\} \oplus \{\nu_5 \circ \bar{\nu}_8\} \oplus \{\nu_5 \circ \varepsilon_8\} \approx Z_8 \oplus Z_2 \oplus Z_2 .$$

$E^2\mu'$ and $-E^2\mu'$ are the only elements of order 4 in π^5_{16}.

Thus

(7.14). $2\zeta_5 = \pm E^2\mu'$ <u>and</u> $4\zeta_n = \eta_n^2 \circ \mu_{n+2}$ <u>for</u> $n \geq 5$.

Consider the exact sequence

$$\pi^6_{17} \xrightarrow{H} \pi^{11}_{17} \xrightarrow{\Delta} \pi^5_{15} \xrightarrow{E} \pi^6_{16} \xrightarrow{H} \pi^{11}_{16}$$

of (4.4). $\pi^{11}_{16} = 0$ by Proposition 5.9. Then E is onto. By Proposition 2.2 and by Lemma 6.2, we have

(7.15). $H(\bar{\nu}_6 \circ \nu_{14}) = \nu^2_{11}$, <u>and</u> $H : \pi^6_{17} \longrightarrow \pi^{11}_{17}$ <u>is onto</u>.

It follows from the exactness of the above sequence that E is an isomorphism onto and $\pi^6_{16} = \{\nu_6 \circ \sigma_9\} \oplus \{\eta_6 \circ \mu_7\} \approx Z_8 \oplus Z_2$.

We shall show the following relations.

(7.16). $2(\nu_5 \circ \sigma_8) = \nu_5 \circ E\sigma'$,

$\Delta(\sigma_9) = x(\nu_4 \circ \sigma') \pm E\varepsilon'$ <u>for an odd</u> x,

$\Delta(\sigma_9 \circ \eta_{16}) = \nu_4 \circ \sigma' \circ \eta_{14} + E\nu' \circ \varepsilon_7$

$\Delta(\sigma_9 \circ \eta^2_{16}) = \nu_4 \circ \sigma' \circ \eta^2_{14} + E\nu' \circ \eta_7 \circ \varepsilon_8$.

By Lemma 5.14, $E(2\nu_5 \circ \sigma_8) = \nu_6 \circ 2\sigma_9 = \nu_6 \circ E^2\sigma' = E(\nu_5 \circ E\sigma')$. Since $E : \pi^5_{15} \longrightarrow \pi^6_{16}$ is an isomorphism, we have that $2(\nu_5 \circ \sigma_8) = \nu_5 \circ E\sigma'$. We know that $\Delta(\sigma_9) = x(\nu_4 \circ \sigma') + yE\varepsilon' + z\eta_4 \circ \mu_5$ for an odd integer x. Applying the homomorphism $E^2 : \pi^4_{14} \longrightarrow \pi^6_{16}$, we have

$$0 = E^2\Delta(\sigma_9) = x\nu_6 \circ E^2\sigma' + yE^3\varepsilon' + z\eta_6 \circ \mu_7$$

$$= x\nu_6 \circ 2\sigma_9 \pm 2y\nu_6 \circ \sigma_9 + z\eta_6 \circ \mu_7 \quad \text{by Lemma 5.14 and (7.10)}$$

$$= 2(x \pm y)\, \nu_6 \circ \sigma_9 + z\eta_6 \circ \mu_7 .$$

Since $\nu_6 \circ \sigma_9$ and $\eta_6 \circ \mu_7$ are independent generators of orders 8 and 2. Thus $x \pm y \equiv 0 \pmod 4$ and $z \equiv 0 \pmod 2$, and we have

$$\Delta(\sigma_9) = x(\nu_4 \circ \sigma') \pm E\varepsilon'$$

for an odd integer x. The remaining two relations in (7.16) follows easily by Proposition 2.5 and (7.12).

In order to compute the group π_{17}^6, consider the exact sequence

$$\pi_{18}^{11} \xrightarrow{\Delta} \pi_{16}^5 \xrightarrow{E} \pi_{17}^6 \xrightarrow{H} \pi_{17}^{11} .$$

By Proposition 2.5, (5.10), (7.4) and by (7.16),

$$\Delta(\sigma_{11}) = \Delta(\iota_{11}) \circ \sigma_9 = \nu_5 \circ \eta_8 \circ \sigma_9$$

$$= \nu_5 \circ E\sigma' \circ \eta_{15} + \nu_5 \circ \bar{\nu}_8 + \nu_5 \circ \varepsilon_8$$

$$= 2(\nu_5 \circ \sigma_8) \circ \eta_{15} + \nu_5 \circ \bar{\nu}_8 + \nu_5 \circ \varepsilon_8$$

$$= \nu_5 \circ \bar{\nu}_8 + \nu_5 \circ \varepsilon_8 .$$

Thus we obtain

(7.17). $\Delta(\sigma_{11}) = \nu_5 \circ \bar{\nu}_8 + \nu_5 \circ \varepsilon_8$ <u>and</u> $\nu_6 \circ \bar{\nu}_9 = \nu_6 \circ \varepsilon_9$.

Since σ_{11} generates π_{18}^{11}, then it follows from the exactness of the above sequence that

$$E\pi_{16}^5 = \{\zeta_6\} \oplus \{\nu_6 \circ \varepsilon_9\} \approx Z_8 \oplus Z_2 .$$

By (7.15), $\pi_{17}^6/E\pi_{16}^5 \approx Z_2$ and it is generated by the class of $\bar{\nu}_6 \circ \nu_{14}$. Then the result $\pi_{17}^6 = \{\zeta_6\} \oplus \{\bar{\nu}_6 \circ \nu_{14}\} \approx Z_8 \oplus Z_4$ follows from the following relation.

(7.18). $$2\bar{\nu}_6 \circ \nu_{14} = \nu_6 \circ \varepsilon_9 = \Delta(\nu_{13}^2) .$$

By (6.1) and Proposition 1.4,

$$\nu_6 \circ \varepsilon_9 \in \nu_6 \circ \{\eta_9, \nu_{10}, 2\nu_{13}\} = \{\nu_6, \eta_9, \nu_{10}\} \circ 2\nu_{14} .$$

By the definition of $\bar{\nu}_6$,

$$2\bar{\nu}_6 \circ \nu_{14} = \bar{\nu}_6 \circ 2\nu_{14} \in \{\nu_6, \eta_9, \nu_{10}\} \circ 2\nu_{14} .$$

The composition $\{\nu_6, \eta_9, \nu_{10}\} \circ 2\nu_{14}$ is a coset of $\nu_6 \circ \pi_{14}^9 \circ 2\nu_{14} = 0$ by (4.7) and Proposition 5.9. Then it consists of a single element which has to be $\nu_6 \circ \varepsilon_9 = 2\bar{\nu}_6 \circ \nu_{14}$.

By Proposition 2.5 and Lemma 6.2, $\Delta(\nu_{13}^2) = \Delta(\nu_{13}) \circ \nu_{14} = \pm 2\bar{\nu}_6 \circ \nu_{14} = \pm \nu_6 \circ \varepsilon_9 = \nu_6 \circ \varepsilon_9$. Thus we have proved the relation (7.18).

Next consider the exact sequence

$$\pi^{13}_{19} \xrightarrow{\Delta} \pi^{6}_{17} \xrightarrow{E} \pi^{7}_{18} \xrightarrow{H} \pi^{13}_{18} \xrightarrow{\Delta} \pi^{6}_{16} \xrightarrow{E} \pi^{7}_{17} \xrightarrow{H} \pi^{13}_{17}$$

of (4.4), where $\pi^{13}_{17} = \pi^{13}_{18} = 0$ and $\pi^{13}_{19} = \{\nu^2_{13}\} \approx Z_2$ by Propositions 5.8, 5.9 and 5.11. It follows that $E : \pi^6_{16} \to \pi^7_{17}$ is an isomorphism onto and that $E : \pi^6_{17} \longrightarrow \pi^7_{18}$ is a homomorphism onto and its kernel is generated by $\Delta(\nu^2_{13}) = 2\bar{\nu}_6 \circ \nu_{14}$, by (7.14). Thus

$$\pi^7_{17} = \{\nu_7 \circ \sigma_{10}\} \oplus \{\eta_7 \circ \mu_8\} \approx Z_8 \oplus Z_2$$

and

$$\pi^7_{18} = \{\zeta_7\} \oplus \{\bar{\nu}_7 \circ \nu_{15}\} \approx Z_8 \oplus Z_2 .$$

By (5.15), Proposition 5.6 and by Proposition 5.8,

$$\pi^8_{18} = \{\sigma_8 \circ \nu_{15}\} \oplus \{\nu_8 \circ \sigma_{11}\} \oplus \{\eta_8 \circ \mu_9\} \approx Z_8 \oplus Z_8 \oplus Z_2$$

and

$$\pi^8_{19} = \{\zeta_8\} \oplus \{\bar{\nu}_8 \circ \nu_{15}\} \approx Z_8 \oplus Z_2 .$$

We shall prove

(7.19). <u>There exists an odd integer</u> x <u>such that</u> $\sigma' \circ \nu_{14} = x\nu_7 \circ \sigma_{10}$ <u>and</u> $\pm\,\Delta(\nu_{17}) = 2\sigma_8 \circ \nu_{15} - x\nu_8 \circ \sigma_{11}$.

Since the group π^7_{17} is generated by $\nu_7 \circ \sigma_{10}$ and $\eta_7 \circ \mu_8$, then $\sigma' \circ \nu_{14} = x\nu_7 \circ \sigma_{10} + y\eta_7 \circ \mu_8$ for some integer x and y. Concerning the composition with the element η_6, we have that

$$\begin{aligned} 0 = \sigma'' \circ \eta_{13} \circ \nu_{14} &= \eta_6 \circ \sigma' \circ \nu_{14} && \text{by (5.9) and (7.4)} \\ &= \eta_6 \circ (x\nu_7 \circ \sigma_{10} + y\eta_7 \circ \mu_8) \\ &= y\eta^2_7 \circ \mu_8 && \text{by (5.9).} \end{aligned}$$

Since $\eta^2_6 \circ \mu_8 \neq 0$, then $y \equiv 0 \pmod 2$ and $y\eta_7 \circ \mu_8 = 0$.

Next we compute that

$$\begin{aligned} E^2(\sigma''' \circ \nu_{12}) &= 4\sigma' \circ \nu_{14} && \text{by Lemma 5.14} \\ &= 4x\,\nu_7 \circ \sigma_{10} = x\eta^2_7 \circ \varepsilon_8 && \text{by (7.10).} \end{aligned}$$

By (7.16), $4(\nu_5 \circ E\sigma') = 8(\nu_5 \circ \sigma_8) = 0$. Then it follows from (6.2) that $\eta^2_7 \circ \varepsilon_8 = E^2(\sigma''' \circ \nu_{12}) = x\eta^2_7 \circ \varepsilon_8$. Since $\eta^2_7 \circ \varepsilon_8 = 4\nu_7 \circ \sigma_{10}$ is of order 2, then $x \not\equiv 0 \pmod 2$ and the first relation of (7.19) is proved. The second relation is a easy consequence of Proposition 2.5 and (5.16).

Now consider the exact sequence

$$\pi^{17}_{21} \xrightarrow{\Delta} \pi^{8}_{19} \xrightarrow{E} \pi^{9}_{20} \xrightarrow{H} \pi^{17}_{20} \xrightarrow{\Delta} \pi^{8}_{18} \xrightarrow{E} \pi^{9}_{19}$$

of (4.4) where $\pi^{17}_{21} = 0$ by Proposition 5.8. $E : \pi^8_{18} \longrightarrow \pi^9_{19}$ is onto by

(7.2) and its kernel is generated by $\Delta(\nu_{17}) = 2\sigma_8 \circ \nu_{15} - x\nu_8 \circ \sigma_{11}$. It follows that $\pi_{19}^9 = \{\sigma_9 \circ \nu_{16}\} \oplus \{\eta_9 \circ \mu_{10}\} \approx Z_8 \oplus Z_2$ and that $\Delta : \pi_{20}^{17} \longrightarrow \pi_{18}^8$ is an isomorphism into. Then the exactness of the above sequence implies that $E : \pi_{19}^8 \longrightarrow \pi_{20}^9$ is an isomorphism onto and

$$\pi_{20}^9 = \{\zeta_9\} \oplus \{\overline{\nu}_9 \circ \nu_{17}\} \approx Z_8 \oplus Z_2 .$$

Consider the reduced joins $\nu' \# \sigma'$, $\nu_4 \# \sigma'$ and $\nu_4 \# \sigma_8$. Then we have from Proposition 3.1, Lemma 5.4 and Lemma 5.14 that

$4\nu_{10} \circ \sigma_{13} = 4\sigma_{10} \circ \nu_{17}$, $2\nu_{11} \circ \sigma_{14} = 2\sigma_{11} \circ \nu_{18}$ and $\nu_{12} \circ \sigma_{15} = \sigma_{12} \circ \nu_{19}$.

Since $E\Delta = 0$, we have from (7.19) that $2\sigma_{10} \circ \nu_{17} = x\nu_{10} \circ \sigma_{13}$ for an odd x. Then we have successively that

$4\sigma_{10} \circ \nu_{17} = 4\nu_{10} \circ \sigma_{13} = 4x\, \nu_{10} \circ \sigma_{13} = 8\sigma_{10} \circ \nu_{17} = 0$,

$2\sigma_{11} \circ \nu_{18} = 2\nu_{11} \circ \sigma_{14} = 2x\, \nu_{11} \circ \sigma_{14} = 4\sigma_{11} \circ \nu_{18} = E(4\sigma_{10} \circ \nu_{17}) = 0$,

and

$\sigma_{12} \circ \nu_{19} = \nu_{12} \circ \sigma_{15} = x\nu_{12} \circ \sigma_{15} = 2\sigma_{12} \circ \nu_{19} = E(2\sigma_{11} \circ \nu_{18}) = 0$.

Thus we have obtained that

(7.20). $4\sigma_{10} \circ \nu_{17} = 2\sigma_{11} \circ \nu_{18} = \sigma_{12} \circ \nu_{19} = 0$,

$4\nu_9 \circ \sigma_{12} = 2\nu_{10} \circ \sigma_{13} = \nu_{11} \circ \sigma_{14} = 0$.

Consider the exact sequences of (4.4)

$$\pi_{n+12}^{2n+1} \xrightarrow{\Delta} \pi_{n+10}^{n} \xrightarrow{E} \pi_{n+11}^{n+1} \xrightarrow{H} \pi_{n+11}^{2n+1} \quad \text{for } n = 9,\ 10 \text{ and } 11.$$

By (7.2) E is onto for $n = 9$ and $n = 10$. Also E is onto for $n = 11$ since $\pi_{22}^{21} = 0$. The kernel of E has at most two elements for $n = 9$ and $n = 10$, since $2\Delta(\eta_{19}^2) = 2\Delta(\eta_{21}) = 0$. This is true for $n = 11$, since $2\Delta(\iota_{23}) = \Delta(2\iota_{25}) = \Delta(\pm H(\Delta(\iota_{25}))) = 0$ by Proposition 2.7. In the other hand, the relations (7.20) show that the kernel of E has at least two elements for $n = 9$, 10 and 11. We obtain consequently the results for π_{20}^{10}, π_{21}^{11} and π_{22}^{12} in Theorem 7.3. The results for stable groups in Theorem 7.3 follows by (4.5).

In the above discussion, we obtain that

(7.21). $\Delta(\eta_{19}^2) = 4\sigma_9 \circ \nu_{16}$, $\Delta(\eta_{21}) = 2\sigma_{10} \circ \nu_{17}$ <u>and</u> $\Delta(\iota_{23}) = \sigma_{11} \circ \nu_{18}$.

Consider the exact sequence

$$\pi_{22}^{19} \xrightarrow{\Delta} \pi_{20}^{9} \xrightarrow{E} \pi_{21}^{10} \xrightarrow{H} \pi_{21}^{19} \xrightarrow{\Delta} \pi_{19}^{9}$$

of (4.4). By (7.21), $\Delta : \pi_{21}^{19} \longrightarrow \pi_{19}^{9}$ is an isomorphism into. It follows that E is onto. We have

(7.22). $$\Delta(\nu_{19}) = \overline{\nu}_9 \circ \nu_{17} .$$

For $\Delta(\nu_{19}) = \Delta(\iota_{19}) \circ \nu_{17}$ by Proposition 2.5

$= \sigma_9 \circ \eta_{16} \circ \nu_{17} + \bar{\nu}_9 \circ \nu_{17} + \varepsilon_9 \circ \nu_{17}$ by (7.1)

$= 0 + \bar{\nu}_9 \circ \nu_{17} + 2\nu_9 \circ \bar{\nu}_{12}$ by (5.9) and (7.12)

$= \bar{\nu}_9 \circ \nu_{17}$.

Then we have from the above exact sequence that

$$\pi^{10}_{21} = \{\zeta_{10}\} \approx Z_8 .$$

In the exact sequence

$$\pi^{21}_{23} \xrightarrow{\Delta} \pi^{10}_{21} \xrightarrow{E} \pi^{11}_{22} \xrightarrow{H} \pi^{21}_{22} \xrightarrow{\Delta} \pi^{10}_{20} ,$$

the last Δ is an isomorphism into by (7.21). $\Delta\pi^{21}_{23}$ is generated by the element $\Delta(\eta^2_{21}) = \Delta(\eta_{21}) \circ \eta_{20} = 2(\sigma_{10} \circ \nu_{17}) \circ \eta_{20} = 0$, by Proposition 2.5 and (7.21). It follows from the exactness of the above sequence that E is an isomorphism onto and $\pi^{11}_{22} = \{\zeta_{11}\} \approx Z_8$.

In the exact sequence

$$\pi^{23}_{24} \xrightarrow{\Delta} \pi^{11}_{22} \xrightarrow{E} \pi^{12}_{23} \xrightarrow{H} \pi^{23}_{23} \xrightarrow{\Delta} \pi^{11}_{21} ,$$

the kernel of the last Δ is generated by $2\iota_{23} = H(\pm \Delta(\iota_{25}))$, by (7.21). Then we have that $\pi^{12}_{23} = \Delta\pi^{25}_{25} \oplus E\pi^{11}_{22} \approx Z \oplus E\pi^{11}_{22}$. By Proposition 2.5, (7.21) and by (5.9),

$$\Delta(\eta_{23}) = \Delta(\iota_{23}) \circ \eta_{21} = \sigma_{11} \circ \nu_{18} \circ \eta_{21} = 0 .$$

Thus $\Delta\ \pi^{23}_{24} = 0$ and E is an isomorphism into. Then we have

$$\pi^{12}_{23} = \{\Delta(\iota_{25})\} \oplus \{\zeta_{12}\} \approx Z \oplus Z_8 .$$

From the exactness of the sequence $\pi^{25}_{25} \xrightarrow{\Delta} \pi^{12}_{23} \xrightarrow{E} \pi^{13}_{24} \xrightarrow{H} \pi^{25}_{24} = 0$ and from (4.5), we have that

$$\pi^n_{n+11} = \{\zeta_n\} \approx Z_8 \quad \text{for} \quad n \geq 13 \quad \text{and} \quad (G_{11};2) = \{\zeta\} \approx Z_8 .$$

Consequently, all the statements of Theorem 7.3 and Theorem 7.4 are established.

iii). <u>The groups</u> π^n_{n+12} <u>and</u> π^n_{n+13} .

Consider the secondary compositions

$$\{\sigma_{12}, \nu_{19}, \eta_{22}\}_1 \quad \text{and} \quad \{\sigma_{11}, 2\nu_{18}, \eta_{21}\}_1 ,$$

which are cosets of

$$\sigma_{12} \circ \pi^{19}_{24} + \pi^{12}_{23} \circ \eta_{23} = \pi^{12}_{23} \circ \eta_{23} = \{\Delta(\iota_{25}) \circ \eta_{23}\} + \{\zeta_{12} \circ \eta_{23}\}$$

and $$\sigma_{11} \circ \pi^{18}_{23} + \pi^{11}_{22} \circ \eta_{22} = \pi^{11}_{22} \circ \eta_{22} = \{\zeta_{11} \circ \eta_{22}\}$$

respectively.

Lemma 7.5: Any element θ of $\{\sigma_{12}, \nu_{19}, \eta_{22}\}_1$ satisfies $H(\theta) = \eta_{23}$ and $2\theta = 0$. Any element θ' of $\{\sigma_{11}, 2\nu_{18}, \eta_{21}\}_1$ satisfies $H(\theta') = \eta_{21}^2$ and $2\theta' = 0$.

Proof: By (7.21) and Proposition 2.6,

$$\eta_{23} = \iota_{23} \circ \eta_{23} \in \Delta^{-1}(\sigma_{11} \circ \nu_{18}) \circ \eta_{23} = H\{\sigma_{12}, \nu_{19}, \eta_{22}\}_1 .$$

$H\{\sigma_{12}, \nu_{19}, \eta_{22}\}_1$ is a coset of $H(\pi_{23}^{12} \circ \eta_{23}) = 2\pi_{23}^{23} \circ \eta_{23} = 0$. Thus $\eta_{23} = H(\theta)$. Similarly we have

$$\eta_{21}^2 = \eta_{21} \circ \eta_{22} \in \Delta^{-1}(\sigma_{10} \circ 2\nu_{17}) \circ \eta_{22} = H\{\sigma_{11}, 2\nu_{18}, \eta_{21}\}_1$$

and from $H(\pi_{22}^{11} \circ \eta_{22}) = 0$ that $\eta_{21}^2 = H(\theta')$.

Next, by Proposition 1.4, (4.7) and by Proposition 5.9,

$$2\theta \in \{\sigma_{12}, \nu_{19}, \eta_{22}\}_1 \circ 2\iota_{24} = \sigma_{12} \circ E\{\nu_{18}, \eta_{21}, 2\iota_{22}\} \subset \sigma_{12} \circ \pi_{24}^{19} = 0$$

and

$$2\theta' \in \{\sigma_{11}, 2\nu_{18}, \eta_{21}\}_1 \circ 2\iota_{23} = \sigma_{11} \circ E\{2\nu_{17}, \eta_{20}, 2\iota_{21}\} \subset \sigma_{11} \circ \pi_{23}^{18} = 0.$$

Then Lemma 7.5 is proved. q. e. d.

By use of these elements θ and θ', our results for the groups π_{n+12}^n and π_{n+13}^n are stated as follows.

Theorem 7.6:
$$\pi_{14}^2 = \{\eta_2 \circ \mu'\} \oplus \{\eta_2 \circ \nu' \circ \bar{\nu}_6\} \oplus \{\eta_2 \circ \nu' \circ \varepsilon_6\} \approx Z_4 \oplus Z_2 \oplus Z_2 ,$$

$$\pi_{15}^3 = \{\nu' \circ \mu_6\} \oplus \{\nu' \circ \eta_6 \circ \varepsilon_7\} \approx Z_2 \oplus Z_2 ,$$

$$\pi_{16}^4 = \{\nu_4 \circ \sigma' \circ \eta_{14}^2\} \oplus \{\nu_4^4\} \oplus \{\nu_4 \circ \mu_7\} \oplus \{\nu_4 \circ \eta_7 \circ \varepsilon_8\} \oplus \{E\nu' \circ \mu_7\} \oplus \{E\nu' \circ \eta_7 \circ \varepsilon_8\} \approx Z_2 \oplus Z_2 \oplus Z_2 \oplus Z_2 \oplus Z_2 \oplus Z_2 ,$$

$$\pi_{17}^5 = \{\nu_5^4\} \oplus \{\nu_5 \circ \mu_8\} \oplus \{\nu_5 \circ \eta_8 \circ \varepsilon_9\} \approx Z_2 \oplus Z_2 \oplus Z_2 ,$$

$$\pi_{18}^6 = \{\Delta(\sigma_{13})\} \approx Z_{16} ,$$

$\pi_{n+12}^n = 0$ for $n = 7, 8$ and 9,

$$\pi_{22}^{10} = \{\Delta(\nu_{21})\} \approx Z_4 ,$$

$$\pi_{23}^{11} = \{\theta'\} \approx Z_2 ,$$

$$\pi_{24}^{12} = \{\theta\} \oplus \{E\theta'\} \approx Z_2 \oplus Z_2 ,$$

$$\pi_{25}^{13} = \{E\theta\} \approx Z_2 ,$$

$\pi_{n+12}^n = (G_{12};2) = 0$ for $n \geq 14$.

Theorem 7.7: $\pi_{15}^2 = \{\eta_2 \circ \nu' \circ \mu_6\} \oplus \{\eta_2 \circ \nu' \circ \eta_6 \circ \varepsilon_7\}$

$$\approx Z_2 \oplus Z_2 ,$$

$\pi_{16}^3 = \{\nu' \circ \eta_6 \circ \mu_7\} \approx Z_2$,

$\pi_{17}^4 = \{\nu_4^2 \circ \sigma_8\} \oplus \{\nu_4 \circ \eta_7 \circ \mu_8\} \oplus \{E\nu' \circ \eta_7 \circ \mu_8\} \approx Z_8 \oplus Z_2 \oplus Z_2$,

$\pi_{18}^5 = \{\nu_5 \circ \sigma_8 \circ \nu_{15}\} \oplus \{\nu_5 \circ \eta_8 \circ \mu_9\} \approx Z_2 \oplus Z_2$,

$\pi_{n+13}^n = \{\nu_n \circ \sigma_{n+3} \circ \nu_{n+10}\} \approx Z_2$ <u>for</u> $n = 6$ <u>and</u> 7,

$\pi_{21}^8 = \{\sigma_8 \circ \nu_{15}^2\} \oplus \{\nu_8 \circ \sigma_{11} \circ \nu_{18}\} \approx Z_2 \oplus Z_2$,

$\pi_{n+13}^n = \{\sigma_n \circ \nu_{n+7}^2\} \approx Z_2$ <u>for</u> $n = 9$ <u>and</u> 10,

$\pi_{24}^{11} = \{\theta' \circ \eta_{23}\} \oplus \{\sigma_{11} \circ \nu_{18}^2\} \approx Z_2 \oplus Z_2$,

$\pi_{25}^{12} = \{\theta \circ \eta_{24}\} \oplus \{E\theta' \circ \eta_{24}\} \approx Z_2 \oplus Z_2$,

$\pi_{26}^{13} = \{E\theta \circ \eta_{25}\} \approx Z_2$,

$\pi_{27}^{14} = \{\Delta(\iota_{29})\} \approx Z$,

$\pi_{n+13}^n = (G_{13};2) = 0$ <u>for</u> $n \geq 15$.

First, by (5.2), Theorem 7.4 and by (7.12), we have

$$\pi_{14}^2 = \{\eta_2 \circ \mu'\} \oplus \{\eta_2 \circ \varepsilon_3 \circ \nu_{11}\} \oplus \{\eta_2 \circ \nu' \circ \varepsilon_6\}$$
$$= \{\eta_2 \circ \mu'\} \oplus \{\eta_2 \circ \nu' \circ \bar{\nu}_6\} \oplus \{\eta_2 \circ \nu' \circ \varepsilon_6\} \approx Z_4 \oplus Z_2 \oplus Z_2 .$$

By Lemma 5.7, $E(\eta_2 \circ \nu' \circ \bar{\nu}_6) = E(\eta_2 \circ \nu' \circ \varepsilon_6) = 0$. By Proposition 2.5 and Lemma 5.2, $\Delta(2\zeta_5) = \Delta(\pm E^2\mu') = \Delta(\iota_5) \circ (\pm \mu') = 2(\eta_2 \circ \mu')$. Thus $\Delta(\zeta_5) \equiv \pm (\eta_2 \circ \mu') \bmod \{\eta_2 \circ \nu' \circ \bar{\nu}_6\} + \{\eta_2 \circ \nu' \circ \varepsilon_6\}$ and $E(\eta_2 \circ \mu') = 0$. Then $E\pi_{14}^2 = 0$ and we have from the exactness of (4.4) that the following sequences are exact.

$$0 \longrightarrow \pi_{15}^3 \xrightarrow{H} \pi_{15}^5 \xrightarrow{\Delta} \pi_{13}^2$$

and
$$\pi_{17}^5 \xrightarrow{\Delta} \pi_{15}^2 \xrightarrow{E} \pi_{16}^3 \xrightarrow{H} \pi_{16}^5 \xrightarrow{\Delta} \pi_{14}^2 \longrightarrow 0$$

By (7.11) and Theorem 7.4, the image of $\nu_5 \circ \sigma_8$ under Δ is an element of order 4 in π_{13}^2. As $\pi_{15}^5 = \{\nu_5 \circ \sigma_8\} \oplus \{\eta_5 \circ \mu_6\} \approx Z_8 \oplus Z_2$, by Theorem 7.3, the kernel of $\Delta : \pi_{15}^5 \longrightarrow \pi_{13}^2$ has at most 4 elements. By Proposition 2.2, (5.3) and (7.10), we have

$$H(\nu' \circ \mu_6) = \eta_5 \circ \mu_6 \quad \text{and} \quad H(\nu' \circ \eta_6 \circ \varepsilon_7) = \eta_5^2 \circ \varepsilon_7 = 4(\nu_5 \circ \sigma_8) .$$

These elements generate the kernel of the homomorphism Δ and it follows from the exactness of the above upper sequence that

$$\pi_{15}^3 = \{\nu' \circ \mu_6\} \oplus \{\nu' \circ \eta_6 \circ \varepsilon_7\} \approx Z_2 \oplus Z_2$$

By (5.2),

$$\pi^2_{15} = \{\eta_2 \circ \nu' \circ \mu_6\} \oplus \{\eta_2 \circ \nu' \circ \eta_6 \circ \varepsilon_7\} \approx Z_2 \oplus Z_2 .$$

We have, by Lemma 5.7,

$$E\pi^2_{15} = \{E(\eta_2 \circ \nu') \circ \mu_7\} \oplus \{E(\eta_2 \circ \nu') \circ \eta_7 \circ \varepsilon_8\} = 0 .$$

It follows from the exactness of the above lower sequence that

(7.23). $\Delta : \pi^5_{17} \longrightarrow \pi^2_{15}$ <u>is onto</u>,

and that the group π^3_{16} is isomorphic to the kernel of $\Delta : \pi^5_{16} \to \pi^2_{14}$ which is a homomorphism onto. As $\pi^5_{16} \approx Z_8 \oplus Z_2 \oplus Z_2$, by Theorem 7.4, then the kernel of this Δ has just two elements 0 and $4\zeta_5$. By Proposition 2.2, (5.3) and by (7.14),

$$H(\nu' \circ \eta_6 \circ \mu_7) = \eta_5^2 \circ \mu_7 = 4\zeta_5 .$$

It follows that

$$\pi^3_{16} = \{\nu' \circ \eta_6 \circ \mu_7\} \approx Z_2 .$$

The results for the groups π^4_{16} and π^4_{17} are obtained from (5.6), Theorem 7.2 and Theorem 7.3.

Consider the homomorphisms $\Delta : \pi^9_{18} \longrightarrow \pi^4_{16}$ and $\Delta : \pi^9_{19} \longrightarrow \pi^4_{17}$. By using Proposition 2.5, we have

$$\Delta(\sigma_9 \circ \eta^2_{16}) = \nu_4 \circ \sigma' \circ \eta^2_{14} + E\nu' \circ \eta_7 \circ \varepsilon_8 \qquad \text{by (7.16)},$$

$$\Delta(\nu^3_9) = \Delta(\nu^2_9) \circ \nu_{13} = \Delta H(\nu_5 \circ \sigma_8) \circ \nu_{13} = 0 \qquad \text{by (7.9)},$$

$$\Delta(\mu_9) = \Delta(\iota_9) \circ \mu_7 = E\nu' \circ \mu_7 \qquad \text{by (5.8)},$$

$$\Delta(\eta_9 \circ \varepsilon_{10}) = \Delta(\iota_9) \circ \eta_7 \circ \varepsilon_8 = E\nu' \circ \eta_7 \circ \varepsilon_8 \qquad \text{by (5.8)},$$

$$\begin{aligned}\Delta(\sigma_9 \circ \nu_{16}) &= \Delta(\sigma_9) \circ \nu_{14} \\ &= x\,(\nu_4 \circ \sigma' \circ \nu_{14}) \pm E\varepsilon' \circ \nu_{14} && \text{by (7.16)} \\ &= x'(\nu^2_4 \circ \sigma_{10}) \pm E(\varepsilon' \circ \nu_{12}) && \text{by (7.19)},\end{aligned}$$

$$\Delta(\eta_9 \circ \mu_{10}) = \Delta(\iota_9) \circ \eta_7 \circ \mu_8 = E\nu' \circ \eta_7 \circ \mu_8 \qquad \text{by (5.8)},$$

where x and x' are odd integers. Then from the exactness of (4.4),

$$E\pi^4_{16} = \{\nu^4_5\} \oplus \{\nu_5 \circ \mu_8\} \oplus \{\nu_5 \circ \eta_8 \circ \varepsilon_9\} \approx Z_2 \oplus Z_2 \oplus Z_2$$

and $$E\pi^4_{17} = \{\nu_5 \circ \eta_8 \circ \mu_9\} \approx Z_2 .$$

By conserning the structure of the groups π^9_{18} and π^9_{19}, we see that $\Delta : \pi^9_{18} \longrightarrow \pi^4_{16}$ has the kernel $\{\nu^3_9\}$ and $\Delta : \pi^9_{19} \longrightarrow \pi^4_{17}$ is an isomorphism into. Then it follows from the exactness of the sequence that

(7.24). $E : \pi^4_{18} \longrightarrow \pi^5_{19}$ <u>is onto</u>

and that the following two sequences are exact:

$$0 \longrightarrow E\pi^4_{16} \longrightarrow \pi^5_{17} \xrightarrow{H} \pi^9_{17} \xrightarrow{\Delta} \pi^4_{15} \quad ,$$
$$0 \longrightarrow E\pi^4_{17} \longrightarrow \pi^5_{18} \xrightarrow{H} \{\nu^3_9\} \longrightarrow 0 \quad .$$

By (7.13), $\Delta : \pi^9_{17} \to \pi^4_{15}$ is an isomorphism into, then it follows that

$$\pi^5_{17} = E\pi^4_{16} = \{\nu^4_5\} + \{\nu_5 \circ \mu_8\} + \{\nu_5 \circ \eta_8 \circ \varepsilon_9\} \approx Z_2 \oplus Z_2 \oplus Z_2 \ .$$

By Proposition 2.2 and (7.9),

$$H(\nu_5 \circ \sigma_8 \circ \nu_{15}) = H(\nu_5 \circ \sigma_8) \circ \nu_{15} = \nu^3_9 \ .$$

By (7.10), $2(\nu_5 \circ \sigma_8 \circ \nu_{15}) = \pm E^2\varepsilon' \circ \nu_{15} \subset E^2\pi^3_{16}$ $= \{E^2\nu' \circ \eta_7 \circ \mu_8\} = E\Delta\{\eta_9 \circ \mu_{10}\} = 0$.

Then it follows from the last exact sequence that

$$\pi^5_{18} = \{\nu_5 \circ \sigma_8 \circ \nu_{15}\} \oplus \{\nu_5 \circ \eta_8 \circ \mu_9\} \approx Z_2 \oplus Z_2 \ .$$

Consider the exact sequence

$$\pi^{11}_{19} \xrightarrow{\Delta} \pi^5_{17} \xrightarrow{E} \pi^6_{18} \xrightarrow{H} \pi^{11}_{18} \xrightarrow{\Delta} \pi^5_{16} \ .$$

By (7.17) and Proposition 5.15, the kernel of $\Delta : \pi^{11}_{18} \to \pi^5_{16}$ is generated by $2\sigma_{11}$ which is $\pm H(\Delta(\sigma_{13}))$ by Proposition 2.2 and 2.7.

By Proposition 2.5, (5.10) and by Lemma 6.3, we have

$$\Delta(\bar{\nu}_{11}) = \nu_5 \circ \eta_8 \circ \bar{\nu}_9 = \nu_5 \circ \nu^3_8 = \nu^4_5$$

and

$$\Delta(\varepsilon_{11}) = \nu_5 \circ \eta_8 \circ \varepsilon_9 \ .$$

Then the result $\pi^6_{18} = \{\Delta(\sigma_{13})\} \approx Z_{16}$ follows from the exact sequence and from the relation

(7.25). $$8\Delta(\sigma_{13}) = \nu_6 \circ \mu_9 \ .$$

This is proved as follows.

$8\Delta(\sigma_{13}) = \Delta(\iota_{13}) \circ 8\sigma_{11}$ by Proposition 2.5

$\in \{\nu_6, \eta_9, 2\iota_{10}\} \circ 8\sigma_{11}$ by Lemma 5.10.

$\nu_6 \circ \mu_9 = \nu_6 \circ -\mu_9$

$\in \nu_6 \circ -E^4\{\eta_5, 2\iota_6, E\sigma'''\}$

$= \{\nu_6, \eta_9, 2\iota_{10}\}_4 \circ E^6\sigma'''$ by Proposition 1.4

$\subset \{\nu_6, \eta_9, 2\iota_{10}\} \circ 8\sigma_{11}$ by (1.15) and Lemma 5.14.

By (4.7), $\{\nu_6, \eta_9, 2\iota_{10}\} \circ 8\sigma_{11}$ is a coset of

$$\nu_6 \circ \pi^9_{11} \circ 8\sigma_{11} = \{\nu_6 \circ \eta^2_9 \circ \sigma_{11}\} = 0 \ .$$

Then we have the equality (7.25).

Next consider the exact sequence

$$\pi^{11}_{20} \xrightarrow{\Delta} \pi^5_{18} \xrightarrow{E} \pi^6_{19} \xrightarrow{H} \pi^{11}_{19} \xrightarrow{\Delta} \pi^5_{17} \quad .$$

The last homomorphism Δ is clarified as above, we see that it is an isomorphism into. It follows from the exactness of the above sequence that E is onto. By Proposition 2.5, (5.10), (5.9), (7.10) and Proposition 5.11, we have

$$\Delta(\nu_{11}^3) = \nu_5 \circ \eta_8 \circ \nu_9^3 = 0 ,$$

$$\Delta(\mu_{11}) = \nu_5 \circ \eta_8 \circ \mu_9$$

and

$$\Delta(\eta_{11} \circ \varepsilon_{12}) = \nu_5 \circ \eta_8^2 \circ \varepsilon_{10} = \nu_5 \circ 4\nu_8 \circ \sigma_{11} = 4\nu_5^2 \circ \sigma_{11} = 0.$$

Then we have that $\pi_{19}^6 = \{\nu_6 \circ \sigma_9 \circ \nu_{16}\} \approx Z_2$ and that

(7.26). <u>the kernel of</u> $\Delta : \pi_{20}^{11} \to \pi_{18}^5$ <u>is generated by</u> ν_{11}^3 <u>and</u> $\eta_{11} \circ \varepsilon_{12}$.

The homomorphism $\Delta : \pi_{20}^{13} \to \pi_{18}^6$ is an isomorphism onto since σ_{13} and $\Delta(\sigma_{13})$ are generators of $\pi_{20}^{13} \approx Z_{16}$ and $\pi_{18}^6 \approx Z_{16}$ respectively. It follows from the exactness of (4.4) that the sequences

$$0 \longrightarrow \pi_{19}^7 \xrightarrow{H} \pi_{19}^{13} \xrightarrow{\Delta} \pi_{17}^6$$

and

$$\pi_{21}^{13} \xrightarrow{\Delta} \pi_{19}^6 \xrightarrow{E} \pi_{20}^7 \longrightarrow 0$$

are exact. By (7.18), the generator ν_{13}^2 of $\pi_{19}^{13} \approx Z_2$ is mapped by Δ onto $2\bar{\nu}_6 \circ \nu_{14} \neq 0$. Thus the homomorphism Δ is an isomorphism into and we have $\pi_{19}^7 = 0$.

By Theorem 7.1, the group $\pi_{21}^{13} \approx Z_2 \oplus Z_2$ is generated by $\bar{\nu}_{13}$ and ε_{13}. Then the result $\pi_{20}^7 = \{\nu_7 \circ \sigma_{10} \circ \nu_{17}\} \approx Z_2$ of Theorem 7.7 is a consequence of

(7.27). $$\Delta(\bar{\nu}_{13}) = \Delta(\varepsilon_{13}) = 0 .$$

This is proved as follows.

$$\begin{aligned}
\Delta(\varepsilon_{13}) &= \Delta(-\varepsilon_{13}) = \Delta(\iota_{13}) \circ (-\varepsilon_{11}) && \text{by Proposition 2.5}\\
&\in \{\nu_6, \eta_9, 2\iota_{10}\} \circ (-\varepsilon_{11}) && \text{by Lemma 5.10}\\
&= \{\nu_6, \eta_9, 2\iota_{10}\}_6 \circ (-\varepsilon_{11}) && \text{since } E^6\pi_5^3 = \pi_{11}^9\\
&= \nu_6 \circ E^6\{\eta_3, 2\iota_4, \varepsilon_4\} && \text{by Proposition 1.4}\\
&\subset \nu_6 \circ E^6\pi_{13}^3 && \text{by (4.7)}\\
&= \{\nu_6 \circ \eta_9 \circ \mu_{10}\} + \{\nu_6 \circ E^6\varepsilon'\} && \text{by Theorem 7.3}\\
&= \{E\Delta(\mu_{11})\} + \{\nu_6 \circ 2\nu_9 \circ \sigma_{12}\} && \text{by (7.10)}\\
&= 0 + \{2\nu_6^2 \circ \sigma_{12}\} = 0 .
\end{aligned}$$

$$\begin{aligned}
\Delta(\bar{\nu}_{13} + \varepsilon_{13}) &= \Delta(\eta_{13} \circ \sigma_{14}) && \text{by Lemma 6.4}\\
&= \Delta(\eta_{13}) \circ \sigma_{12} && \text{by Proposition 2.5}\\
&= \Delta(H(\sigma')) \circ \sigma_{12} = 0 && \text{by Lemma 5.14.}
\end{aligned}$$

Thus $\Delta(\varepsilon_{13}) = \Delta(\bar{\nu}_{13} + \varepsilon_{13}) - \Delta(\bar{\nu}_{13}) = 0$ and (7.27) is proved.

By (5.15), Proposition 5.9 and Proposition 5.11, we have

$$\pi^8_{20} = 0$$

$$\pi^8_{21} = \{\sigma_8 \circ \nu^2_{15}\} \oplus \{\nu_8 \circ \sigma_{11} \circ \nu_{18}\} \approx Z_2 \oplus Z_2 .$$

In the exact sequence $0 = \pi^8_{20} \xrightarrow{E} \pi^9_{21} \xrightarrow{H} \pi^{17}_{21}$ of (4.4), $\pi^{17}_{21} = 0$ by Proposition 5.8. Thus we have $\pi^9_{21} = 0$.

Consider the exact sequence

$$\pi^8_{22} \xrightarrow{E} \pi^9_{23} \xrightarrow{H} \pi^{17}_{23} \xrightarrow{\Delta} \pi^8_{21} \xrightarrow{E} \pi^9_{22} \xrightarrow{H} \pi^{17}_{22} ,$$

of (4.4), where $\pi^{17}_{22} = 0$ and $\pi^{17}_{23} = \{\nu^2_{17}\} \approx Z_2$ by Proposition 5.9 and Proposition 5.11. By Proposition 2.5 and (7.19), we have for an odd integer x that

$$\Delta(\nu^2_{17}) = \Delta(\nu_{17}) \circ \nu_{18} = 2\sigma_8 \circ \nu^2_{15} - x\nu_8 \circ \sigma_{11} \circ \nu_{18}$$
$$= \nu_8 \circ \sigma_{11} \circ \nu_{18} .$$

It follows from the exactness of the above sequence that

$$\pi^9_{22} = \{\sigma_9 \circ \nu^2_{16}\} \approx Z_2$$

and that

(7.28). $E : \pi^8_{22} \to \pi^9_{23}$ <u>is a homomorphism onto</u>.

$\pi^{19}_{22} = \{\nu_{19}\} \approx Z_8$ and $\pi^{19}_{24} = \pi^{19}_{23} = 0$ by Proposition 5.6, 5.8 and 5.9. It follows from the exactness of (4.4) that

$$E : \pi^9_{22} \approx \pi^{10}_{23} = \{\sigma_{10} \circ \nu^2_{17}\} \approx Z_2$$

and the sequence

$$0 = \pi^9_{21} \xrightarrow{E} \pi^{10}_{22} \xrightarrow{H} \pi^{19}_{22} \xrightarrow{\Delta} \pi^9_{20}$$

is exact. By (7.22), the kernel of the homomorphism Δ is generated by $2\nu_{19} = \pm H(\Delta(\nu_{21}))$. It follows that $\pi^{10}_{22} = \{\Delta(\nu_{21})\} \approx Z_4$.

Obviously the homomorphism $\Delta : \pi^{21}_{24} \to \pi^{10}_{22}$ is onto and its kernel is generated by $4\nu_{21} = \eta^3_{21}$. It follows from the exactness of (4.4) that the following two sequences are exact:

$$0 \longrightarrow \pi^{11}_{23} \xrightarrow{H} \pi^{21}_{23} = \{\eta^2_{21}\} ,$$

$$\pi^{21}_{25} \xrightarrow{\Delta} \pi^{10}_{23} \xrightarrow{E} \pi^{11}_{24} \xrightarrow{H} \{\eta^3_{21}\} ,$$

where $\pi^{21}_{25} = 0$ by Proposition 5.8. By Lemma 7.5 and Proposition 2.2, we see that the elements θ' and $\theta' \circ \eta_{23}$ are mapped onto η^2_{21} and η^3_{21} respectively and they are of order 2.. It follows that

$$\pi^{11}_{23} = \{\theta'\} \approx Z_2$$

and

$$\pi_{24}^{11} = \{\theta' \circ \eta_{23}\} \oplus \{\sigma_{11} \circ \nu_{18}^2\} \approx Z_2 \oplus Z_2 .$$

Next consider the exact sequence

$$\pi_{26}^{23} \xrightarrow{\Delta} \pi_{24}^{11} \xrightarrow{E} \pi_{25}^{12} \xrightarrow{H} \pi_{25}^{23} \xrightarrow{\Delta} \pi_{23}^{11} \xrightarrow{E} \pi_{24}^{12} \xrightarrow{H} \pi_{24}^{23}$$

of (4.4). By Proposition 2.5 and (7.21) we have

$$\Delta(\nu_{23}) = \Delta(\iota_{23}) \circ \nu_{21} = \sigma_{11} \circ \nu_{18}^2 .$$

Thus we have that $E\pi_{24}^{11} = \{E\theta' \circ \eta_{24}\} \approx Z_2$ and that

(7.29). <u>the kernel of</u> $\Delta : \pi_{26}^{23} \to \pi_{24}^{11}$ <u>is generated by</u> $2\nu_{23}$.

By Lemma 7.5 and Proposition 2.2, we see that the elements θ and $\theta \circ \eta_{24}$ are mapped by H onto the generators of π_{24}^{23} and π_{25}^{23} respectively and they are of order 2. Then it follows from the exactness of the above sequence that

$$\pi_{24}^{12} = \{\theta\} \oplus \{E\theta'\} \approx Z_2 \oplus Z_2$$

and

$$\pi_{25}^{12} = \{\theta \circ \eta_{24}\} \oplus \{E\theta' \circ \eta_{24}\} \approx Z_2 \oplus Z_2 .$$

By use of Proposition 1.2 and Proposition 1.4, we have

$$E^2\theta' \in E^2\{\sigma_{11}, 2\nu_{18}, \eta_{21}\}_1 \subset \{\sigma_{13}, 2\nu_{20}, \eta_{23}\}_3$$

and

$$0 \in \{\sigma_{13}, \nu_{20}, 0\}_3 = \{\sigma_{13}, \nu_{20}, 2\eta_{23}\}_3 \subset \{\sigma_{13}, 2\nu_{20}, \eta_{23}\}_3 .$$

The secondary composition $\{\sigma_{13}, 2\nu_{20}, \eta_{23}\}_3$ is a coset of

$$\begin{aligned} &\sigma_{13} \circ E^3\pi_{22}^{17} + \pi_{24}^{13} \circ \eta_{24} && \text{by (4.7)}\\ &= 0 + \{\zeta_{13} \circ \eta_{24}\} && \text{by Proposition 5.9 and Theorem 7.4}\\ &\subset E^6\pi_{19}^7 = 0 . \end{aligned}$$

Thus we have proved that $E^2\theta' = 0$ and that $E\theta'$ is in the kernel of $E : \pi_{24}^{12} \to \pi_{25}^{13}$. By the exactness of the sequence (4.4), we have that $E\theta' = \Delta(\eta_{25})$.

By Proposition 2.5 we have also $E\theta' \circ \eta_{24} = \Delta(\eta_{25}^2)$.

Then we see that the homomorphisms $\Delta : \pi_{27}^{25} \to \pi_{25}^{12}$ and $\Delta : \pi_{26}^{25} \to \pi_{24}^{12}$ are isomorphisms into. Also we see easily in the last part of the proof of Theorem 7.4 that $\Delta : \pi_{25}^{25} \to \pi_{23}^{12}$ is an isomorphism into. It follows from the exactness of the sequence (4.4) that

$$\pi_{25}^{13} = \{E\theta\} \approx Z_2$$

and

$$\pi_{26}^{13} = \{E\theta \circ \eta_{25}\} \approx Z_2 .$$

Next consider the stable element $E^\infty\theta \in G_{12}$. $E^\infty\theta$ is an element of the secondary composition

$$< \sigma, \nu, \eta > \ \in \ G_{12} / (\sigma \circ G_5 + G_{11} \circ \eta) .$$

By (4.7), Proposition 5.8 and Theorem 7.4,

$$\sigma \circ G_5 + G_{11} \circ \eta = 0 + \{\zeta \circ \eta\} = E^\infty(\zeta_7 \circ \eta_{18}) = 0 .$$

Thus $< \sigma, \nu, \eta >$ consists of the element $E^\infty\theta$ only. By i) of (3.9),

$$< \sigma, \nu, \eta > = < \eta, \nu, \sigma > .$$

Since $\eta_{11} \circ \nu_{12} = 0$ by (5.9) and since $\nu_{11} \circ \sigma_{14} = 2\sigma_{11} \circ \nu_{18} = 0$ by (7.19) and (7.20), then the secondary composition $\{\eta_{11}, \nu_{12}, \sigma_{19}\}_1$ is defined and its image under E^∞ is $< \eta, \nu, \sigma >$. By (4.7), there exists an element of $\pi_{23}^{11} = \{\theta'\}$ belongs to this secondary composition. Since $E^2\pi_{23}^{11} = \{E^2\theta'\} = 0$, we have that $< \eta, \nu, \sigma >$ contains 0. Thus we conclude that $E^\infty\theta = 0$.

Since $E^\infty : \pi_{26}^{14} \rightarrow (G_{12};2)$ is an isomorphism onto by (4.5), then it follows that $E^2\theta = 0$.

From the exactness of the sequence (4.4), we have that $\Delta : \pi_{27}^{27} \rightarrow \pi_{25}^{13} = \{E\theta\}$ is onto, $\Delta(\iota_{27}) = E\theta$ and

$$\pi_{26}^{14} = \pi_{n+12}^{n} = (G_{12};2) = 0, \quad \text{for} \quad n \geq 14.$$

By Proposition 2.5, we have

$$\Delta(\eta_{27}) = \Delta(\iota_{27}) \circ \eta_{25} = E\theta \circ \eta_{25} .$$

Then it follows from the exactness of the sequence (4.4) that $E\pi_{26}^{13} = 0$ and the sequence

$$0 \rightarrow \pi_{27}^{14} \xrightarrow{H} \pi_{27}^{27} \xrightarrow{\Delta} \pi_{25}^{13}$$

is exact. The kernel of Δ is generated by $2\iota_{27} = \pm H(\Delta(\iota_{29}))$. Thus we have

$$\pi_{27}^{14} = \{\Delta(\iota_{29})\} \approx Z .$$

Finally by (4.5) and by the exact sequence (4.4), it is computed easily that

$$\pi_{n+13}^{n} = (G_{13};2) = 0 \quad \text{for} \quad n \geq 15.$$

We see in the above discussion that

(7.30). $\Delta(\iota_{27}) = E\theta$, $\Delta(\eta_{25}) = E\theta'$ <u>and the homomorphisms</u> $\Delta : \pi_{n+15}^{2n+1} \rightarrow \pi_{n+13}^{n}$ <u>are isomorphisms into for</u> $n = 12, 13$ <u>and</u> 14.

PART 2.

CHAPTER VIII

Squaring Operations

Let

$$Sq^i : H^n(X, Z_2) \to H^{n+i}(X, Z_2)$$

be Steenrod's squaring operation. In his paper [18], Steenrod gives a homomorphism of $\pi_{n+i-1}(S^n)$ into Z_2 in terms of the functional squaring operation. The homomorphism will be denoted by

$$H_2 : \pi_{n+i-1}(S^n) \to Z_2, \quad i > 1.$$

H_2 may be defined as follows. Let α be an element of $\pi_{n+i-1}(S^n$ and consider a cell complex $K_\alpha = S^n \cup_\alpha CS^{n+i-1}$ of Chapter I which is uniquely determined by α up to homotopy type. Then

$$H_2(\alpha) \not\equiv 0 \pmod{2}$$

if and only if the squaring operation $Sq^i : H^n(K_\alpha, Z_2) \to H^{n+i}(K_\alpha, Z_2)$ is an isomorphism onto.

For the case $i = 1$, H_2 may be defined for the elements of $2\pi_n(S^n)$ and we have a homomorphism

$$H_2 : 2\pi_n(S^n) \to Z_2 .$$

Since Sq^1 is the Bockstein homomorphism associated with the exact sequence $0 \to Z_2 \to Z_4 \to Z_2 \to 0$ of the coefficient groups, then it is easily verified that

(8.1) $\qquad H_2(2r\iota_n) \equiv r \pmod{2}$ <u>for an integer</u> r .

It follows from properties of the squaring operation that [18]

(8.2) $\quad$ i). $H_2(\alpha) = 0$ <u>if</u> $\alpha \in \pi_{n+i-1}(S^n)$ <u>and</u> $i > n$.

ii). $H_2 \circ E = H_2$.

It is known [1] that H_2 is trivial unless $i = 0, 1, 3, 7$.

<u>Proposition</u> 8.1. <u>Let</u> $k = 1$, (3 <u>or</u> 7 <u>respectively</u>), <u>then</u> $H_2 : \pi_{n+k}(S^n) \to Z_2$ <u>is onto if and only if</u> $n > k$. <u>Let</u> $n > k+1$, <u>then</u> $H_2(\alpha) \not\equiv 0 \pmod{2}$ <u>if and only if</u> $\alpha \equiv \eta_n$, (ν_n <u>or</u> σ_n <u>respectively</u>)

mod $2\pi_{n+k}(S^n)$. Let $n = k+1$, then $H_2(\alpha) \neq 0$ (mod.2) if and only if $\alpha \equiv \eta_2$, (ν_4 or σ_8 respectively) mod $2\pi_{2k+1}(S^{k+1}) + E\pi_{2k}(S^k)$.

Proof. Consider projective plane of complex numbers, quaternions or Cayley numbers. This is a cell complex of the type $K_\alpha = S^{k+1} \cup_\alpha CS^{2k+1}$ such that $Sq^{k+1} \neq 0$. Thus $H_2(\alpha) \neq 0$ for the class α of the attaching map, in fact, the Hopf class. We have that $H_2 : \pi_{n+k}(S^n) \to Z_2$ is a homomorphism onto for $n = k+1$. It follows from ii) of (8.2) that H_2 is onto for $n \geq k+1$. It follows from i) of (8.2) that H_2 is trivial for $n \leq k$. Thus the first statement is proved.

Next consider the case $n > k+1$. Then the group $\pi_{n+k}(S^n)$ is finite. By Propositions 5.1, 5.6 and 5.15, η_n, (ν_n or σ_n respectively) generates the 2-primary component of $\pi_{n+k}(S^n)$. Since H_2 is onto, then the kernel of H_2 is $2\pi_{n+k}(S^n)$. Thus we have that $H_2(\alpha) \not\equiv 0$ (mod.2) if and only if $\alpha \equiv \eta_n$, (ν_n or σ_n respectively) mod $2\pi_{n+k}(S^n)$.

Consider the case $n = k+1$. By Hopf fibering we have that $\pi_{2k+1}(S^{k+1})$ is a direct sum of $E\pi_{2k}(S^k)$ and an infine cyclic subgroup generated by an element β such that $H(\beta) = \pm\iota_{2k+1}$. Here β may be chosen arbitrary under the condition $H(\beta) = \pm\iota_{2k+1}$. Then, we choose that $\beta = \eta_2$, ν_4 or σ_8 respectively. By (8.2), $H_2(E\pi_{2k}(S^k)) = H_2(\pi_{2k}(S^k)) = 0$. Since H_2 is onto, it follows that $H_2(\beta) \not\equiv 0$ (mod.2) and the kernel of H_2 is $2\pi_{2k+1}(S^{k+1}) + E\pi_{2k}(S^k)$. Thus $H_2(\alpha) \not\equiv 0$ if and only if $\alpha \equiv \beta$ mod $2\pi_{2k+1}(S^{k+1}) + E\pi_{2k}(S^k)$. q.e.d.

Let α be an element of $\pi_n(X)$ and consider the space $X \cup_\alpha CS^n$ of Chapter I. Let e^{n+1} be the interior of CS^n, then e^{n+1} is an open $(n+1)$-cell and $X \cup_\alpha CS^n = X \cup e^{n+1}$. Let $p : X \cup e^{n+1} \to S^{n+1}$ be the shrinking map of (1.17). Denote by

$$u_{n+1} \in H^{n+1}(X \cup e^{n+1}, Z_2)$$

the image of the fundamental class of S^{n+1} under the homomorphism p^* induced by p. Now consider an element β of $\pi_{n+i-1}(S^n)$ such that

$$\alpha \circ \beta = 0 .$$

According to the definition of the coextension, we have a mapping

$$f : S^{n+i} \to X \cup e^{n+1}$$

which represents a coextension of β. By (1.18), the composition $p \circ f$ represents $E\beta$.

Lemma 8.2. Assume that $H_2(\beta) \neq 0$, then $Sq^i(u_{n+1}) = u_{n+i+1}$ in $H^*(X \cup e^{n+1} \cup_f CS^{n+i}, Z_2)$, where u_{n+i+1} is represented by CS^{n+i} as same as u_{n+1}.

Proof. By shrinking the subset X of $X \cup e^{n+1} \cup_f CS^{n+i}$ to a point, we have a mapping $p : X \cup e^{n+1} \cup_f CS^{n+i} \to S^{n+1} \cup_{p \circ f} CS^{n+i}$. Since $H_2(E\beta) = H_2(\beta) \neq 0$, then $Sq^i \neq 0$ in $S^{n+1} \cup_{p \circ f} CS^{n+i}$. It follows by the induced homomorphism $H^*(S^{n+1} \cup_{p\circ f} CS^{n+i}, Z_2) \to H^*(X \cup e^{n+1} \cup_f CS^{n+i}, Z_2)$ that $Sq^i u_{n+1} = u_{n+i+1}$. q.e.d.

We shall see in the following examples how the lemma is applied, in which Adem's relations in iterated squaring operations are essentially useful.

Example 1. Since $H_2 : \pi_{n+k}(S^n) \to Z_2$ is onto for $n \geq k+1$ and for $k = 1, 3, 7$, then the elements η_n, ν_n and σ_n are not divisible by 2 and thus they are not zero.

Example 2. The secondary composition $\{2\iota_n, \eta_n, 2\iota_{n+1}\} \not\ni 0$. Since $\pi_{n+2}(S^n) = \{\eta_n \circ \eta_{n+1}\} \approx Z_2$, it follows that the secondary composition consists of a single element $\eta_n \circ \eta_{n+1}$.

To prove this we assume that $\{2\iota_n, \eta_n, 2\iota_{n+1}\}$ contains 0. Then by Proposition 1.7, there exists an extension $\alpha \in \pi(K \to S^n)$ of $2\iota_n$ and a coextension $\beta \in \pi(S^{n+2} \to K)$ of $2\iota_{n+1}$ such that the composition $\alpha \circ \beta$ is zero, where $K = S^n \cup e^{n+2}$ has η_n as the class of the attaching map and thus $Sq^2 \neq 0$ in K. Let $L = S^n \cup_\alpha CK$ which is a cell complex of the form $S^n \cup e^{n+1} \cup e^{n+3}$. Since $S^n \cup e^{n+1}$ has $2\iota_n$ as the attaching class, then $Sq^1 \neq 0$ in $S^n \cup e^{n+1}$. It is easy to see that $Sq^2Sq^1 \neq 0$ in L. Let $\tilde{\beta} \in \pi(S^{n+3} \to L)$ be a coextension of β. $\tilde{\beta}$ exists since $\alpha \circ \beta = 0$. Construct a cell complex $M = L \cup_{\tilde{\beta}} CS^{n+3} = S^n \cup e^{n+1} \cup e^{n+3} \cup e^{n+4}$, then by use of the above lemma we have that $Sq^1Sq^2Sq^1 \neq 0$ in M. On the other hand, we have an Adem's relation $Sq^2Sq^2 = Sq^1Sq^2Sq^1$. Since there is no cell of dimension $n+2$, $Sq^2Sq^2 = 0$ in M, but this is a contradiction. Thus we conclude that $\{2\iota_n, \eta_n, 2\iota_{n+1}\} \not\ni 0$.

Example 3. The compositions $\eta_n^2, \nu_n^2, \eta_n \circ \sigma_{n+1}$ and σ_n^2 are not divisible by 2 and thus they are not trivial (Adem). We shall show for σ_n^2, the other cases are proved similarly. Assume that $\sigma_n^2 = 2\alpha$ for some element α of $\pi_{n+14}(S^n)$. Then we may construct a cell complex $K = S^n \cup e^{n+8} \cup e^{n+15} \cup e^{n+16}$

such that e^{n+8} and e^{n+15} are attached by σ_n and α respectively, and that e^{n+16} is attached by a coextension of $\sigma_{n+7} - 2\iota_{n+14}$.

Let u_n, u_{n+8} and u_{n+16} be cohomology classes mod 2 which are represented by S^n, e^{n+8} and e^{n+16} respectively. Then we have that

$$Sq^8 Sq^8 u_n = Sq^8 u_{n+8} = u_{n+16} .$$

Now we have a relation $Sq^8Sq^8 = Sq^{12}Sq^4 + Sq^{14}Sq^2 + Sq^{15}Sq^1$ of Adem. Obviously, $Sq^1u_n = Sq^2u_n = Sq^4u_n = 0$. Thus $Sq^8Sq^8u_n = 0$, but this is a contradiction. Therefore we have proved that σ_n^2 cannot be divisible by 2.

Example 4. The elements of the secondary composition $\{\eta_n, \nu_{n+1}, \eta_{n+4}\}$, $\{\nu_n, \eta_{n+3}, \nu_{n+4}\}$, $\{\nu_n, \sigma_{n+3}, \nu_{n+10}\}$ and $\{\sigma_n, \nu_{n+7}, \sigma_{n+10}\}$ are not divisible by 2 and thus they are not trivial.

We shall show this for the last secondary composition. For the other secondary compositions, the proofs are similar.

Assume that there exists an element α of $\pi_{n+18}(S^n)$ such that 2α is contained in $\{\sigma_n, \nu_{n+7}, \sigma_{n+10}\}$. By Proposition 1.7, 2α is a composition $\beta \circ \gamma$ of an extension $\beta \in \pi(K \to S^n)$ of σ_n and a coextension γ of σ_{n+10}, where $K = S^{n+7} \cup e^{n+11} = S^{n+7} \cup_\delta CS^{n+10}$ for $\delta = \nu_{n+7}$. Let $K \vee S^{n+18}$ be the union of K and S^{n+18} with the base point in common. Let $F : K \vee S^{n+18} \to S^n$ represent β on K and represent γ on S^{n+18}. Let $G : S^{n+18} \to K \vee S^{n+18}$ be a mapping which represents the sum $\gamma + (-2\iota_{n+18})$. Since $\beta \circ \gamma$ equals to 2α, then it follows that the composition $F \circ G$ is homotopic to zero. By a similar way to Example 2, we construct a cell complex $L = S^n \cup_F C(K \vee S^{n+18}) = S^n \cup e^{n+8} \cup e^{n+12} \cup e^{n+19}$ and a coextension $\lambda \in \pi(S^{n+19} \to L)$ of $\{G\}$. Let $M = L \cup_\lambda CS^{n+19} = L \cup e^{n+20}$. Then it is verified that $Sq^8Sq^4Sq^8 \neq 0$ in M. By use of Adem's relation, we have a relation

$$Sq^8Sq^4Sq^8 = Sq^4Sq^{16} + Sq^{16}Sq^4 + (Sq^{18} + Sq^{14}Sq^4)Sq^2 + Sq^{15}Sq^4Sq^1 .$$

Since there is no cell of dimension $(n+1)$, $(n+2)$, $(n+4)$ or $(n+16)$, in M, the right side of the above equation vanishes in M, but this is a contradiction. Thus we have proved that any element of $\{\sigma_n, \nu_{n+7}, \sigma_{n+10}\}$ cannot be divisible by 2.

Next we shall consider about the operation Sq^{16}.

Lemma 8.3. Let $n \geq 16$, then there exists a cell complex $K = S^n \cup e^{n+8} \cup e^{n+16}$ such that, for a generator u of $H^n(K, Z_2)$, $Sq^{16}u \neq 0$ and *that the attaching map of* e^{n+8} *represents* σ_n *and the attaching map of* e^{n+16} *represents a coextension of* $2\sigma_{n+7}$.

Proof. Let $M = S^8 \cup e^{16}$ be given by attaching e^{16} by σ_8. Then $Sq^8 u_8 = u_{16}$ for the classes u_8 and u_{16} represented by S^8 and e^{16}. By Cartan's formula, $Sq^{16}(u_8 \times u_8) = u_{16} \times u_{16}$ in the product $M \times M$. Consider the reduced product $M \# M$ of two copies of M and let $\Phi : M \times M \to M \# M$ be a shrinking map which defines $M \# M$. Then $M \# M$ is a cell complex of the form $S^{16} \cup e_1^{24} \cup e_2^{24} \cup e^{32}$ such that $S^{16} \cup e_1^{24} = M \# S^8 = E^8 M$ and $S^{16} \cup e_2^{24} = S^8 \# M$. The cohomology mod 2 of S^{16} and e^{32} are mapped by Φ^* onto $u_8 \times u_8$ and $u_{16} \times u_{16}$ respectively. Then it follows that $Sq^{16} \neq 0$ in $M \times M$. The attaching maps of e_1^{24} and e_2^{24} represent $E^8\sigma_8 = \sigma_{16}$ and $\iota_8 \# \sigma_8$ respectively. By Proposition 3.1, $\iota_8 \# \sigma_8 = E^8\sigma_8 = \sigma_{16}$. Thus there exists an extension

$$f : S^{16} \cup e_2^{24} \to S^{16} \cup e_1^{24}$$

of the identity of S^{16} such that e_2^{24} is mapped onto e_1^{24} by mapping-degree 1. Identifying each point of e_2^{24} with its image under f, we obtain a cell complex

$$K = S^{16} \cup e^{24} \cup e^{32} .$$

We shall prove that this complex K satisfies the condition of the lemma for $n = 16$. There is no changement for S^{16} and e^{32} in the above identification. Thus $Sq^{16} \neq 0$ still holds. Since $S^{16} \cup e^{24} = E^8 M$, then we have that e^{24} is attached to S^{16} by a representative of σ_{16}. Next consider the class of the attaching map of e^{32}, which depends on the choice of the orientation of e^{32}. Let $F : M \# M \to K$ be the above identification of f. Let p and p' be mappings which shrink S^{16} of K and $M \# M$ respectively. Then there exists a mapping F' such that the following diagram is commutative.

$$\begin{array}{ccc} M \# M & \xrightarrow{\quad F \quad} & K \\ \Big\downarrow p' & & \Big\downarrow p \\ (S_1^{24} \vee S_2^{24}) \cup e^{32} & \xrightarrow{\quad F' \quad} & S^{24} \cup e^{32} \end{array} .$$

The mapping F' maps e^{32} homeomorphically onto e^{32} and maps S_1^{24} and S_2^{24} onto S^{24} by degree 1. Let $L = (S_1^{24} \vee S_2^{24}) \cup e^{32}$. By shrinking S_1^{24}

to a point, we get from L the reduced join $M \# S^{16} = S_2^{24} \cup e^{32}$, in which e^{32} is attached by $E^{16}\sigma_8 = \sigma_{24}$. By shrinking S_2^{24} to a point we get $S^{16} \# M = S_1^{24} \cup e^{32}$, in which e^{32} is attached to S_1^{24} by $\iota_{16} \# \sigma_8 = \sigma_{24}$ (Proposition 3.1). Remark that these discussions are allowed under suitable choice of orientations in S_1^{24}, S_2^{24}, and e^{32} such that the shrinking maps preserve the orientations. Now, we see that the attaching map of e^{32} in L represents the sum of the classes of the attaching maps in $M \# S^{16}$ and $S^{16} \# M$. Then, by the mapping F', we see that the class of the attaching map of e^{32} represents $2\sigma_{24}$ by a suitable choice of the orientation of e^{32}. Let $p_0 : S^{16} \cup e^{24} \to S^{24}$ be the restriction of p on $S^{16} \cup e^{24}$, and consider the homomorphism

$$p_{0*} : \pi_{31}(S^{16} \cup e^{24}) \to \pi_{31}(S^{24})$$

induced by p_0. $p_{0*} : \pi_{31}(S^{16} \cup e^{24}, S^{16}) \approx \pi_{31}(S^{24})$ by Theorem 2 of [6]. By use of (1.18), we have that $p_{0*}(\gamma) = 2\sigma_{24}$ if and only if γ is a coextension of $2\sigma_{23}$. Then it follows that the attaching map of e^{32} in K represents $\pm\gamma$. By changing the orientation of e^{32}, if it is necessary, we have proved that K satisfies the lemma for $n = 16$.

The case that $n > 16$ is proved by considering $E^{n-16}K$.

q.e.d.

<u>Theorem</u> 8.4. i). <u>The secondary composition</u> $\{\sigma_n, 2\sigma_{n+7}, \eta_{n+14}\}$ <u>contains an element</u> α <u>which can not be decomposed into the sum of a form</u> $\alpha = 2\alpha' + \Sigma\beta_i \circ \gamma_i$, <u>where</u> $\alpha' \in \pi_{n+16}(S^n)$, $\beta_i \in \pi_{n+k_i}(S^n)$, $\gamma_i \in \pi_{n+16}(S^{n+k_i})$ <u>and</u> $1 < k_i < 15$.

ii). <u>The secondary composition</u> $\{\sigma_n, 2\sigma_{n+7}, \nu_{n+14}\}$ <u>contains an element</u> α <u>which can not be decomposed into the sum of a form</u> $\alpha = 2\alpha' + \Sigma\beta_i \circ \gamma_i$, <u>where</u> $\alpha' \in \pi_{n+18}(S^n)$, $\beta_i \in \pi_{n+k_i}(S^n)$, $\gamma_i \in \pi_{n+18}(S^{n+k_i})$, $0 < k_i < 18$ <u>and</u> $k_i \neq 3, 15$.

<u>Proof</u>. We shall prove i). The proof of ii) is similar.

Consider the complex $K = S^n \cup e^{n+8} \cup e^{n+16}$ of Lemma 8.3. Let $\delta \in \pi_{n+15}(S^n \cup e^{n+8})$ be the class of the attaching map of e^{n+16}. By Lemma 8.3, δ is a coextension of $2\sigma_{n+7}$. Then it follows from Proposition 2.8, that there exists an element α of $\{\sigma_n, 2\sigma_{n+7}, \eta_{n+14}\}$ such that

$$i_*(\alpha) = -\delta \circ \eta_{n+15} = \delta \circ \eta_{n+15}.$$

Now assume that α is decomposed into $\alpha = 2\alpha' + \Sigma \beta_i \circ \gamma_i$. By attaching cells of appropriate dimensions to S^n by representatives of α' and β_i, we have a cell complex L such that its $(n+2)$-skeleton is S^n. By the assumption, we have that $\delta \circ \eta_{n+15}$ is inessential in $L_0 = L \cup_\delta e^{n+16}$. Thus there exists a coextension $\delta' \in \pi_{n+17}(L_0 \cup_\delta e^{n+16}) = \pi_{n+17}(L)$ of η_{n+15}. Consider the complex

$$M = L \cup_{\delta'} e^{n+18} = L_0 \cup_\delta e^{n+16} \cup_{\delta'} e^{n+18} .$$

By Lemma 8.2, $Sq^2 u_{n+16} = u_{n+18}$. In the inessentiality of $\delta \circ \eta_{n+15}$, we use the relation $\alpha = 2\alpha' + \Sigma \beta_i \circ \gamma_i$. Then we get the coextension δ' by use of coextensions $2\iota_{n+16}$ and γ_i for corresponding cells and by combining them in S^n by use of the relation. In particular, e^{n+18} cover e^{n+17} by degree 2, where e^{n+17} is the only $(n+17)$-cell and it corresponds to α'. Thus, we have that $u_{n+18} \neq 0$.

Since e^{n+16} is the only $(n+16)$-cell of M, then it follows from Lemma 8.3 that

$$Sq^2 Sq^{16} u_n = Sq^2 u_{n+16} = u_{n+18} \neq 0.$$

By shrinking S^n of L to a point, we get a complex L/S^n which is the union of $K/S^n = S^{n+8} \cup_{2\sigma} e^{n+16}$ and some spheres having a point in common. Since the squaring operations in K/S^n are trivial, then the squaring operations in L/S^n are trivial. It follows that

$$Sq^a Sq^b Sq^c = 0 \text{ in } M = L \cup e^{n+18}$$

if a, b and c are positive. Now consider the relation

$$Sq^2 Sq^{16} = Sq^8 Sq^2 Sq^8 + (Sq^2 Sq^{12} + Sq^{13} Sq^1) Sq^4$$
$$+ (Sq^{16} + Sq^{15} Sq^1) Sq^2 + (Sq^{15} Sq^2 + Sq^{14} Sq^3 + Sq^{13} Sq^4) Sq$$

of Adem. Since there is no $(n+2)$-cell, then is follows that

$$Sq^2 Sq^{16} u_n = 0.$$

But this is a contradiction. Thus α can not be decomposed into a form $2\alpha' + \Sigma \beta_i \circ \gamma_i$. q.e.d.

Remark that the theorem 8.4 still holds even if we allow to take $k_i = 1, 15$ in i) and $k_i = 3, 15$ in ii), since there is no element of $H_2 \neq 0$ in $\pi_{n+16}(S^n)$, i.e., there is no complex $S^n \cup e^{n+16}$ of $Sq^{16} \neq 0$.

Then we see that the theorem 8.4 holds for arbitrary element of $\{\sigma_n, 2\sigma_{n+7}, \eta_{n+14}\}$ and $\{\sigma_n, 2\sigma_{n+7}, \nu_{n+1}\}$.

CHAPTER IX

Lemmas for Generators of $\pi_{n+11}(S^n;2)$.

By Theorem 7.4, the 2-primary component $(G_{11};2)$ of the stable group G_{11} is isomorphic to Z_8 and it is generated by $\zeta = E^\infty \zeta_5$. Consider the secondary composition

$$\langle \nu, 16\iota, \sigma \rangle$$

which is a coset of

$$\nu \circ G_8 + \sigma \circ G_4 = \nu \circ (G_8;2) + \sigma \circ (G_4;2)$$

$$= \{\nu \circ \varepsilon\} + \{\nu \circ \bar{\nu}\} + 0 \qquad \text{By Theorem 7.1 and Proposition 5.8}$$

$$= 0 \qquad \text{by (7.17) and (7.18).}$$

Thus $\langle \nu, 16\iota, \sigma \rangle$ consists of a single element. By the definitions of ζ and ζ_5, we have

$$\zeta = E^\infty \zeta_5 \in E^\infty\{\nu_5, 8\iota_8, E\sigma'\}$$

$$\subset \langle E^\infty \nu_5, E^\infty 8\iota_8, E^\infty \sigma' \rangle$$

$$= \langle \nu, 8\iota, 2\sigma \rangle \qquad \text{by Lemma 5.14}$$

$$\subset \langle \nu, 16\iota, \sigma \rangle \qquad \text{by (3.5).}$$

Thus we have obtained

(9.1). $\langle \nu, 8\iota, 2\sigma \rangle$ <u>and</u> $\langle \nu, 16\iota, \sigma \rangle$ <u>consist of a single element</u> ζ.

By i) of (3.9),

(9.2) $\langle 2\sigma, 8\iota, \nu \rangle = \langle \sigma, 16\iota, \nu \rangle$ <u>consists of a single element</u> ζ.

By ii) of (3.9), we have a relation

$$-\langle \nu, 8\iota, 2\sigma \rangle + \langle 8\iota, 2\sigma, \nu \rangle + \langle 2\sigma, \nu, 8\iota \rangle \equiv 0$$

mod. $\nu \circ G_8 + 2\sigma \circ G_4 + 8\iota \circ G_{11} = 8G_{11}$. Since $(G_{11};2) \approx Z_8$, then $8G_{11}$ coincides with the odd component of G_{11}. By (3.5),

$$\langle 2\sigma, \nu, 8\iota \rangle \equiv 2\langle \sigma, \nu, 8\iota \rangle \mod 8G_{11} = 16G_{11}.$$

Then it follows that

$$\langle \nu, 8\iota, 2\sigma \rangle \subset \langle 8\iota, 2\sigma, \nu \rangle + \langle 2\sigma, \nu, 8\iota \rangle + 8G_{11}$$
$$\subset \langle 8\iota, 2\sigma, \nu \rangle + 2G_{11}.$$

Since $2G_{11}/8G_{11}$ is generated by the class of 2ζ, then we have that the secondary composition $\langle 8\iota, 2\sigma, \nu \rangle$ contains $x\zeta$ for an odd integer x. By i) of (3.9), $\langle \nu, 2\sigma, 8\iota \rangle = \langle 8\iota, 2\sigma, \nu \rangle$. By (3.5), $\langle 8\iota, 2\sigma, \nu \rangle \supset \langle 16\iota, \sigma, \nu \rangle$, where the secondary compositions are cosets of the same subgroup $8G_{11} = 16G_{11}$. Thus $\langle 8\iota, 2\sigma, \nu \rangle = \langle 16\iota, \sigma, \nu \rangle$. By i) of (3.9), $\langle \nu, \sigma, 16\iota \rangle = \langle 16\iota, \sigma, \nu \rangle$. Consequently we have obtained

(9.3). $\langle 8\iota, 2\sigma, \nu \rangle = \langle 16\iota, \sigma, \nu \rangle = \langle \nu, \sigma, 16\iota \rangle = \langle \nu, 2\sigma, 8\iota \rangle$
$= x\zeta + 8G_{11}$ <u>for an odd integer</u> x.

Next we prove

(9.4). <u>The stable secondary compositions</u> $\langle \sigma, \eta^3, 2\iota \rangle$, $\langle 2\iota, \eta^3, \sigma \rangle$, $\langle \eta \circ \varepsilon, \eta, 2\iota \rangle$, $\langle 2\iota, \eta, \eta \circ \varepsilon \rangle$, $\langle \varepsilon, \eta^2, 2\iota \rangle$, $\langle 2\iota, \eta^2, \varepsilon \rangle$, $\langle \eta, \eta \circ \varepsilon, 2\iota \rangle$, $\langle 2\iota, \eta \circ \varepsilon, \eta \rangle$, $\langle 2\iota, \eta \circ \varepsilon, \eta \rangle$, $\langle \eta^2, \varepsilon, 2\iota \rangle$, $\langle 2\iota, \varepsilon, \eta^2 \rangle$ <u>are the same coset of the subgroup</u> $2G_{11}$. $\langle \nu^3, \eta, 2\iota \rangle = \langle 2\iota, \eta, \nu^3 \rangle = 2G_{11}$.

First remark that

$$\sigma \circ G_4 = \sigma \circ (G_4;2) = 0,$$
$$\varepsilon \circ G_2 = \{\varepsilon \circ \eta^2\} = 0 \qquad \text{by (7.10) and (7.20)},$$
$$\varepsilon \circ G_3 = \{\varepsilon \circ \nu\} = 0, \qquad \text{by (7.18)},$$
$$\eta \circ G_{10} = \{\eta^2 \circ \mu\} = \{4\zeta\} \subset 2G_{11},$$
$$\eta^2 \circ G_9 = \{\eta^2 \circ \varepsilon\} + \{\eta^2 \circ \nu^3\} + \{\eta^2 \circ \mu\},$$
$$= 0 + 0 + \{4\zeta\} \subset 2G_{11},$$

and $$\nu^3 \circ G_2 = \{\nu^3 \circ \eta^2\} = 0.$$

Then the secondary compositions in (9.4) are cosets of $2G_{11}$. Then, by use of (3.5),

$$\langle \eta \circ \varepsilon, \eta, 2\iota \rangle = \langle \varepsilon, \eta^2, 2\iota \rangle$$

and $\langle \eta \circ \varepsilon, \eta, 2\iota \rangle = \langle \eta, \eta \circ \varepsilon, 2\iota \rangle = \langle \eta^2, \varepsilon, 2\iota \rangle$.

By i) of (3.9),

$$\langle \sigma, \eta^3, 2\iota \rangle = \langle 2\iota, \eta^3, \sigma \rangle, \ \langle \eta \circ \varepsilon, \eta, 2\iota \rangle = \langle 2\iota, \eta, \eta \circ \varepsilon \rangle,$$
$$\langle \varepsilon, \eta^2, 2\iota \rangle = \langle 2\iota, \eta^2, \varepsilon \rangle, \ \langle \eta, \eta \circ \varepsilon, 2\iota \rangle = \langle 2\iota, \eta \circ \varepsilon, \eta \rangle,$$
$$\langle \eta^2, \varepsilon, 2\iota \rangle = \langle 2\iota, \varepsilon, \eta^2 \rangle \text{ and } \langle \nu^3, \eta, 2\iota \rangle = \langle 2\iota, \eta, \nu^3 \rangle.$$

Since $(G_5;2) = 0$, then

$$\nu^2 \circ < \nu, \eta, 2\iota > \subset \nu^2 \circ G_5 = \nu^2 \circ (G_5;2) = 0 .$$

By (3.5),

$$< \nu^3, \eta, 2\iota > \supset \nu^2 \circ < \nu, \eta, 2\iota > = 0.$$

Thus $< \nu^3, \eta, 2\iota > = 2G_{11}$. Finally, we have

$$\begin{aligned} < \eta \circ \varepsilon, \eta, 2\iota > &= < \eta \circ \varepsilon, \eta, 2\iota > + 2G_{11} \\ &= < \eta \circ \varepsilon, \eta, 2\iota > + < \nu^3, \eta, 2\iota > \\ &= < \eta \circ \varepsilon + \nu^3, \eta, 2\iota > \qquad \text{by (3.8)} \\ &= < \sigma \circ \eta^2, \eta, 2\iota > \quad \text{by Lemma 6.4 and Lemma 6.3} \\ &= < \sigma, \eta^3, 2\iota >. \end{aligned}$$

Consequently we see that (9.4) is proved.

<u>Lemma</u> 9.1. $< \sigma, \eta^3, 2\iota > = < 2\iota, \eta^3, \sigma > = \zeta + 2G_{11}$,

$$< \eta \circ \varepsilon, \eta, 2\iota > = < 2\iota, \eta, \eta \circ \varepsilon > = \zeta + 2G_{11} ,$$

$$< \varepsilon, \eta^2, 2\iota > = < 2\iota, \eta^2, \varepsilon > = \zeta + 2G_{11} ,$$

$$< \eta, \eta \circ \varepsilon, 2\iota > = < 2\iota, \eta \circ \varepsilon, \eta > = \zeta + 2G_{11} ,$$

$$< \eta^2, \varepsilon, 2\iota > = < 2\iota, \varepsilon, \eta^2 > = \zeta + 2G_{11}.$$

<u>Proof</u>. By (9.4), it is sufficient to prove that

$$< \eta \circ \varepsilon, \eta, 2\iota > \not\equiv 0 \mod 2G_{11}.$$

We shall show that the assumption $< \eta \circ \varepsilon, \eta, 2\iota > \equiv 0 \mod 2G_{11}$ leads us to a contradiction, then the lemma is proved.

Let n be sufficiently large, for example $n > 13$. Assume that $< \eta \circ \varepsilon, \eta, 2\iota > \equiv 0 \mod 2G_{11}$, then $< \eta_n \circ \varepsilon_{n+1}, \eta_{n+9}, 2\iota_{n+10} >$ contains 0. By Proposition 1.7, the trivial element 0 is the composition of an extension $\alpha_1 \in \pi(K_1 \to S^n)$ of $\eta_n \circ \varepsilon_{n+1}$ and a coextension $\beta_1 \in \pi_{n+11}(K_1)$ of $2\iota_{n+10}$, where $K_1 = S^{n+9} \cup e^{n+11}$ has η_{n+9} as the class of the attaching map of e^{n+11}. Since $H_2(\eta_{n+9}) \neq 0$, then $Sq^2 \neq 0$ in K_1. Let $L_1 = K_1 \cup e^{n+12}$ be constructed by attaching e^{n+12} by a representative of β_1. Since $\alpha_1 \circ \beta_1 = 0$, then there exists an extension $\tilde{\alpha} \in \pi(L_1 \to S^n)$ of α_1. Let

$$f_1 : L_1 \to S^n$$

be a representative of $\tilde{\alpha}$. By Lemma 8.1, we see that

$$Sq^3 = Sq^1Sq^2 \neq 0 \text{ in } L_1.$$

Similarly, from the relation $< \nu^3, \eta, 2\iota > \equiv 0$ of (9.4) we construct a cell complex $L_2 = S^{n+9} \cup e^{n+11} \cup e^{n+12}$ such that

$$Sq^3 = Sq^1 Sq^2 \neq 0 \quad \text{in} \quad L_2$$

and a mapping

$$f_2 : L_2 \to S^n$$

such that $f_2 \mid S^{n+9}$ represents ν_n^3. Let $f_3 : S^{n+9} \to S^n$ be a representative of μ_n. Then the mappings f_1, f_2 and f_3 define a mapping

$$f : L_1 \vee L_2 \vee S^{n+9} \to S^n ,$$

where $L_1 \vee L_2 \vee S^{n+9}$ is the union of L_1, L_2 and S^{n+9} having the base point in common. Since $\eta_n \circ \varepsilon_{n+1}$, ν_n^3 and μ_n generate the 2-primary component of $\pi_{n+9}(S^n)$, by Theorem 7.2, then

$$f_* : \pi_{n+9}(L_1 \vee L_2 \vee S^{n+9}) \to \pi_{n+9}(S^n)$$

is a homomorphism onto the 2-primary component of $\pi_{n+9}(S^n)$.

Next, we introduce a result of [23]. Let K_9 be a CW-complex such that its $(n+9)$-skeleton is S^n and $\pi_i(K_9) = 0$ for $i \geq n+9$. Then Proposition 4.9 of [23] states that

(9.5). $H^{n+10}(K_9, Z_2) \approx Z_2 + Z_2 + Z_2$. <u>There are generators</u> h_9, i_9 <u>and</u> j_9 of $H^{n+10}(K_9, Z_2)$ <u>such that</u> $Sq^2 h_9 = Sq^3 i_9 = 0$.

Consider a space S of paths in K_9 which start in K_9 and end in S^n. By associating to each path the starting point, we have a fibering

$$p' : S \to K_9$$

in the sense of Serre [13], such that E has S^n as its deformation retract and p' is equivalent to the injection of S^n into K_9 with respect to the retraction. The fibre $Y = p'^{-1}(e_0)$ of this fibering consists of the paths starting at e_0. It is verified easily from the homotopy exact sequence of the fibering that the injection homomorphism

$$i_* : \pi_i(Y) \to \pi_i(S)$$

is an isomorphism onto for $i > n+8$ and $\pi_i(Y) = 0$ for $i \leq n+8$. Then Y is an $(n+8)$-connective fibre space over S^n. In fact, by associating the end point to each path, we have a fibering

$$p : Y \to S^n$$

and this is equivalent to the injection $i : Y \subset S$ by the retraction. Thus

$$p_* : \pi_i(Y) \to \pi_i(S^n)$$

is an isomorphism onto for $i > n+8$.

By Proposition 5, Chapter III of [13], we have the following exact sequence of cohomology groups associated with the fibering $p' : S \to K_9$.

$$\cdots \to H^i(S, Z_2) \xrightarrow{i^*} H^i(Y, Z_2) \xrightarrow{\Sigma} H^{i+1}(K_9, Z_2) \xrightarrow{p'^*} H^{i+1}(S, Z_2) \to \cdots$$

for $i < 2n + 8$, where Σ is a suspension homomorphism. Since $H^i(S, Z_2) = 0$ for $i > n$, then we have that Σ is an isomorphism for $n < i < 2n+8$. Since Σ commutes with the squaring operations, then it follows from (9.5) that

(9.6). $H^{n+9}(Y, Z_2) \approx Z_2 + Z_2 + Z_2$ <u>and</u> $H^{n+9}(Y, Z_2)$ <u>contains a subgroup isomorphic to</u> $Z_2 + Z_2$ <u>which vanishes under the operation</u> $Sq^3 = Sq^1Sq^2$.

Now consider the mapping f. Since the $(n+8)$-skeleton of the cell complex $L_1 \vee L_2 \vee S^{n+9}$ is trivial, there exists a mapping $F : L_1 \vee L_2 \vee S^{n+9} \to Y$ such that

$$p \circ F = f.$$

Consider the following commutative diagram.

$$\begin{array}{ccc} \pi_{n+9}(L_1 \vee L_2 \vee S^{n+9}) & \xrightarrow{F_*} & \pi_{n+9}(Y) \\ & \searrow^{f_*} & \downarrow p_* \\ & & \pi_{n+9}(S^n). \end{array}$$

Since p_* is an isomorphism onto and since f_* is a homomorphism onto the 2-primary component, then F_* is a homomorphism onto the 2-primary component. Since Y and $L_1 \vee L_2 \vee S^{n+9}$ are $(n+8)$-connected, then it follows that $F_* : H_{n+9}(L_1 \vee L_2 \vee S^{n+9}, Z_2) \to H_{n+9}(Y, Z_2)$ is onto. By the duality,

$$F^* : H^{n+9}(Y, Z_2) \to H^{n+9}(L_1 \vee L_2 \vee S^{n+9}, Z_2)$$

is an isomorphism into. Obviously $H^{n+9}(L_1 \vee L_2 \vee S^{n+9}, Z_2) \approx Z_2 + Z_2 + Z_2$. Thus the homomorphism F^* is an isomorphism onto. Since $Sq^3 \neq 0$ in L_1 and L_2 and since $Sq^3 = 0$ in S^{n+9}, then we have that

$$Sq^3H^{n+9}(L_1 \vee L_2 \vee S^{n+9}, Z_2) \approx Z_2 + Z_2 .$$

It follows from (9.6) that

$$Sq^3H^{n+9}(Y, Z_2) \approx Z_2 \text{ or } 0.$$

But these results contradict to the commutativity of the following diagram.

$$\begin{array}{ccc} H^{n+9}(Y, Z_2) & \xrightarrow{F^*} & H^{n+9}(L_1 \vee L_2 \vee S^{n+9}, Z_2) \\ \downarrow Sq^3 & & \downarrow Sq^3 \\ H^{n+12}(Y, Z_2) & \xrightarrow{F^*} & H^{n+12}(L_1 \vee L_2 \vee S^{n+9}, Z_2). \end{array}$$

Consequently we get a contradiction from the assumption $\langle \eta \circ \varepsilon, \eta, 2\iota \rangle \equiv 0$. Thus the lemma is proved. q.e.d.

Lemma 9.2. $\nu_7 \circ \zeta_{10} = E^2\sigma''' \circ \sigma_{14}$.

Proof. Consider the homomorphism $E^\infty : \pi_{22}(S^{11}) \to G_{11}$. This is an isomorphism of the 2-primary components, by Theorem 7.4. Moreover this is an isomorphism onto, since $\pi_{22}(S^{11})$ is stable in odd promary components (cf. (13.5)). By (7.20), $\nu_{11} \circ \sigma_{14} = 0$. Then the secondary composition $\{8\iota_{11}, \nu_{11}, \sigma_{14}\}$ is defined and it is mapped onto $< 8\iota, \nu, \sigma >$ by E^∞. $< 8\iota, \nu, \sigma >$ is a coset of $8G_{11} + \sigma \circ G_4 = 8G_{11}$. By (3.5), $< 8\iota, \nu, \sigma > \subset < 2\iota, 4\nu, \sigma > = < 2\iota, \eta^3, \sigma >$. Then it follows from Lemma 9.1 that

$$x\zeta \in < 8\iota, \nu, \sigma >$$

for an odd integer x. Thus, $x\zeta_{11} \in \{8\iota_{11}, \nu_{11}, \sigma_{14}\}$. We have

$$E(x\nu_7 \circ \zeta_{10}) = \nu_8 \circ x\zeta_{11} \in \nu_8 \circ \{8\iota_{11}, \nu_{11}, \sigma_{14}\}$$

$$= -\{\nu_8, 8\iota_{11}, \nu_{11}\} \circ \sigma_{15} \qquad \text{by Proposition 1.4}$$

and $$E(E^2\sigma''' \circ \sigma_{14}) \in E^3\{\nu_5, 8\iota_8, \nu_8\} \circ \sigma_{15} \qquad \text{by Lemma 5.13}$$

$$\subset -\{\nu_8, 8\iota_{11}, \nu_{11}\} \circ \sigma_{15} \qquad \text{by Proposition 1.3.}$$

The composition $-\{\nu_8, 8\iota_{11}, \nu_{11}\} \circ \sigma_{15}$ is a coset of

$$\nu_8 \circ \pi_{15}(S^{11}) \circ \sigma_{15} = \nu_8 \circ \pi_{15}^{11} \circ \sigma_{15} \qquad \text{by (4.7)}$$

$$= 0 \qquad \text{by Proposition 5.8.}$$

Thus $E(x\nu_7 \circ \zeta_{10}) = E(E^2\sigma''' \circ \sigma_{14})$. Since $E : \pi_{21}(S^7) \to \pi_{22}(S^8)$ is an isomorphism into, by Lemma 4.5, then it follows that

$$x\nu_7 \circ \zeta_{10} = E^2\sigma''' \circ \sigma_{14}.$$

Since $2\sigma''' = 0$ and since x is odd, then we have

$$E^2\sigma''' \circ \sigma_{14} = \nu_7 \circ \zeta_{10}. \qquad \text{q.e.d.}$$

CHAPTER X.

2-Primary Components of $\pi_{n+k}(S^n)$ for $k = 14$ and 15.

i). The elements κ_n and $\bar{\varepsilon}_n$.

Consider secondary compositions

$\{\eta_9, 2\iota_{10}, \bar{\nu}_{10}\}_1$, $\{\eta_9, 2\iota_{10}, \varepsilon_{10}\}_1$ and $\{\eta_9, 2\iota_{10}, \eta_{10} \circ \sigma_{11}\}_1$.

Since π_{11}^9 is generated by $\eta_9 \circ \eta_{10}$, then the subgroups $\pi_{11}^9 \circ \bar{\nu}_{11}$, $\pi_{11}^9 \circ \varepsilon_{11}$ and $\pi_{11}^9 \circ \eta_{11} \circ \sigma_{12}$ are contained in $\eta_9 \circ E\pi_{18}^9$.

Thus the above secondary compositions are cosets of the same subgroup

$$\eta_9 \circ E\pi_{18}^9 .$$

By Lemma 6.4, $\eta_9 \circ \sigma_{10} + \varepsilon_9 = \bar{\nu}_9$. Then it follows from Proposition 1.6 that

$$\{\eta_9, 2\iota_{10}, \bar{\nu}_{10}\}_1 = \{\eta_9, 2\iota_{10}, \varepsilon_{10}\}_1 + \{\eta_9, 2\iota_{10}, \eta_{10} \circ \sigma_{11}\}_1 .$$

We shall show

(10.1). $$0 \in \{\eta_9, 2\iota_{10}, \bar{\nu}_{10}\}_1 .$$

First remark that the subgroup $\eta_9 \circ E\pi_{18}^9$ contains $\eta_9 \circ \mu_{10}$ and that $\eta_9^2 \circ \varepsilon_{11} = 4\nu_9 \circ \sigma_{12} = 8\sigma_9 \circ \nu_{16} = 0$, by Theorem 7.3, (7.10) and (7.20). By Proposition 1.2 and Proposition 1.3,

$$2\nu_9 \circ \sigma_{12} = E^6\nu' \circ \sigma_{12} \in E^6 \{\eta_3, 2\iota_4, \eta_4\} \circ E\sigma_{11}$$

$$\subset \{\eta_9, 2\iota_{10}, \eta_{10} \circ \sigma_{11}\}_1 ,$$

$$E^6\{\eta_3, 2\iota_4, \varepsilon_4\}_1 \subset \{\eta_9, 2\iota_{10}, \varepsilon_{10}\}_1 .$$

By Lemma 5.2, $H\{\eta_3, 2\iota_4, \varepsilon_4\}_1 = \varepsilon_5$. It is seen in Theorem 7.3 and its proof that $H(\alpha) = \varepsilon_5$ implies $\alpha \equiv \varepsilon'$ mod $\eta_3 \circ E\pi_{12}^3$. By use of (7.10), we have that $2\nu_9 \circ \sigma_{12} = \pm E^6\varepsilon' = E^6\varepsilon'$ is contained in $\{\eta_9, 2\iota_{10}, \varepsilon_{10}\}_1$. Therefore the secondary composition $\{\eta_9, 2\iota_{10}, \bar{\nu}_{10}\}_1$ contains $2\nu_9 \circ \sigma_{12} + 2\nu_9 \circ \sigma_{12} = 4\nu_9 \circ \sigma_{12} = \eta_9^2 \circ \varepsilon_{11} = 0$. Then (10.1) is proved.

By Proposition 1.7,

(10.2) $$0 = \alpha \circ E\beta$$

for an extension $\alpha \in \pi(EK \to S^9)$ of η_9 and a coextension $\beta \in \pi_{18}(K)$ of $\bar{\nu}_9$, where $K = S^9 \cup CS^9$ is given by attaching the base S^9 of CS^9 by $2\iota_9$. Next we have

(10.3) $$\nu_7 \circ E\alpha = 0.$$

By Proposition 5.9, $\pi_{12}(S^7)$ has no 2-primary component, then $2\pi_{12}(S^7) = \pi_{12}(S^7) = E\pi_{11}(S^6)$, (in fact $\pi_{12}(S^7) = 0$). Thus $\{\nu_7, \eta_{10}, 2\iota_{11}\}_1$ is a subset of $\pi_{12}(S^7) \circ 2\iota_{12}$. Then it follows from Proposition 1.9 that $\nu_7 \circ E\alpha = (E\,p)^* 0 = 0$.

By virtue of the relations (10.2) and (10.3), we can define a secondary composition

$$\{\nu_7, E\alpha, E^2\beta\}_1 \subset \pi_{21}(S^7).$$

Choose an element κ_7 from this secondary composition and denote that

$$\kappa_n = E^{n-7}\kappa_7 \quad \text{for} \quad n \geq 7 \quad \text{and} \quad \kappa = E^\infty \kappa_7 .$$

<u>Lemma</u> 10.1. $2\kappa_7 \equiv \bar{\nu}_7 \circ \nu_{15}^2$ mod. $\nu_7 \circ \zeta_{10}$. $\eta_6 \circ \kappa_7 \in \{\nu_6^2, 2\iota_{12}, \bar{\nu}_{12}\}_3$.

<u>Proof</u>. By Proposition 1.2,

$$2\kappa_7 \in \{\nu_7, E\alpha, E^2\beta\}_1 \circ 2\iota_{21} \subset \{\nu_7, E\alpha, E^2(\beta \circ 2\iota_{18})\}_1 .$$

By Proposition 1.8,

$$\beta \circ 2\iota_{18} \in i_*\{2\iota_9, \bar{\nu}_9, 2\iota_{17}\}$$

for the homomorphism i_* induced by the injection $i : S^9 \subset K$, which annihilates $2\iota_9 \circ \pi_{18}(S^9)$. It follows then from Corollary 3.7 and Lemma 6.3 that

$$E(\beta \circ 2\iota_{18}) = Ei_*\bar{\nu}_{10} \circ \eta_{18} = Ei_*\nu_{10}^3 .$$

By Proposition 1.2,

$$2\kappa_7 \in \{\nu_7, E\alpha, E^2(\beta \circ 2\iota_{18})\}_1 \subset \{\nu_7, E^2i^*E\alpha, \nu_{11}^3\}_1 = \{\nu_7, \eta_{10}, \nu_{11}^3\}_1$$

and $\bar{\nu}_7 \circ \nu_{15}^2 \in \{\nu_7, \eta_{10}, \nu_{11}\}_1 \circ \nu_{15}^2 \subset \{\nu_7, \eta_{10}, \nu_{11}^3\}_1$.

By (4.7), Proposition 5.9 and by Theorem 7.4, the secondary composition $\{\nu_7, \eta_{10}, \nu_{11}^3\}_1$ is a coset of

$$\nu_7 \circ E\pi_{20}^9 + \pi_{12}^7 \circ \nu_{12}^3 = \{\nu_7 \circ \zeta_{10}\} .$$

It follows that $2\kappa_7 \equiv \bar{\nu}_7 \circ \nu_{15}^2$ mod. $\nu_7 \circ \zeta_{10}$.

Next by Proposition 1.4,

$$\eta_6 \circ \kappa_7 \in \eta_6 \circ \{\nu_7, E\alpha, E^2\beta\} = \{\eta_6, \nu_7, E\alpha\} \circ -E^3\beta .$$

Let γ be an element of $\{\eta_6, \nu_7, E\alpha\} \subset \pi(E^3K \to S^6)$ such that $\eta_6 \circ \kappa_7 = \gamma \circ -E^3\beta$. The restriction of γ on S^{12} is an element of $\{\eta_6, \nu_7, \eta_{10}\}$, which coincides with ν_6^2 by Lemma 5.12. Thus γ is an extension of ν_6^2. It follows from Proposition 1.7 that

$$\eta_6 \circ \kappa_7 = \gamma \circ -E^3\beta \in \{\nu_6^2, 2\iota_{12}, \bar{\nu}_{12}\}_3. \qquad \text{q.e.d.}$$

Next consider the following secondary composition

$$\{\varepsilon_3, 2\iota_{11}, \nu_{11}^2\}_6.$$

Choose an element $\bar{\varepsilon}_3$ from this secondary composition and denote that

$$\bar{\varepsilon}_n = E^{n-3}\,\bar{\varepsilon}_3 \text{ for } n \geq 3 \text{ and } \bar{\varepsilon} = E^\infty\,\bar{\varepsilon}_3 .$$

<u>Lemma</u> 10.2. $H(\bar{\varepsilon}_3) \equiv \nu_5 \circ \sigma_8 \circ \nu_{15}$ mod. $\nu_5 \circ \eta_8 \circ \mu_9$ <u>and</u> $2\bar{\varepsilon}_n = 2\bar{\varepsilon} = 0$ <u>for</u> $n \geq 3$.

<u>Proof</u>. By Proposition 2.3, Lemma 6.1, Proposition 2.6 and by (5.13),

$$H(\bar{\varepsilon}_3) \in H\{\varepsilon_3, 2\iota_{11}, \nu_{11}^2\}_6$$
$$\subset \{H(\varepsilon_3), 2\iota_{11}, \nu_{11}^2\}_6 = \{\nu_5^2, 2\iota_{11}, \nu_{11}^2\}_6.$$

and $H(H(\bar{\varepsilon}_3)) \in H\{\nu_5^2, 2\iota_{11}, \nu_{11}^2\}_1 = \Delta^{-1}(\nu_4^2 \circ 2\iota_{10}) \circ \nu_{12}^2 = \pm\,\nu_9 \circ \nu_{12}^2 = \nu_9^3$.

It was seen in the proof of Theorem 7.7 that π_{18}^5 has two generators $\nu_5 \circ \sigma_6 \circ \nu_{15}$ and $E(\nu_4 \circ \eta_7 \circ \mu_8)$ and that $H(\nu_5 \circ \sigma_8 \circ \nu_{15}) = \nu_9^3$. Since $HE = 0$, we have that

$$H(\bar{\varepsilon}_3) \equiv \nu_5 \circ \sigma_8 \circ \nu_{15} \text{ mod. } \nu_5 \circ \eta_7 \circ \mu_8 .$$

Next, by Proposition 1.4, (5.9) and Corollary 3.7,

$$2\bar{\varepsilon}_3 \in \{\varepsilon_3, 2\iota_{11}, \nu_{11}^2\}_1 \circ 2\iota_{18} = \varepsilon_3 \circ \{2\iota_{11}, \nu_{11}^2, 2\iota_{17}\}$$

and $$0 = \varepsilon_3 \circ \nu_{11}^2 \circ \eta_{17} \in \varepsilon_3 \circ \{2\iota_{11}, \nu_{11}^2, 2\iota_{17}\} .$$

Since $\varepsilon_3 \circ \{2\iota_{11}, \nu_{11}^2, 2\iota_{17}\}$ is a coset of $\varepsilon_3 \circ \pi_{18}^{11} \circ 2\iota_{11} = \{\varepsilon_3 \circ 2\sigma_{11}\} = \{2\varepsilon_3 \circ \sigma_{11}\} = 0$, then it follows that

$$2\bar{\varepsilon}_3 = 0 \text{ and } 2\bar{\varepsilon}_n = 2\bar{\varepsilon} = 0 \text{ for } n \geq 3. \qquad \text{q.e.d.}$$

ii). <u>The Groups</u> π_{n+14}^n.

We shall prove

<u>Theorem</u> 10.3. $\pi_{16}^2 = \{\eta_2 \circ \nu' \circ \eta_6 \circ \mu_7\} \approx Z_2$,

$$\pi_{17}^3 = \{\varepsilon_3 \circ \nu_{11}^2\} \approx Z_2 ,$$

$$\pi^4_{18} = \{\varepsilon_4 \circ \nu^2_{12}\} \oplus \{\nu_4 \circ \zeta_7\} \oplus \{\nu_4 \circ \bar{\nu}_7 \circ \nu_{15}\} \approx Z_2 \oplus Z_8 \oplus Z_2 ,$$
$$\pi^5_{19} = \{\nu_5 \circ \zeta_8\} \oplus \{\nu_5 \circ \bar{\nu}_8 \circ \nu_{16}\} \approx Z_2 \oplus Z_2 ,$$
$$\pi^6_{20} = \{\sigma'' \circ \sigma_{13}\} \oplus \{\bar{\nu}_6 \circ \nu^2_{14}\} \approx Z_4 \oplus Z_2 ,$$
$$\pi^7_{21} = \{\sigma' \circ \sigma_{14}\} \oplus \{\kappa_7\} \approx Z_8 \oplus Z_4 ,$$
$$\pi^8_{22} = \{\sigma^2_8\} \oplus \{E\sigma' \circ \sigma_{15}\} \oplus \{\kappa_8\} \approx Z_{16} \oplus Z_8 \oplus Z_4 ,$$
$$\pi^9_{23} = \{\sigma^2_9\} \oplus \{\kappa_9\} \approx Z_{16} \oplus Z_4,$$
$$\pi^n_{n+14} = \{\sigma^2_n\} + \{\kappa_n\} \approx Z_{16} \oplus Z_2 \qquad \underline{\text{for}}\ n = 10, 11, 13,$$
$$\pi^{12}_{26} = \{\sigma^2_{12}\} + \{\kappa_{12}\} \oplus \{\Delta(\nu_{25})\} \approx Z_{16} \oplus Z_2 \oplus Z_4 ,$$
$$\pi^{14}_{28} = \{\sigma^2_{14}\} \oplus \{\kappa_{14}\} \approx Z_8 \oplus Z_2 ,$$
$$\pi^{15}_{29} = \{\sigma^2_{15}\} \oplus \{\kappa_{15}\} \approx Z_4 \oplus Z_2 ,$$
$$\pi^n_{n+14} = \{\sigma^2_n\} \oplus \{\kappa_n\} \approx Z_2 \oplus Z_2 \qquad \underline{\text{for}}\ n \geq 16 ,$$
$$(G_{14};2) = \{\sigma^2\} \oplus \{\kappa\} \approx Z_2 \oplus Z_2 ,$$

<u>where</u> $\sigma^2_n = \sigma_n \circ \sigma_{n+7}$ <u>and</u> $\sigma^2 = \sigma \circ \sigma$.

First $\pi^2_{16} = \{\eta_2 \circ \nu' \circ \eta_6 \circ \mu_7\} \approx Z_2$, by Theorem 7.7 and by (5.2). By Lemma 5.7 and by Proposition 2.5, we have

(10.4) $\Delta(\nu_5 \circ \eta_8 \circ \mu_9) = \eta_2 \circ \nu' \circ \eta_6 \circ \mu_7$ and $E\pi^2_{16} = 0$. It follows from the exactness of the sequence (4.4) that $H : \pi^3_{17} \to \pi^5_{17}$ maps π^3_{17} isomorphically into the kernel of $\Delta : \pi^5_{17} \to \pi^2_{15}$ which is onto by (7.23). By Theorem 7.6 and Theorem 7.7, the groups π^5_{17} and π^2_{15} have 8 and 4 elements respectively. Thus the kernel of Δ is isomorphic to Z_2. By Proposition 2.2 and Lemma 6.1,

$$H(\varepsilon_3 \circ \nu^2_{11}) = H(\varepsilon_3) \circ \nu^2_{11} = \nu^4_5 ,$$

and $\nu^4_5 \neq 0$ by Theorem 7.6. Consequently we have obtained that

$$\pi^3_{17} = \{\varepsilon_3 \circ \nu^2_{11}\} \approx Z_2 .$$

By (5.6) and Theorem 7.4,

$$\pi^4_{18} = \{\nu_4 \circ \zeta_7\} \oplus \{\nu_4 \circ \bar{\nu}_7 \circ \nu_{15}\} \oplus \{\varepsilon_4 \circ \nu^2_{12}\} \approx Z_8 \oplus Z_2 \oplus Z_2 .$$

By the exactness of (4.4) and by (7.24),

$$E : \pi^4_{18}/\Delta\pi^9_{20} \approx \pi^5_{19} .$$

The group $\pi^9_{20} \approx Z_8 + Z_2$ is generated by ζ_9 and $\bar{\nu}_9 \circ \nu_{17}$, by Theorem 7.4.

We have

$$\Delta(\bar{\nu}_9 \circ \nu_{17}) = \Delta(\bar{\nu}_9) \circ \nu_{15} = \varepsilon_4 \circ \nu_{12}^2 \quad \text{by Proposition 2.5 and (7.13)}$$

$$\Delta(\zeta_9) = \Delta(\iota_9) \circ \zeta_7 \quad \text{by Proposition 2.5}$$

$$= \pm (2\nu_4 \circ \zeta_7 - E\nu' \circ \zeta_7) \text{ by (5.8).}$$

Since $E\nu' \circ \zeta_7 \in E\pi_{17}^3 = \{\varepsilon_4 \circ \nu_{12}^2\}$, then $\Delta\pi_{20}^9$ is generated by $\varepsilon_4 \circ \nu_{12}^2$ and $2\nu_4 \circ \zeta_7$. Thus

$$\pi_{19}^5 = \{\nu_5 \circ \zeta_9\} \oplus \{\nu_5 \circ \bar{\nu}_8 \circ \nu_{16}\} \approx Z_2 \oplus Z_2$$

and

(10.5) <u>the kernel of</u> $\Delta : \pi_{20}^9 \to \pi_{18}^4$ <u>is</u> $\{4\zeta_9\} = \{\eta_9^2 \circ \mu_{11}\}$.

Next we prove

<u>Lemma</u> 10.4. $H\{\sigma''', 16\iota_{12}, \sigma_{12}\}_1 = 4\zeta_9 = \eta_9^2 \circ \mu_{11}$,

$$H\{\sigma'', 16\iota_{13}, \sigma_{13}\}_1 = \eta_{11} \circ \mu_{12},$$

$$H\{\sigma', 16\iota_{14}, \sigma_{14}\}_1 = \mu_{13} + \{\nu_{13}^3\} + \{\eta_{13} \circ \varepsilon_{14}\}$$

<u>and</u> $H\{\sigma_9, 16\iota_{16}, \sigma_{16}\}_1 = 8\sigma_{17}$.

<u>Proof</u>. By Proposition 2.6,

$$H\{\sigma_9, 16\iota_{16}, \sigma_{16}\}_1 = \Delta^{-1}(\sigma_8 \circ 16\iota_{15}) \circ \sigma_{17}.$$

By (5.16), $\Delta^{-1}(\sigma_8 \circ 16\iota_{15})$ consists of the elements $8(2x+1)\iota_{17}$ for the integers x. Then $\Delta^{-1}(\sigma_8 \circ 16\iota_{15}) \circ \sigma_{17} = (16x + 8)\sigma_{17} = 8\sigma_{17}$ and thus $H\{\sigma_9, 16\iota_{16}, \sigma_{16}\}_1 = 8\sigma_{17}$.

Next, by Proposition 2.3 and by Lemma 5.14,

$$H\{\sigma', 16\iota_{14}, \sigma_{14}\}_1 \subset \{H(\sigma'), 16\iota_{14}, \sigma_{14}\}_1 = \{\eta_{13}, 16\iota_{14}, \sigma_{14}\}_1,$$

where $\{\eta_{13}, 16\iota_{14}, \sigma_{14}\}_1$ is a coset of

$$\eta_{13} \circ E\pi_{21}^{13} + \pi_{15}^{13} \circ \sigma_{15} = \{\eta_{13} \circ \bar{\nu}_{14}\} + \{\eta_{13} \circ \varepsilon_{14}\} + \{\eta_{13}^2 \circ \sigma_{15}\}$$

$$= \{\nu_{13}^3\} + \{\eta_{13} \circ \varepsilon_{14}\},$$

by (4.7), Theorem 7.1 and Lemma 6.4. By Lemma 6.5 and Proposition 1.2,

$$\mu_{13} \in \{\eta_{13}, 2\iota_{14}, E^9\sigma'''\}_9 + \{\nu_{13}^3\}$$

$$= \{\eta_{13}, 2\iota_{14}, 8\sigma_{14}\}_9 \subset \{\eta_{13}, 16\iota_{14}, \sigma_{14}\}_1.$$

Thus we conclude that $H\{\sigma', 16\iota_{14}, \sigma_{14}\}_1$ is a coset containing μ_{13}.

We have

$$H\{\sigma'', 16\iota_{13}, \sigma_{13}\}_1 \subset \{H(\sigma''), 16\iota_{13}, \sigma_{13}\}_1 \quad \text{by Proposition 2.3}$$

$$= \{\eta_{11}^2, 16\iota_{13}, \sigma_{13}\}_1 \quad \text{by Lemma 5.14}$$

and $\eta_{11} \circ \mu_{12} \in \eta_{11} \circ \{\eta_{12}, 16\iota_{13}, \sigma_{13}\}_1$

$\subset \{\eta_{11}^2, 16\iota_{13}, \sigma_{13}\}_1$ by Proposition 1.2.

Since $\{\eta_{11}^2, 16\iota_{13}, \sigma_{13}\}_1$ is a coset of

$$\eta_{11}^2 \circ E\pi_{20}^{12} + \pi_{14}^{11} \circ \sigma_{14} = \{\eta_{11}^2 \circ \bar{\nu}_{13}\} + \{\eta_{11}^2 \circ \varepsilon_{13}\} + \{\nu_{11} \circ \sigma_{14}\} = 0$$

by (4.7), Theorem 7.1, Lemma 6.3, (5.9), (7.10) and (7.20), then

$$H\{\sigma'', 16\iota_{13}, \sigma_{13}\}_1 = \eta_{11} \circ \mu_{12}.$$

Similarly,

$$H\{\sigma''', 16\iota_{12}, \sigma_{12}\}_1 \subset \{H(\sigma'''), 16\iota_{12}, \sigma_{12}\}_1$$

$$= \{\eta_9^3, 16\iota_{12}, \sigma_{12}\}_1$$

and $4\zeta_9 = \eta_9^2 \circ \mu_{11} \in \{\eta_9^3, 16\iota_{12}, \sigma_{12}\}_1$, where $\{\eta_9^3, 16\iota_{12}, \sigma_{12}\}_1$ is a coset of

$$\eta_9^3 \circ E\pi_{19}^{11} + \pi_{13}^9 \circ \sigma_{13} = \{\eta_9^3 \circ \bar{\nu}_{12}\} + \{\eta_9^3 \circ \varepsilon_{12}\} = 0 .$$

Thus

$$H\{\sigma''', 16\iota_{12}, \sigma_{12}\}_1 = \eta_9^2 \circ \mu_{11} .$$

q.e.d.

Now consider homomorphisms

$$\Delta : \pi_{21}^{11} \to \pi_{19}^5 \quad \text{and} \quad \Delta : \pi_{22}^{13} \to \pi_{20}^6$$

By use of Proposition 2.5, we have

$\Delta(\eta_{11} \circ \mu_{12}) = \Delta\, H\{\sigma'', 16\iota_{13}, \sigma_{13}\}_1 = 0$ by Lemma 10.4,

$\Delta(\sigma_{11} \circ \nu_{18}) = \Delta(\sigma_{11}) \circ \nu_{16}$

$= \nu_5 \circ \bar{\nu}_8 \circ \nu_{16} + \nu_5 \circ \varepsilon_8 \circ \nu_{16}$ by (7.17)

$= \nu_5 \circ \bar{\nu}_8 \circ \nu_{16} + \nu_5 \circ E^4\Delta(\bar{\nu}_9)$ by (7.13)

$= \nu_5 \circ \bar{\nu}_8 \circ \nu_{16}$,

$\Delta(\nu_{13}^3) = \Delta(\eta_{13} \circ \bar{\nu}_{14}) = \Delta(\eta_{13}) \circ \bar{\nu}_{12}$ by Lemma 6.3

$= \Delta H(\sigma') \circ \bar{\nu}_{12} = 0$, by Lemma 5.14,

$\Delta(\eta_{13} \circ \varepsilon_{14}) = \Delta H(\sigma') \circ \varepsilon_{12} = 0$,

and $\Delta(\mu_{13}) \in \Delta(H\{\sigma', 16\iota_{14}, \sigma_{14}\}_1 + \{\nu_{13}^3\} + \{\eta_{13} \circ \varepsilon_{14}\}) = 0$ by Lemma 10.4.

It follows from Theorem 7.2, Theorem 7.3 and from the exactness of the sequence (4.4) that

(10.6) <u>the kernel of</u> $\Delta : \pi_{21}^{11} \to \pi_{19}^5$ <u>is</u> $\{\eta_{11} \circ \mu_{12}\}$, $E\pi_{19}^5 = \{\nu_6 \circ \zeta_9\} \approx Z_2$ <u>and that</u> $E : \pi_{20}^6 \to \pi_{21}^7$ <u>is an isomorphism into</u>.

By Lemma 9.2, $E(\nu_6 \circ \zeta_9) = E(E\sigma''' \circ \sigma_{13}) = E(2\sigma'' \circ \sigma_{13})$. It follows from the last assertion of (10.6) that

(10.7) $\nu_6 \circ \zeta_9 = 2\sigma'' \circ \sigma_{13}$.

It follows from (7.26) and from the exactness of the sequence (4.4) the following sequence is exact.

$$0 \to E\pi^5_{19} \longrightarrow \pi^6_{20} \xrightarrow{H} \{\nu^3_{11}\} + \{\eta_{11} \circ \varepsilon_{12}\} \to 0.$$

We have

$$\begin{aligned} H(\sigma'' \circ \sigma_{13}) &= H(\sigma'') \circ \sigma_{13} && \text{by Proposition 2.2} \\ &= \eta^2_{11} \circ \sigma_{13} && \text{by Lemma 5.14} \\ &= \eta_{11} \circ \bar{\nu}_{12} + \eta_{11} \circ \varepsilon_{12} && \text{by Lemma 6.4} \\ &= \nu^3_{11} + \eta_{11} \circ \varepsilon_{12} && \text{by Lemma 6.3} \\ H(\bar{\nu}_6 \circ \nu^2_{14}) &= H(\bar{\nu}_6) \circ \nu^2_{14} && \text{by Proposition 2.2} \\ &= \nu_{11} \circ \nu^2_{14} = \nu^3_{11} && \text{by Lemma 6.2 .} \end{aligned}$$

Then we have the result

$$\pi^6_{20} = \{\sigma'' \circ \sigma_{13}\} \oplus \{\bar{\nu}_6 \circ \nu^2_{14}\} \approx Z_4 \oplus Z_2 \quad .$$

It follows from (10.6) and the exactness of the sequence (4.4), the sequence

$$0 \longrightarrow \pi^6_{20} \xrightarrow{E} \pi^7_{21} \xrightarrow{H} \pi^{13}_{21}$$

is exact. Since $\pi^{13}_{21} \approx Z_2 + Z_2$, then $\pi^7_{21}/E\pi^6_{20}$ has at most 4 elements. Consider the elements κ_7 and $\sigma' \circ \sigma_{14}$, then

$$2\sigma' \circ \sigma_{14} = E(\sigma'' \circ \sigma_{13}) \qquad \text{by Lemma 5.14}$$

and

$$2\kappa_7 = E(\bar{\nu}_6 \circ \nu^2_{14}) + 2xE(\sigma'' \circ \sigma_{13}) \qquad \text{by Lemma 10.1 and (10.7),}$$

where $x = 0$ or 1. This shows that $\sigma' \circ \sigma_{14}$ and κ_7 are elements of orders 8 and 4 respectively. Thus we have

$$\pi^7_{21} = \{\sigma' \circ \sigma_{14}\} \oplus \{\kappa_7\} \approx Z_8 \oplus Z_4 \quad .$$

By (5.15) and Proposition 5.15,

$$\pi^8_{22} = \{\sigma_8 \circ \sigma_{15}\} \oplus \{E\sigma' \circ \sigma_{15}\} \oplus \{\kappa_8\} \approx Z_{16} \oplus Z_8 \oplus Z_4 \quad .$$

It follows from (7.28) and the exactness of the sequence (4.4) that

$$E : \pi^8_{22}/\Delta\pi^{17}_{24} \approx \pi^9_{23}.$$

By Proposition 2.5 and by (5.16),

$$\begin{aligned} \Delta(\sigma_{17}) = \Delta(\iota_{17}) \circ \sigma_{15} &= \pm (2\sigma_8 - E\sigma') \circ \sigma_{15} \\ &= \pm (2\sigma^2_8 - E\sigma' \circ \sigma_{15}) \; . \end{aligned}$$

Thus we have

$$\pi^9_{23} = \{\sigma^2_9\} \oplus \{\kappa_9\} \approx Z_{16} \oplus Z_4$$

and

(10.8) <u>the kernel of</u> $\Delta : \pi^{17}_{24} \to \pi^8_{22}$ <u>is</u> $\{8\sigma_{17}\} \approx Z_2$.

In the exact sequence

$$\pi^{19}_{25} \xrightarrow{\Delta} \pi^9_{23} \xrightarrow{E} \pi^{10}_{24} \xrightarrow{H} \pi^{19}_{24}$$

of (4.4), $\pi_{24}^{19} = 0$ by Proposition 5.9. Thus E is onto and its kernel is generated by $\Delta(\nu_{19}^2)$, since $\pi_{25}^{19} = \{\nu_{19}^2\} \approx Z_2$ by Proposition 5.11. By Proposition 2.5 and (7.22),

$$\Delta(\nu_{19}^2) = \Delta(\nu_{19}) \circ \nu_{20} = \bar{\nu}_9 \circ \nu_{17}^2 .$$

Then we have that

$$\pi_{24}^{10} = \{\sigma_{10}^2\} + \{\kappa_{10}\} \approx Z_{16} \oplus Z_2$$

where $2\kappa_{10} = 0$ or $2\kappa_{10} = 8\sigma_{10}^2$. We have also that $\Delta : \pi_{25}^{19} \to \pi_{23}^{9}$ is an isomorphism into. It follows from the exactness of (4.4) that

(10.9). $E : \pi_{24}^{9} \to \pi_{25}^{10}$ <u>is onto</u>.

It follows from Proposition 5.8, Proposition 5.9 and from the exactness of (4.4) that $E : \pi_{24}^{10} \to \pi_{25}^{11}$ is an isomorphism onto and that the sequence

$$0 \to \pi_{25}^{11} \xrightarrow{E} \pi_{26}^{12} \xrightarrow{H} \pi_{26}^{23} \xrightarrow{\Delta} \pi_{24}^{11}$$

is exact. Thus

$$\pi_{25}^{11} = \{\sigma_{11}^2\} + \{\kappa_{11}\} \approx Z_{16} \oplus Z_2$$

and

$$E\pi_{25}^{11} = \{\sigma_{12}^2\} + \{\kappa_{12}\} \approx Z_{16} \oplus Z_2$$

By (7.29), the kernel of $\Delta : \pi_{26}^{23} \to \pi_{24}^{11}$ is generated by $2\nu_{23}$, which is $\pm\,\Delta(\nu_{25})$ by Proposition 2.5 and Proposition 2.7. Then the result

$$\pi_{26}^{12} = \{\sigma_{12}^2\} + \{\kappa_{12}\} \oplus \{\Delta(\nu_{25})\} \approx Z_{16} \oplus Z_2 \oplus Z_4$$

is obtained by use of the first relation of the following

(10.10).
$$\Delta(\eta_{25}^3) = 4\Delta(\nu_{25}) = 0 ,$$
$$\Delta(\eta_{27}^2) = 8\sigma_{13}^2 ,$$
$$\Delta(\eta_{29}) = 4\sigma_{14}^2 ,$$
<u>and</u>
$$\Delta(\iota_{31}) = 2\sigma_{15}^2 .$$

Apply Corollary 3.3 for the case $\alpha = \sigma''$ and $\beta = \sigma'$, then we have, by Lemma 5.14,

$$\begin{aligned}\Delta(\eta_{25}^3) = \Delta(E^{14}\eta_{11}^2 \circ E^{14}\eta_{13}) &= \Delta(E^{14}H(\sigma'') \circ E^{14}H(\sigma'))\\ &= E^6\sigma'' \circ E^{12}\sigma' + E^5\sigma' \circ E^{13}\sigma''\\ &= 4\sigma_{12} \circ 2\sigma_{19} + 2\sigma_{12} \circ 4\sigma_{19}\\ &= 16\sigma_{12}^2 = 0.\end{aligned}$$

Similarly, we have

$$\begin{aligned}\Delta(\eta_{27}^2) &= \Delta(E^{14}H(\sigma') \circ E^{15}H(\sigma'))\\ &= E^6\sigma' \circ E^{13}\sigma' + E^6\sigma' \circ E^{13}\sigma'\\ &= 8\sigma_{13}^2 ,\end{aligned}$$

$$\Delta(\eta_{29}) = \Delta(E^{14}H(\sigma_8) \circ E^{16}H(\sigma')) = E^6\sigma_8 \circ E^{14}\sigma' + E^7\sigma' \circ E^{13}\sigma_8 = 4\sigma_{14}^2$$

and

$$\Delta(\iota_{31}) = \Delta(E^{16}H(\sigma_8) \circ E^{16}H(\sigma_8)) = E^7\sigma_8 \circ E^{14}\sigma_8 + E^7\sigma_8 \circ E^{14}\sigma_8 = 2\sigma_{15}^2 .$$

Thus (10.10) is proved.

Consider the exact sequences

$$\pi_{n+16}^{2n+1} \xrightarrow{\Delta} \pi_{n+14}^n \xrightarrow{E} \pi_{n+15}^{n+1} \xrightarrow{H} \pi_{n+15}^{2n+1} \xrightarrow{\Delta} \pi_{n+13}^n$$

of (4.4) for $n = 12, 13, 14$ and 15. It follows from (7.30) that the above E are homomorphisms onto for $n = 12, 13$ and 14. It is also onto for $n = 15$, since $\pi_{30}^{31} = 0$. Then, by use of (10.10), we have

$$\pi_{27}^{13} = \{\sigma_{13}^2\} + \{\kappa_{13}\} \approx Z_{16} \oplus Z_2 ,$$

$$\pi_{28}^{14} = \{\sigma_{14}^2\} \oplus \{\kappa_{14}\} \approx Z_8 \oplus Z_2 ,$$

$$\pi_{29}^{15} = \{\sigma_{15}^2\} \oplus \{\kappa_{15}\} \approx Z_4 \oplus Z_2 ,$$

and

$$\pi_{30}^{16} = \{\sigma_{16}^2\} \oplus \{\kappa_{16}\} \approx Z_2 \oplus Z_2 .$$

By (4.5), $\pi_{n+14}^n = \{\sigma_n^2\} \oplus \{\kappa_n\} \approx Z_2 \oplus Z_2$ for $n \geq 16$ and $(G_{14};2) = \{\sigma^2\} \oplus \{\kappa\} \approx Z_2 \oplus Z_2$.

Consequently Theorem 10.3 is proved.

We have from (10.10) and from the exactness of the sequence (4.4) that

(10.11). $H\pi_{28}^{13} = \{\eta_{25}^3\} \approx Z_2$, $H\pi_{31}^{16} = \{2\iota_{13}\} \approx Z$ <u>and that</u> $E : \pi_{n+15}^n \to \pi_{n+16}^{n+1}$ <u>are homomorphisms onto for</u> $n = 13$ <u>and</u> $n = 14$.

iii). <u>The groups</u> π_{n+15}^n <u>for</u> $n \leq 12$.

Choose elements $\rho^{IV} \in \pi_{20}^5$, $\rho''' \in \pi_{21}^6$, $\rho'' \in \pi_{22}^7$ and $\rho' \in \pi_{24}^9$ as follows.

$$\rho^{IV} \in \{\sigma''', 2\iota_{12}, 8\sigma_{12}\}_1 ,$$

$$\rho''' \in \{\sigma'' , 4\iota_{13}, 4\sigma_{13}\}_1 ,$$

$$\rho'' \in \{\sigma' , 8\iota_{14}, 2\sigma_{14}\}_1 ,$$

$$\rho' \in \{\sigma_9 , 16\iota_{16}, \sigma_{16}\}_1 .$$

By Lemma 10.4, we have

(10.12). $H(\rho^{IV}) = \eta_9^2 \circ \mu_{11} = 4\zeta_9$, $H(\rho''') = \eta_{11} \circ \mu_{12}$, $H(\rho'') \equiv \mu_{13}$

mod $\{\nu_{13}^3\} + \{\eta_{13}\circ\varepsilon_{14}\}$ and $H(\rho') = 8\sigma_{17}$.

Theorem 10.5. $\pi_{17}^2 = \{\eta_2\circ\varepsilon_3\circ\nu_{11}^2\} \approx Z_2$,

$$\pi_{n+15}^n = \{\bar\varepsilon_n\} \approx Z_2 \quad \text{for } n = 3 \text{ and } 4 \; ,$$

$$\pi_{20}^5 = \{\rho^{IV}\} \oplus \{\bar\varepsilon_5\} \approx Z_2 \oplus Z_2 \; ,$$

$$\pi_{21}^6 = \{\rho'''\} \oplus \{\bar\varepsilon_6\} \approx Z_4 \oplus Z_2$$

$$\pi_{22}^7 = \{\rho''\} \oplus \{\sigma'\circ\bar\nu_{14}\} \oplus \{\sigma'\circ\varepsilon_{14}\} \oplus \{\bar\varepsilon_7\} \approx Z_8 \oplus Z_2 \oplus Z_2 \oplus Z_2 \; ,$$

$$\pi_{23}^8 = \{\sigma_8\circ\bar\nu_{15}\} \oplus \{\sigma_8\circ\varepsilon_{15}\} \oplus \{E\rho''\} \oplus \{E\sigma'\circ\bar\nu_{15}\} \oplus \{E\sigma'\circ\varepsilon_{15}\} \; ,$$
$$\oplus \{\bar\varepsilon_8\} \approx Z_2 \oplus Z_2 \oplus Z_8 \oplus Z_2 \oplus Z_2 \oplus Z_2 \; ,$$

$$\pi_{24}^9 = \{\rho'\} \oplus \{\sigma_9\circ\bar\nu_{16}\} \oplus \{\sigma_9\circ\varepsilon_{16}\} \oplus \{\bar\varepsilon_9\} \approx Z_{16} \oplus Z_2 \oplus Z_2 \oplus Z_2 \; ,$$

$$\pi_{25}^{10} = \{E\rho'\} \oplus \{\sigma_{10}\circ\overline{\nu_{17}}\} \oplus \{\bar\varepsilon_{10}\} \approx Z_{16} \oplus Z_2 \oplus Z_2 \; ,$$

$$\pi_{26}^{11} = \{E^2\rho'\} \oplus \{\bar\varepsilon_{11}\} \approx Z_{16} \oplus Z_2 \; ,$$

$$\pi_{27}^{12} = \{E^3\rho'\} \oplus \{\bar\varepsilon_{12}\} \approx Z_{16} \oplus Z_2 \; .$$

First we have from (5.2) and Theorem 10.3 that

$$\pi_{17}^2 = \{\eta_2\circ\varepsilon_3\circ\nu_{11}^2\} \approx Z_2 \; .$$

By (7.12) and Lemma 5.7,

$$E(\eta_2\circ\varepsilon_3\circ\nu_{11}^2) = E(\eta_2\circ\nu'\circ\bar\nu_6\circ\nu_{14}) = 0 \; .$$

Thus $E\pi_{17}^2 = 0$. It follows from the exactness of the sequence (4.4) that the following sequences are exact.

$$(10.13). \qquad \pi_{19}^5 \xrightarrow{\Delta} \pi_{17}^2 \longrightarrow 0 \; ,$$

$$0 \longrightarrow \pi_{18}^3 \xrightarrow{H} \pi_{18}^5 \xrightarrow{\Delta} \pi_{16}^2 \; .$$

By (10.4), $\Delta(\nu_5\circ\eta_8\circ\mu_9) \neq 0$. Since $\pi_{18}^5 = \{\nu_5\circ\eta_8\circ\mu_9\} \oplus \{\nu_5\circ\sigma_8\circ\nu_{15}\} \approx Z_2 \oplus Z_2$, by Theorem 7.7, then the image $H\pi_{18}^3$ has at most two elements. By Lemma 10.2, $H(\bar\varepsilon_3) \neq 0$. Thus we have

$$\pi_{18}^3 = \{\bar\varepsilon_3\} \approx Z_2 \; .$$

It follows from (5.6) and Theorem 7.6 that

$$\pi_{19}^4 = \{\bar\varepsilon_4\} \approx Z_2 \; .$$

Consider the exact sequence

$$\pi_{21}^9 \longrightarrow \pi_{19}^4 \xrightarrow{E} \pi_{20}^5 \xrightarrow{H} \pi_{20}^9 \xrightarrow{\Delta} \pi_{18}^4$$

of (4.4). $\pi_{21}^9 = 0$ by Theorem 7.6. Then E is an isomorphism into. By (10.5) and (10.12), the kernel of Δ is generated by $H(\rho^{IV}) = \eta_9^2\circ\mu_{11}$. Thus π_{20}^5 has 4 elements and it is generated by $\bar\varepsilon_5$ and ρ^{IV}. Then the result

$$\pi^5_{20} = \{\rho^{IV}\} + \{\bar{\varepsilon}_5\} \approx Z_2 \oplus Z_2$$

follows from the first relation of the following lemma.

Lemma 10.6. $2\rho^{IV} = 0$,

$$2\rho''' \equiv E\,\rho^{IV} \quad \text{mod } \sigma'' \circ \pi^{13}_{21} ,$$

$$2\rho'' \equiv E\,\rho''' \quad \text{mod } \sigma' \circ \pi^{14}_{22} ,$$

and

$$2\rho' \equiv E^2\rho'' \quad \text{mod } \sigma_9 \circ \pi^{16}_{24} .$$

Proof. By Proposition 1.4,

$$2\rho^{IV} \in \{\sigma''', 2\iota_{13}, 8\sigma_{12}\}_1 \circ 2\iota_{20} = \sigma''' \circ E\{2\iota_{11}, 8\sigma_{11}, 2\iota_{18}\} .$$

By Corollary 3.7, $\sigma''' \circ E\{2\iota_{11}, 8\sigma_{11}, 2\iota_{18}\}$ contains

$$\sigma''' \circ 8\sigma_{12} \circ \eta_{19} = 8\sigma''' \circ \sigma_{12} \circ \eta_{19} = 0 .$$

Since $\sigma''' \circ E\{2\iota_{11}, 8\sigma_{11}, 2\iota_{18}\}$ is a coset of

$$\sigma''' \circ E\pi^{11}_{19} \circ 2\iota_{20} = 2\sigma''' \circ E\pi^{11}_{19} = 0 ,$$

then it follows that $2\rho^{IV} = 0$.

Next we have

$2\rho''' \in \{\sigma'', 4\iota_{13}, 4\sigma_{13}\}_1 \circ 2\iota_{21} \subset \{\sigma'', 4\iota_{13}, 8\sigma_{13}\}_1$ by Proposition 1.2,

$E\rho^{IV} \in E\{\sigma''', 2\iota_{12}, 8\sigma_{12}\}_1$

$\subset \{E\sigma''', 2\iota_{12}, 8\sigma_{12}\}_1 = \{2\sigma'', 2\iota_{13}, 8\sigma_{13}\}_1$ by Proposition 1.3,

$\subset \{\sigma'', 4\iota_{13}, 8\sigma_{13}\}_1$ by Proposition 1.2.

The secondary composition $\{\sigma'', 4\iota_{13}, 8\sigma_{13}\}_1$ is a coset of

$$\sigma'' \circ E\pi^{12}_{20} + \pi^6_{14} \circ 8\sigma_{14} = \sigma'' \circ E\pi^{12}_{20} + 8\pi^6_{14} \circ \sigma_{14}$$

$$= \sigma'' \circ \pi^{13}_{21}$$

by (4.7) and Theorem 7.1. Therefore we have

$$2\rho''' \equiv E\rho^{IV} \quad \text{mod } \sigma'' \circ \pi^{13}_{21} .$$

The last two relations of the lemma are proved similarly.

q.e.d.

Next we prove

(10.14). $E : \pi^n_{n+15} \to \pi^{n+1}_{n+16}$ are isomorphisms into for $n = 5$ and 6.

It is sufficient to prove that $\Delta\pi^{11}_{22} = \Delta\pi^{13}_{23} = 0$, by the exactness of (4.4). π^{11}_{22} is generated by ζ_{11}, since Theorem 7.4. Then

$\Delta(\zeta_{11}) = \Delta(\iota_{11}) \circ \zeta_9 = \nu_5 \circ \eta_8 \circ \zeta_9$ by Proposition 2.5 and (5.10)

$\in \nu_5 \circ \pi^8_{20} = 0$ by Theorem 7.6.

Thus $\Delta\pi^{11}_{22} = 0$. π^{13}_{23} is generated by $\eta_{13} \circ \mu_{14}$, since Theorem 7.3. Then, by Proposition 2.5, and by Lemma 5.14,

$$\Delta(\eta_{13} \circ \mu_{14}) = \Delta(\eta_{13}) \circ \mu_{12} = \Delta H(\sigma') \circ \mu_{12} = 0 .$$

Thus $\Delta\pi^{13}_{23} = 0$ and (10.14) is proved.

Since $2\pi^{13}_{21} = 0$, by Theorem 7.1, then $E(\sigma'' \circ \pi^{13}_{21}) = 2\sigma' \circ E\pi^{13}_{21} = \sigma' \circ E2\pi^{13}_{21} = 0$. It follows from (10.14) that $\sigma'' \circ \pi^{13}_{21} = 0$. Then, by Lemma 10.6, we have

(10.15). $$E\rho^{IV} = 2\rho''' .$$

We have an exact sequence

$$0 \longrightarrow \pi^5_{20} \xrightarrow{E} \pi^6_{21} \xrightarrow{H} \{\eta_{11} \circ \mu_{12}\} \longrightarrow 0$$

from the exactness of (4.4) and from (10.6) and (10.14). Then it follows, by use of (10.15) and (10.12), that

$$\pi^6_{21} = \{\rho'''\} \oplus \{\bar{\varepsilon}_6\} \approx Z_4 \oplus Z_2$$

We have also an exact sequence

$$0 \longrightarrow \pi^6_{21} \xrightarrow{E} \pi^7_{22} \xrightarrow{H} \pi^{13}_{22} ,$$

where $\pi^{13}_{22} = \{\nu^3_{13}\} \oplus \{\mu_{13}\} \oplus \{\eta_{13} \circ \varepsilon_{14}\} \approx Z_2 \oplus Z_2 \oplus Z_2$ by Theorem 7.2. We have

$$\begin{aligned} H(\sigma' \circ \bar{\nu}_{14}) &= H(\sigma') \circ \bar{\nu}_{14} && \text{by Proposition 2.2,} \\ &= \eta_{13} \circ \bar{\nu}_{14} = \nu^3_{13} && \text{by Lemma 5.14 and Lemma 6.3,} \\ H(\sigma' \circ \varepsilon_{14}) &= H(\sigma') \circ \varepsilon_{14} = \eta_{13} \circ \varepsilon_{14} \end{aligned}$$

and $H(\rho'') \equiv \mu_{13} \mod \{H(\sigma' \circ \bar{\nu}_{14})\} + \{H(\sigma' \circ \varepsilon_{14})\}$ by (10.12).

Thus π^7_{22} is generated by $E\pi^6_{21}$, $\sigma' \circ \bar{\nu}_{14}$, $\sigma' \circ \varepsilon_{14}$ and ρ''. Obviously $2(\sigma' \circ \bar{\nu}_{14}) = 2(\sigma' \circ \varepsilon_{14}) = 0$. By Lemma 10.6 and Theorem 7.1,

$$2\rho'' = E\rho''' + a\sigma' \circ \bar{\nu}_{14} + b\sigma' \circ \varepsilon_{14}$$

for some integers a and b. Applying the homomorphism H to this equation, we have that

$$\begin{aligned} a\nu^3_{13} + b\eta_{13} \circ \varepsilon_{14} &= H(E\rho''' + a\sigma' \circ \bar{\nu}_{14} + b\sigma' \circ \varepsilon_{14}) \\ &= 2H(\rho'') \in 2\pi^{13}_{22} = 0 . \end{aligned}$$

It follows that $a \equiv b \equiv 0$ (mod.2) and

(10.16). $$E\rho''' = 2\rho'' .$$

By the exactness of the above sequence, we have the result

$$\pi^7_{22} = \{\rho''\} \oplus \{\sigma' \circ \bar{\nu}_{14}\} \oplus \{\sigma' \circ \varepsilon_{14}\} \oplus \{\bar{\varepsilon}_7\} \approx Z_8 \oplus Z_2 \oplus Z_2 \oplus Z_2 .$$

By (5.15) and by Theorem 7.1, we have the result of π^8_{23} in Theorem 10.5.

Next consider the homomorphism $\Delta : \pi^{17}_{25} \to \pi^8_{23}$. By Proposition 2.5 and by (5.16),

$$\begin{aligned} \Delta(\bar{\nu}_{17}) &= \Delta(\iota_{17}) \circ \bar{\nu}_{15} = \pm (2\sigma_8 - E\sigma') \circ \bar{\nu}_{15} = E\sigma' \circ \bar{\nu}_{15} \\ \Delta(\varepsilon_{17}) &= \Delta(\iota_{17}) \circ \varepsilon_{15} = \pm (2\sigma_8 - E\sigma') \circ \varepsilon_{15} = E\sigma' \circ \varepsilon_{15} . \end{aligned}$$

Since $\bar{\nu}_{17}$ and ε_{17} generate $\pi^{17}_{25} \approx Z_2 \oplus Z_2$, by Theorem 7.1, then it follows from the exactness of (4.4) that

(10.17). $E : \pi^8_{24} \to \pi^9_{25}$ <u>is onto</u>

and that

$$E\pi^8_{23} = \{E^2\rho''\} \oplus \{\sigma_9 \circ \bar{\nu}_{16}\} \oplus \{\sigma_9 \circ \varepsilon_{16}\} \oplus \{\bar{\varepsilon}_9\} \approx Z_8 \oplus Z_2 \oplus Z_2 \oplus Z_2 .$$

We have an exact sequence

$$0 \longrightarrow E\pi^8_{23} \longrightarrow \pi^9_{24} \xrightarrow{H} \{8\sigma_{17}\} \longrightarrow 0$$

from (10.8). The element ρ' satisfies $H(\rho') = 8\sigma_{17}$ and $2\rho' \equiv E^2\rho''$ mod $\{\sigma_9 \circ \bar{\nu}_{16}\} + \{\sigma_9 \circ \varepsilon_{16}\}$, by (10.11), Lemma 10.6 and Theorem 7.1. Then we conclude that

$$\pi^9_{24} = \{\rho'\} \oplus \{\sigma_9 \circ \bar{\nu}_{16}\} \oplus \{\sigma_9 \circ \varepsilon_{16}\} \oplus \{\bar{\varepsilon}_9\} \approx Z_{16} \oplus Z_2 \oplus Z_2 \oplus Z_2 .$$

<u>Remark</u>. It can be proved that $2\rho' = E^2\rho''$ by use of the methods in the next chapter.

<u>Lemma</u> 10.7. $\varepsilon_n \circ \sigma_{n+8} = 0$ <u>for</u> $n \geq 3$, $\sigma_n \circ \varepsilon_{n+7} = 0$ <u>for</u> $n \geq 11$ <u>and</u> $\bar{\nu}_n \circ \sigma_{n+8} = 0$ <u>for</u> $n \geq 6$.

<u>Proof</u>. $\varepsilon_3 \circ \sigma_{11}$ is an element of $\pi^3_{18} = \{\bar{\varepsilon}_3\} \approx Z_2$. Thus $\varepsilon_3 \circ \sigma_{11} = x\bar{\varepsilon}_3$ for an integer x. Apply $H \circ H : \pi^3_{18} \to \pi^5_{18} \to \pi^9_{18}$ to this relation. We have for some integer y

$$\begin{aligned} HH(\bar{\varepsilon}_3) &= H(\nu_5 \circ \sigma_8 \circ \nu_{15} + yE(\nu_4 \circ \eta_7 \circ \mu_8)) && \text{by Lemma 10.2,} \\ &= H(\nu_5 \circ \sigma_8) \circ \nu_{15} + 0 && \text{by Proposition 2.2} \\ &= \nu_9^2 \circ \nu_{15} = \nu_9^3 && \text{by (7.9)} \end{aligned}$$

and

$$\begin{aligned} HH(\varepsilon_3 \circ \sigma_{11}) &= H(H(\varepsilon_3) \circ \sigma_{11}) = HH(\varepsilon_3) \circ \sigma_{11} && \text{by Proposition 2.2} \\ &= H(\nu_2^5) \circ \sigma_{11} = HE(\nu_2^4) \circ \sigma_{11} = 0 && \text{by Lemma 6.1.} \end{aligned}$$

Since ν_9^3 is an element of order 2, then the relation $HH(\varepsilon_3 \circ \sigma_{11}) = x\, HH(\bar{\varepsilon}_3)$ implies that $x \equiv 0$ (mod.2) and $\varepsilon_3 \circ \sigma_{11} = 0$. Obviously $\varepsilon_n \circ \sigma_{n+8} = E^{n-3}(\varepsilon_3 \circ \sigma_{11}) = 0$ for $n \geq 3$. By Proposition 3.1,

$$\sigma_n \circ \varepsilon_{n+7} = E^{n-11}(\sigma_8 \# \varepsilon_3) = \varepsilon_n \circ \sigma_{n+8} = 0$$

for $n \geq 11$. Next

$$\begin{aligned} \bar{\nu}_8 \circ \sigma_{16} &\in \{\nu_8, \eta_{11}, \nu_{12}\}_1 \circ \sigma_{16} && \text{by Lemma 6.2,} \\ &= \nu_8 \circ E\{\eta_{10}, \nu_{11}, \sigma_{14}\} && \text{by Proposition 1.4,} \\ &\subset \nu_8 \circ E\pi^{10}_{22} = \{\nu_8 \circ E\Delta(\nu_{21})\} && \text{by Theorem 7.6} \\ &= 0 . \end{aligned}$$

Since $E^2 : \pi_{21}^6 \to \pi_{23}^8$ is an isomorphism into, then $E^2(\bar{\nu}_6 \circ \sigma_{14}) = \bar{\nu}_8 \circ \sigma_{16} = 0$ implies that $\bar{\nu}_6 \circ \sigma_{14} = 0$. Thus $\bar{\nu}_n \circ \sigma_{n+8} = E^{n-6}(\bar{\nu}_6 \circ \sigma_{14}) = 0$ for $n \geq 6$. q.e.d.

As a corollary, we have

(10.18). $\Delta(\sigma_{19}) = \sigma_9 \circ \bar{\nu}_{16} + \sigma_9 \circ \varepsilon_{16}$, $\sigma_{10} \circ \bar{\nu}_{17} = \sigma_{10} \circ \varepsilon_{17}$ <u>and</u> $\sigma_n \circ \bar{\nu}_{n+7} = 0$ <u>for</u> $n \geq 11$.

For, by use of Lemma 6.4, Proposition 2.5 and (7.1),

$$\begin{aligned}\Delta(\sigma_{19}) &= \Delta(\iota_{19}) \circ \sigma_{17} = (\sigma_9 \circ \eta_{16} + \bar{\nu}_9 + \varepsilon_9) \circ \sigma_{17}\\ &= \sigma_9 \circ \eta_{16} \circ \sigma_{17} + \bar{\nu}_9 \circ \sigma_{17} + \varepsilon_9 \circ \sigma_{17}\\ &= \sigma_9 \circ (\bar{\nu}_{16} + \varepsilon_{16}) = \sigma_9 \circ \bar{\nu}_{16} + \sigma_9 \circ \varepsilon_{16}\,.\end{aligned}$$

Thus $\sigma_{10} \circ \bar{\nu}_{17} = \sigma_{10} \circ \varepsilon_{17} + E\Delta(\sigma_{19}) = \sigma_{10} \circ \varepsilon_{17}$ and then $\sigma_n \circ \bar{\nu}_{n+7} = \sigma_n \circ \varepsilon_{n+7} = 0$ for $n \geq 11$.

Now, in the exact sequence

$$\pi_{26}^{10} \xrightarrow{H} \pi_{26}^{19} \xrightarrow{\Delta} \pi_{24}^{9} \xrightarrow{E} \pi_{25}^{10}$$

of (4.4), the homomorphism E is onto, by (10.9), and its kernel is generated by $\sigma_9 \circ \bar{\nu}_{16} + \sigma_9 \circ \varepsilon_{16}$. Then we have that

$$\pi_{25}^{10} = \{E\rho'\} \oplus \{\sigma_{10} \circ \bar{\nu}_{17}\} \oplus \{\bar{\varepsilon}_{10}\} \approx Z_{16} \oplus Z_2 \oplus Z_2 \quad ,$$

and

(10.19). <u>The image of H is generated by</u> $2\sigma_{19}$.

In the exact sequence

$$\pi_{27}^{21} \xrightarrow{\Delta} \pi_{25}^{10} \xrightarrow{E} \pi_{26}^{11} \xrightarrow{H} \pi_{26}^{21}$$

of (4.4), $\pi_{26}^{21} = 0$ by Proposition 5.9 and $\pi_{27}^{21} = \{\nu_{21}^2\} \approx Z_2$ by Proposition 5.11. $\sigma_{10} \circ \bar{\nu}_{17} \neq 0$, but $E(\sigma_{10} \circ \bar{\nu}_{17}) = \sigma_{11} \circ \bar{\nu}_{18} = 0$ by (10.18). Thus E has a non-trivial kernel. It follows from the exactness of the above sequence that

$$\pi_{26}^{11} = \{E^2\rho'\} \oplus \{\bar{\varepsilon}_{11}\} \approx Z_{16} \oplus Z_2$$

and that

(10.20). $\Delta : \pi_{27}^{21} \to \pi_{25}^{10}$ <u>is an isomorphism into and</u> $\Delta(\nu_{21}^2) = \sigma_{10} \circ \bar{\nu}_{17} = \sigma_{10} \circ \varepsilon_{17}$.

Since $\pi_{27}^{23} = \pi_{28}^{23} = 0$ by Proposition 5.8 and Proposition 5.9, then we have from the exactness of the sequence (4.4) that $E : \pi_{26}^{11} \to \pi_{27}^{12}$ is an isomorphism onto and that

$$\pi_{27}^{12} = \{E^3\rho'\} \oplus \{\bar{\varepsilon}_{12}\} \approx Z_{16} \oplus Z_2 \quad .$$

Consequently all the assertions of Theorem 10.5 are proved.

iv). <u>The groups</u> π^n_{n+15} <u>for</u> $n \geq 13$.

We shall continue the computation of the groups π^n_{n+15}.

<u>Lemma</u> 10.8. i). $E : \pi^{12}_{27} \to \pi^{13}_{28}$ <u>is an isomorphism into and</u> $H : \pi^{13}_{28} / E\pi^{12}_{27} \approx Z_2$.

ii). $E^{n-13} : \pi^{13}_{28} \to \pi^n_{n+15}$ <u>are isomorphisms for</u> $n \geq 13$ <u>and</u> $n \neq 16$.

iii). $E : \pi^{15}_{30} \to \pi^{16}_{31}$ <u>is an isomorphism into and</u> $\pi^{16}_{31} = E\pi^{15}_{30} \oplus \{\Delta(\iota_{33})\} \approx \pi^{15}_{30} \oplus Z$.

<u>Proof</u>. The assertion of i) is proved by the exactness of the sequence (4.4), and by the results $H\pi^{13}_{28} = \{\eta^3_{25}\} \approx Z_2$ of (10.11) and $\pi^{25}_{29} = 0$ of Proposition 5.8. Next we prove

(10.21). $\Delta(\nu_{27}) = 0$, $\Delta(\eta^2_{29}) = 0$ <u>and</u> $\Delta(\eta_{31}) = 0$.

For, by use of Proposition 2.5,

$$\begin{aligned}
\Delta(\nu_{27}) &= \Delta(\iota_{27}) \circ \nu_{25} = E\theta \circ \nu_{25} && \text{by (7.30)}\\
&= E(\{\sigma_{12}, \nu_{19}, \eta_{22}\} \circ \nu_{24})\\
&= - E(\sigma_{12} \circ \{\nu_{19}, \eta_{22}, \nu_{23}\}) && \text{by Proposition 1.4}\\
&= - E(\sigma_{12} \circ \bar{\nu}_{19}) && \text{by Lemma 6.2}\\
&= 0 && \text{by (10.18)},\\
\Delta(\eta^2_{29}) &= \Delta(\eta_{29}) \circ \eta_{28} = 4\sigma^2_{14} \circ \eta_{28} && \text{by (10.10)}\\
&= \sigma^2_{14} \circ 4\eta_{28} = 0,\\
\text{and} \quad \Delta(\eta_{31}) &= \Delta(\iota_{31}) \circ \eta_{29} = 2\sigma^2_{15} \circ \eta_{29} && \text{by (10.10)}\\
&= \sigma^2_{15} \circ 2\eta_{29} = 0,
\end{aligned}$$

where we have to remark that the composition $\{\sigma_{12}, \nu_{19}, \eta_{22}\} \circ \nu_{24}$ consists of a single element, but it is proved easily.

Apply this (10.21) and (10.11) to the exact sequence (4.4), then we have exact sequences

$$0 \longrightarrow \pi^{13}_{28} \xrightarrow{E} \pi^{14}_{29} \longrightarrow 0,$$

$$0 \longrightarrow \pi^{14}_{29} \xrightarrow{E} \pi^{15}_{30} \longrightarrow 0,$$

$$0 \longrightarrow \pi^{15}_{30} \xrightarrow{E} \pi^{16}_{31} \longrightarrow 2\pi^{31}_{31} \longrightarrow 0$$

and

$$\pi^{33}_{33} \xrightarrow{\Delta} \pi^{16}_{31} \xrightarrow{E} \pi^{17}_{32} \longrightarrow \pi^{33}_{32} = 0.$$

By Proposition 2.7, we may replace $2\pi^{31}_{31}$ by $\Delta\pi^{33}_{33}$. Then the lemma is proved by these exact sequences and by (4.5). q.e.d.

Next we prove

Lemma 10.9. *There exists an element* ρ_{13} *of* π_{28}^{13} *such that* $2\rho_{13} = E^4\rho'$ *and that* $E^\infty\rho_{13} \in \langle \sigma, 2\sigma, 8\iota \rangle$.

Proof. Since $\rho' \in \{\sigma_9, 16\iota_{16}, \sigma_{16}\}$, then $E^\infty\rho' \in \langle \sigma, 16\iota, \sigma \rangle$. $\langle \sigma, 16\iota, \sigma \rangle$ is a coset of $G_8 \circ \sigma = (G_8;2) \circ \sigma$. $(G_8;2)$ is generated by $\bar{\nu}$ and ε (Theorem 7.1). By Lemma 10.7, $\varepsilon \circ \sigma = \bar{\nu} \circ \sigma = 0$. It follows that $G_8 \circ \sigma = 0$ and that $\langle \sigma, 16\iota, \sigma \rangle$ consists of a single element $E^\infty\rho'$. Apply (3.10) to $\alpha = \sigma$ and $\beta = 16\iota$, then we have that

$$E^\infty\rho' \in \langle \sigma, 2\sigma, 16\iota \rangle .$$

By (3.5),

$$2 \langle \sigma, 2\sigma, 8\iota \rangle = \langle \sigma, 2\sigma, 8\iota \rangle \circ 2\iota \subset \langle \sigma, 2\sigma, 16\iota \rangle .$$

Since $\langle \sigma, 2\sigma, 16\iota \rangle$ is a coset of $G_8 \circ \sigma + G_{15} \circ 16\iota = 16G_{15}$, then we have that there exist elements $\alpha \in \langle \sigma, 2\sigma, 8\iota \rangle$ and $\alpha \in G_{15}$ such that

$$E^\infty\rho' - 2\alpha = 16\alpha' .$$

Set $\rho = \alpha + 8\alpha'$, then $\rho \in \langle \sigma, 2\sigma, 8\iota \rangle$ and $2\rho = E^\infty\rho'$. By Lemma 10.8, $E^\infty : \pi_{28}^{13} \to (G_{15};2)$ is an isomorphism into. Obviously, $\rho \in (G_{15};2)$. Then it follows that there exists an element ρ_{13} of π_{28}^{13} such that $2\rho_{13} = E^4\rho'$ and $E^\infty\rho_{13} = \rho \in \langle \sigma, 2\sigma, 8\iota \rangle$. q.e.d.

Since the order of $2\rho_{13} = E^4\rho'$ is 16, then the order of ρ_{13} is 32. It follows from i) of Lemma 10.8 that

$$\pi_{28}^{13} = \{\rho_{13}\} \oplus \{\bar{\varepsilon}_{13}\} \approx Z_{32} \oplus Z_2 .$$

We have also that

(10.22). $$H(\rho_{13}) = 4\nu_{25} = \eta_{25}^3 .$$

Denote that

$$\rho_n = E^{n-13}\rho_{13} \quad \text{for} \quad n \geq 13 \quad \text{and} \quad \rho = E^\infty\rho_{13} ,$$

then it follows from Lemma 10.8 the following theorem.

Theorem 10.10. $\pi_{31}^{16} = \{\rho_{16}\} \oplus \{\bar{\varepsilon}_{16}\} \oplus \{\Delta(\iota_{33})\} \approx Z_{32} \oplus Z_2 \oplus Z$,

$\pi_{n+15}^n = \{\rho_n\} \oplus \{\bar{\varepsilon}_n\} \approx Z_{32} \oplus Z_2$ *for* $n \geq 13$ *and* $n \neq 16$,

and $(G_{15};2) = \{\rho\} \oplus \{\bar{\varepsilon}\} \approx Z_{32} \oplus Z_2$.

Next we shall prove the following relations.

(10.23). $\eta_n \circ \kappa_{n+1} = \bar{\varepsilon}_n$ *for* $n \geq 6$ *and* $\kappa_n \circ \eta_{n+14} = \bar{\varepsilon}_n$ *for* $n \geq 9$.

By Lemma 10.1, $\eta \circ \kappa \in E^\infty\{\nu_6^2, 2\iota_{12}, \bar{\nu}_{12}\} \subset -\langle \nu^2, 2\iota, \bar{\nu} \rangle$. By the definition of $\bar{\varepsilon}_3$, $\bar{\varepsilon} \in E^\infty\{\varepsilon_3, 2\iota_{11}, \nu_{11}^2\} \subset \langle \varepsilon, 2\iota, \nu^2 \rangle$. By i) of (3.9), $\langle \varepsilon, 2\iota, \nu^2 \rangle = -\langle \nu^2, 2\iota, \varepsilon \rangle$. The stable secondary compositions $< \nu^2, 2\iota, \bar{\nu} >$, $< \nu^2, 2\iota, \varepsilon >$ and $< \nu^2, 2\iota, \eta \circ \sigma >$ are cosets of 0, since $G_9 \circ \nu^2 = G_7 \circ \bar{\nu} = G_7 \circ \varepsilon = 0$ by Theorem 7.6 and Lemma 10.7. By (3.5), (3.8) and by Lemma 6.4,

$$< \nu^2, 2\iota, \bar{\nu} > + < \nu^2, 2\iota, \varepsilon > = < \nu^2, 2\iota, \eta > \circ \sigma \in G_8 \circ \sigma = 0 .$$

Thus we conclude that $\eta \circ \kappa = \bar{\varepsilon}$. Since $E^\infty : \pi_{21}^6 \to (G_{15};2)$ is an isomorphsim into, it follows from $E^\infty(\eta_6 \circ \kappa_7) = \eta \circ \kappa = \bar{\varepsilon} = E^\infty(\bar{\varepsilon}_6)$ that $\eta_6 \circ \kappa_7 = \bar{\varepsilon}_6$. Obviously, $\eta_n \circ \kappa_{n+1} = \bar{\varepsilon}_n$ for $n \geq 6$. By Proposition 3.1, $\kappa_n \circ \eta_{n+14} = E^{n-9}(\eta_2 \circ \kappa_7) = \eta_n \circ \kappa_{n+1} = \bar{\varepsilon}_n$ for $n \geq 9$.

CHAPTER XI.

Relative J-homomorphisms.

The homomorphism

$$J : \pi_i(SO(n)) \to \pi_{n+i}(S^n)$$

of G. W. Whitehead was defined as follows (cf.[26]). Let $f : S^i \to SO(n)$ be a representative of an element α of $\pi_i(SO(n))$. Define a mapping

$$F : S^i \times S^{n-1} \to S^{n-1}$$

by the formula $F(x,y) = f(x)(y)$, $x \in S^i$, $y \in S^{n-1}$. Let

$$G(F) : S^i * S^{n-1} \to S^n$$

be the Hopf construction of F. By a suitable homeomorphism of $S^i * S^{n-1}$ with S^{n+i}, $G(F)$ represents an element

$$J(\alpha) \in \pi_{i+n}(S^n) .$$

The Hopf construction

$$G(h) : A * B \to EC$$

of a mapping $h : A \times B \to C$ is defined as follows, where $A * B$ denotes a join of A and B. We consider that $A * B$ is obtained from the product $A \times I^1 \times B$ by identifying with the relations : $(a, 0, b) = (a, 0, b')$ and $(a, 1, b) = (a', 1, b)$ for every $a, a' \in A$ and $b, b' \in B$. We represent each point of $A * B$ by a symbol (a,t,b) with the above relations. Then $G(h)$ is defined by the formula

$$G(h)(a, t, b) = d_C(h(a, b), t),$$

where $d_C : C \times I^1 \to EC$ is a shrinking map which defines EC.

For two mappings $f : A \to A'$ and $g : B \to B'$, their join

$$f * g : A * B \to A' * B'$$

is defined by the formula $(f * g)(a, t, b) = (f(a), t, g(b))$.

Consider a mapping

$$p : A * B \to EA \# B$$

which is defined by the formula $p(a, t, b) = \Phi(d_A(a, t), b)$, where $\Phi : EA \times B \to EA \# B$ is a shrinking map which defines $EA \# B$. It is verified easily that p shrinks the subset $A * b_0 \cup a_0 * B$ to a point and maps homeomorphically elsewhere. If A and B are finite cell complexes, then the subset $A * b_0 \cup a_0 * B$ is a contractible subcomplex of $A * B$, and thus p is a homotopy equivalence. In particular, we denote by

$$p_n : SO(n) * S^{n-1} \to E^n SO(n) = ESO(n) \# S^{n-1}$$

the homotopy equivalence of the case that $A = SO(n)$ and $B = S^{n-1}$. In the case that $A = S^i$ and $B = S^{n-1}$, the join $S^i * S^{n-1}$ is homeomorphic to S^{n+i} and $p : S^i * S^{n-1} \to S^{n+i}$ is a mapping of degree ± 1. Thus p is homotopic to a homeomorphism p_0.

In general, we have the following commutative diagram.

$$\text{(11.1)} \qquad \begin{array}{ccc} A * B & \xrightarrow{\ f * g\ } & A' * B' \\ \downarrow p & & \downarrow p \\ EA \# B & \xrightarrow{\ Ef \# g\ } & EA' \# B' \end{array}$$

Now let

$$r_n : SO(n) \times S^{n-1} \to S^{n-1}$$

be the action of $SO(n)$ as the rotations of S^{n-1}. Then the representative $G(F)$ of $J(\alpha)$ satisfies the formula

$$G(F) = G(r_n) \circ (f * i_{n-1}) ,$$

where i_{n-1} is the identity of S^{n-1}. By taking orientations of $S^i * S^{n-1}$ and S^{n+i} such that p_0 preserves the orientations, we have that $J(\alpha)$ is the class of $G(r_n) \circ (f * i_{n-1}) \circ p_0$. It follows from the commutativity of the diagram (11.1) that $G(r_n) \circ (f * i_{n-1}) \circ p_0$ is homotopic to the composition

$$G(r_n) \circ q_n \circ Ef \# i_{n-1} = G(r_n) \circ q_n \circ E^n f ,$$

where q_n is a homotopy-inverse of p_n. We denote that

$$G_n = G(r_n) \circ q_n : E^n SO(n) \to S^n .$$

We remark that G_n is independent of the choice of the inverse q_n up to homotopy. Then the homomorphism J is defined by the following formula.

(11.2) $J = G_{n*} \circ E^n : \pi_i(SO(n)) \to \pi_{n+i}(E^n SO(n)) \to \pi_{n+i}(S^n)$.

Next consider the natural injection i of $SO(n-1)$ into $SO(n)$, which is given by considering that each rotation of $SO(n-1)$ is a rotation of $SO(n)$ leaving the last coordinate fixed.

<u>Lemma</u> 11.1. <u>The restriction</u> $G_n \mid E^n SO(n-1)$ <u>is homotopic to</u> $- EG_{n-1}$.

<u>Proof</u>. We use the notations $C_+(A) = d_A(A \times [\frac{1}{2}, 1])$ and $C_-(A) = d_A(A \times [0, \frac{1}{2}])$. We identify A with $C_+(A) \cap C_-(A)$ by the correspondence $a \to d_A(a, \frac{1}{2})$. Then the restriction of r_n on $SO(n-1) \times S^{n-1}$ satisfies the conditions

$$r_n \mid SO(n-1) \times S^{n-2} = r_{n-1}, \quad r_n(SO(n-1) \times C_+(S^{n-2})) \subset C_+(S^{n-2})$$

and $r_n(SO(n-1) \times C_-(S^{n-2})) \subset C_-(S^{n-2})$.

It follows from the definition of the Hopf construction that $G(r_n) \mid SO(n-1) * S^{n-2} = G(r_{n-1})$, $G(r_n)(SO(n-1) * C_+(S^{n-2})) \subset EC_+(S^{n-2})$ and $G(r_n)(SO(n-1) * C_-(S^{n-2})) \subset EC_-(S^{n-2})$.

We have also similar properties on p_n :

$p_n \mid SO(n-1) * S^{n-2} = p_{n-1}$, $p_n(SO(n-1) * C_+(S^{n-2})) \subset C_+(E^{n-1}SO(n-1))$ and $p_n(SO(n-1) * C_-(S^{n-2})) \subset C_-(E^{n-1}SO(n-1))$.

We show that there are homotopy inverses q_n and q_{n-1} of p_{n-1}, respectively, such that

$$q_n \mid E^{n-1}SO(n-1) = q_{n-1}, \quad q_n(C_+E^{n-1}SO(n-1)) \subset SO(n-1) * C_+(S^{n-2})$$

and $q_n(C_-E^{n-1}SO(n-1)) \subset SO(n-1) * C_-(S^{n-2})$.

Let U be the closures of a regular neighbourhood of $SO(n) * e_0 \cup e_0 * S^{n-1}$ in $SO(n) * S^{n-1}$ such that U and its intersections $U_0 = U \cap SO(n-1) * S^{n-2}$, $U_+ = U \cap SO(n-1) * C_+S^{n-2}$ and $U_- = U \cap SO(n-1) * C_-S^{n-2}$ are all contractible to a point. This is possible if we take a suitable simplicial decomposition of $SO(n) * S^{n-1}$ such that $SO(n-1) * S^{n-2}$, $SO(n-1) * C_+S^{n-2}$ and $SO(n-1) * C_-S^{n-2}$ are subcomplexes. Since $SO(n) * e_0 \cup e_0 * S^{n-1}$ is contractible to a point and it is a deformation retract of U, then U is contractible to a point. Similarly, U_0, U_+ and U_- are contractible to a point. Further, applying the identification p_n, we have that the images $V = p_n(U)$, $V_0 = p_n(U_0)$, $V_+ = p_n(U_+)$ and $V_- = p_n(U_-)$ are contractible to a point.

The mapping p_n maps the outside of U homeomorphically onto the outside of V. Then we obtain a mapping q_n by setting $q_n = p_n^{-1}$ on $E^nSO(n) - V$ and by extending over V into U. Since U is contractible, such q_n exists and it is unique up to homotopy. Further, q_n is a homotopy inverse of p_n.

Now, we can choose the extension of q_n over V such that $q_n(V_0) \subset U_0$, $q_n(V_+) \subset U_+$ and $q_n(V_-) \subset V_-$, since U_0, U_+ and U_- are contractible. Set $q_{n-1} = q_n \mid E^{n-1}SO(n-1)$, then $q_{n-1} = p_{n-1}^{-1}$ on $E^{n-1}SO(n-1) - V_0$ and thus q_{n-1} is a homotopy inverse of p_{n-1}. Then the inverses q_n and q_{n-1} satisfy the required condition.

Consider mappings G_n and G_{n-1} which are defined by use of these q_n and q_{n-1}, then the following conditions are satisfied.

$G_n \mid E^{n-1}SO(n-1) = G_{n-1}$, $G_n(C_+E^{n-1}SO(n-1)) \subset EC_+(S^{n-2})$ and $G_n(C_-E^{n-1}SO(n-1)) \subset EC_-(S^{n-2})$.

Let $\sigma : S^n \to S^n$ and $\rho : S^n \to S^n$ be defined by the formulas $\sigma(\psi_n(t_1, \ldots, t_{n-2}, t_{n-1}, t_n)) = \psi_n(t_1, \ldots, t_{n-2}, t_n, t_{n-1})$ and $\rho(d_n(x, t)) = d_n(x, 1-t)$. Then we have that σ and ρ are homotopic to each other, $\rho \circ EG_{n-1} = - EG_{n-1}$ and that σ maps C_+S^{n-1} and C_-S^{n-1} homeomorphically onto $EC_+(S^{n-2})$ and $EC_-(S^{n-2})$ respectively. It is easily verified that $\sigma \circ EG_{n-1}$ satisfies the same conditions as G_n. Since $EC_+(S^{n-2})$ is contractible to a point, the restrictions of G_n and $\sigma \circ EG_{n-1}$ on $C_+E^{n-1}SO(n-1)$ are homotopic to each other fixing the points of $E^{n-1}SO(n-1)$. A similar statement is true for C_-. Then $G_n \mid E^nSO(n-1)$ is homotopic to $\sigma \circ EG_{n-1}$ and $\sigma \circ EG_{n-1}$ is homotopic to $\rho \circ EG_{n-1} = - EG_{n-1}$. Thus we have obtained the lemma. q.e.d.

Corollary 11.2. The diagram

$$\begin{array}{ccc} \pi_i(SO(n-1)) & \xrightarrow{-J} & \pi_{n+i-1}(S^{n-1}) \\ \downarrow i_* & & \downarrow E \\ \pi_i(SO(n)) & \xrightarrow{J} & \pi_{n+i}(S^n) \end{array}$$

is commutative.

This is a direct consequence of (11.2) and Lemma 11.1.

Corollary 11.3. There exists a sequence of mappings $F_n : E^nSO(n) \to S^n$ for $n = 2, 3, \ldots$ such that $F_{n+1} \mid E^{n+1}SO(n) = EF_n$ and $F_{n*} \circ E^n = (-1)^nJ$.

Proof. By use of homotopy extension theorem and Lemma 11.1, we have a sequence of mappings G_n of Lemma 11.1 such that $G_{n+1} \mid E^{n+1}SO(n) = - EG_n$. Let σ_n be a homeomorphism of $E^nSO(n)$ on itself given by the formula $\sigma_n(\phi(r, \psi_n(t_1, \ldots, t_n))) = \phi(r, \psi_n(1-t_1, \ldots, 1-t_n))$, where ϕ is a shrinking map which defines $E^nSO(n) = SO(n) \# S^n$. Define F_n by setting $F_n = G_n \circ \sigma_n$. Then it is easily verified that $F_{n+1} \mid E^{n+1}SO(n) =$

F_n and that F_n is homotopic to $(-1)^n G_n$. Thus $F_{n*} \circ E^n = (-1)^n G_{n*} \circ E^n = (-1)^n J$. q.e.d.

Consider the canonical injection $i : X \to \Omega(EX)$ of Chapter I. Then canonical injection of EX into $\Omega(E(EX))$ defines an injection of $\Omega(EX)$ into $\Omega^2(E^2X)$. As the composition of these injections, we have an injection of X into $\Omega^2(E^2X)$. Repeating the process, we have an injection

$$i : X \to \Omega^k(E^kX)$$

for $k = 1, 2, 3, \ldots$, where Ω^k indicates the k-fold iteration of Ω. It follows from the commutativity of (1.9) that the diagram

$$\begin{array}{ccc} X & \xrightarrow{\ f\ } & Y \\ \downarrow i & & \downarrow i \\ \Omega^k(E^kX) & \xrightarrow{\ \Omega^k(E^kf)\ } & \Omega^k(E^kY) \end{array}$$

is commutative. As k-fold iteration of Ω_0 of (1.10), we have a one-to-one correspondence

$$\Omega_0^k : \pi(E^kX \to Y) \approx \pi(X \to \Omega^k(E^kY))$$

and a commutative diagram

$$\begin{array}{ccccc} \pi(X \to Y) & & \xrightarrow{\ E^k\ } & & \pi(E^kX \to E^kY) \\ & \searrow i_* & & \swarrow \Omega_0^k & \\ & & \pi(X \to \Omega^k(E^kY)). & & \end{array}$$

The mappings F_n of Corollary 11.3 satisfy $\Omega_0 F_{n+1} \mid E^n SO(n) = F_n$ and thus

(11.3). $$\Omega_0^k F_{n+k} \mid E^n SO(n) = F_n,$$

where $\Omega_0^k F_{n+k}$ is a mapping of $E^n SO(n+k)$ into $\Omega^k(S^{n+k})$.

The mapping $\Omega_0^k F_{n+k}$ induces a homomorphism of the homotopy exact sequence of the pair $(E^n SO(n+k), E^n SO(n))$ into that of the pair $(\Omega^k(S^{n+k}), S^n)$. Combining the suspension homomorphisms E^n, we have a diagram

(11.4) $$\begin{array}{ccccccc} \ldots \to \pi_i(SO(n)) & \xrightarrow{\ i_*\ } & \pi_i(SO(n+k)) & \xrightarrow{\ i_*\ } & \pi_i(SO(n+k), SO(n)) & \xrightarrow{\partial} & \ldots \\ \downarrow J & & \downarrow J & & \downarrow (\Omega_0^k F_{n+k})_* \circ E^n & & \\ \ldots \to \pi_{n+i}(S^n) & \xrightarrow{\ E^k\ } & \pi_{n+k+i}(S^{n+k}) & \longrightarrow & \pi_{n+i}(\Omega^k(S^{n+k}), S^n) & \xrightarrow{\partial} & \ldots \end{array}$$

which is commutative up to sign.

Lemma 11.4. *$(\Omega_0 F_{n+1})_* \circ E^n : \pi_n(SO(n+1), SO(n)) \to \pi_{2n}(\Omega(S^{n+1}), S^n)$ is an isomorphism onto. The groups are infinite cyclic.*

Proof. We introduce the following result. For a generator α of $\pi_n(SO(n+1), SO(n))$, we have the relation

$$J(\partial(\alpha)) = \pm [\iota_n, \iota_n].$$

This is proved in the section 9 of [26]. By the projection of the fibering $SO(n+1)/SO(n) = S^n$, the group $\pi_n(SO(n+1), SO(n))$ is isomorphic to $\pi_n(S^n) \approx Z$. By Theorem 2.4, $\pi_{2n}(\Omega(S^{n+1}), S^n)$ is isomorphic to $\pi_{2n}((S^{2n})_\infty) \approx Z$. The group $\pi_{2n}(\Omega(S^{n+1}), S^n)$ is replaced by $\pi_{2n+1}(S^{2n+1})$, and the lower sequence of (11.4) becomes (2.11). First, consider the case that n is even. By Proposition 2.7, $[\iota_n, \iota_n] = \pm \Delta(\iota_{2n+1})$ has Hopf invariant ± 2. Then $J(\partial(\alpha))$ is of infinite order. Thus $\Delta : \pi_{2n+1}(S^{2n+1}) \to \pi_{2n-1}(S^n)$ is an isomorphism into and its image coincides with the image of $J \circ \partial$. It follows from the commutativity of (11.4) that $(\Omega_0 F_{n+1})_* \circ E^n$ is onto, and thus it is an isomorphism onto.

Next, consider the case that n is odd. It is known that the boundary homomorphism $\partial' : \pi_{n+1}(SO(n+2), SO(n+1)) \to \pi_n(SO(n+1), SO(n))$ of the triple $(SO(n+2), SO(n+1), SO(n))$ is a homomorphism of degree 2 (cf. [17]). Let α be a generator of $\pi_{n+1}(SO(n+2), SO(n+1))$, and let β be its image in $\pi_n(SO(n+1))$ under the boundary homomorphism. Then $\partial'\alpha = j_* \beta$ and $J(\beta) = \pm [\iota_{n+1}, \iota_{n+1}]$. By Proposition 2.7, $H(J(\beta)) = \pm 2\iota_{2n+1}$. By the commutativity of the diagram (11.4), we have that $j_*(\beta)$ is mapped to a generator of $2(\pi_{2n}(\Omega(S^{n+1}), S^n))$ under the homomorphism $(\Omega_0 F_{n+1})_* \circ E^n$. Since $\partial'\alpha = j_*(\beta)$ generates $2\pi_n(SO(n+1), SO(n))$, it follows that $(\Omega_0 F_{n+1})_* \circ E^n$ is an isomorphism onto. q.e.d

According to [29], we embed an n-dimensional real projective space P^n into $SO(n+1)$ such that an $(n-k)$-dimensional subspace P^{n-k} of P^n is embedded in the subgroup $SO(n-k+1)$ and that the projection $p : SO(n+k-1) \to S^{n-k}$ maps $P^{n-k} - P^{n-k-1}$ homeomorphically onto $S^{n-k} - e_0$.

Let ι be an element of $\pi_n(P^n, P^{n-1})$ which is represented by a characteristic map of the cell e^n. Then the injection homomorphism $i_* : \pi_n(P^n, P^{n-1}) \to \pi_n(SO(n+1), SO(n))$ maps ι to a generator of $\pi_n(SO(n+1), SO(n))$, since i_* is mapped onto a generator of $\pi_n(S^n)$ by the projection of $SO(n+1)$ onto S^n. We have also that $E^n : \pi_n(P^n, P^{n-1}) \to$

$\to \pi_{2n}(E^nP^n, E^nP^{n-1})$ maps ι to a generator of $\pi_{2n}(E^nP^n, E^nP^{n-1})$ which is infinite cyclic. It follows from the commutativity of the diagram

$$\begin{array}{ccc}
\pi_n(P^n, P^{n-1}) & \xrightarrow{i_*} & \pi_n(SO(n+1), SO(n)) \\
\downarrow E^n & & \downarrow E^n \\
\pi_{2n}(E^nP^n, E^nP^{n-1}) & \xrightarrow{E^n i_*} & \pi_{2n}(E^nSO(n+1), E^nSO(n)) \\
\downarrow (\Omega_0 F_{n+1})_* & \swarrow (\Omega_0 F_{n+1})_* & \\
\pi_{2n}(\Omega(S^{n+1}), S^n) & &
\end{array}$$

that $E^n\iota$ is mapped onto a generator of $\pi_{2n}(\Omega(S^{n+1}), S^n)$. Thus we have the following Corollary.

<u>Corollary</u> 11.5. $(\Omega_0F_{n+1})_* : \pi_{2n}(E^nP^n, E^nP^{n-1}) \to \pi_{2n}(\Omega(S^{n+1}), S^n)$ <u>is an isomorphism onto</u>.

Next, we use the following notation of the space of paths.

$$\Omega(X, A) = \{\ell : I^1 \to X \mid \ell(0) \in A \text{ and } \ell(1) = x_0\} .$$

For a mapping $g : (CS^i, S^i, e_0) \to (X, A, x_0)$, we define a mapping

$$\Omega'g : (S^i, e_0) \to (\Omega(X, A), \ell_0), \quad \ell_0(I^1) = x_0,$$

by the formula $(\Omega'g(x))(t) = g(d'(x, t)), (x, t) \in S^i \times I^1$, where $d' : S^i \times I^1 \to CS^i$ is a mapping which defines the cone CS^i.

Then it is verified that Ω' is independent of homotopy and it defines an isomorphism

$$\Omega' : \pi_{i+1}(X, A) \approx \pi_i(\Omega(X, A)).$$

Let $f : (X, A) \to (Y, B)$ be a mapping, then the diagram

$$\begin{array}{ccc}
\pi_{i+1}(X, A) & \xrightarrow{\Omega'} & \pi_i(\Omega(X, A)) \\
\downarrow f_* & & \downarrow \Omega f_* \\
\pi_{i+1}(Y, B) & \xrightarrow{\Omega'} & \pi_i(\Omega(Y, B))
\end{array}$$

is commutative, where $\Omega f : \Omega(X, A) \to \Omega(Y, B)$ is a mapping defined by the formula $(\Omega f(\ell))(t) = f(\ell(t))$ for $\ell \in \Omega(X, A)$, $t \in I^1$.

Applying the commutativity for the mapping $\Omega_0F_{n+1} : (E^nP^n, E^nP^{n-1}) \to (\Omega(S^{n+1}), S^n)$, we have that

(11.5) $\Omega(\Omega_0F_{n+1})_* : \pi_{2n-1}(\Omega(E^nP^n, E^nP^{n-1})) \to \pi_{2n-1}(\Omega(\Omega(S^{n+1}), S^n))$ <u>is an isomorphism onto</u>.

Consider the union $X \cup CA$ of X and CA in which A and the base of CA are identified to each other. Then the canonical injection $i : X \to \Omega(EX)$ is extended to an injection

$$i : X \cup CA \to \Omega(EX, EA)$$

by the formula $i(d_A'(a, t))(s) = d_A(a, (1-t)s + t)$.

In particular, we have an injection

(11.6) $i : E^{n-1}(P^{n+k-1} \cup CP^{n-1}) \to \Omega(E^n P^{n+k-1}, E^n P^{n-1})$, _where we use the identification_ $E^{n-1}(P^{n+k-1} \cup CP^{n-1}) = E^{n-1}P^{n+k-1} \cup CE^{n-1}P^{n-1}$ _of_ (1.16).

By shrinking the subcomplex P^{n-1} of P^{n+k-1}, we obtain a cell complex

$$P_n^{n+k-1} = S^n \cup e^{n+1} \cup \dots \cup e^{n+k-1}.$$

In particular P_n^n is an n-sphere. The complex P_n^{n+k-1} is also obtained from $P^{n+k-1} \cup CP^{n+1}$ by shrinking CP^{n-1}. Since CP^{n-1} is contractible to a point, then we have

(11.7). _The shrinking map of_ $P^{n+k-1} \cup CP^{n-1}$ _onto_ P_n^{n+k-1} _is a homotopy equivalence_.

Next, for a mapping $g : (ECS^{i-1}, ES^{i-1}, e_0) \to (X, A, x_0)$, we associate the mapping $\Omega_0 g : (CS^{i-1}, S^{i-1}, e_0) \to (\Omega(X), \Omega(A), \ell_0)$. Then, by taking their homotopy classes, we have an isomorphism

$$\Omega_0 : \pi_{i+1}(X, A) \approx \pi_i(\Omega(X), \Omega(A)).$$

The commutativity of the diagram

$$\begin{array}{ccc} \pi_{i+1}(X, A) & \xrightarrow{\Omega_0} & \pi_i(\Omega(X), \Omega(A)) \\ \downarrow f_* & & \downarrow \Omega f_* \\ \pi_{i+1}(Y, B) & \xrightarrow{\Omega_0} & \pi_i(\Omega(Y), \Omega(B)) \end{array}$$

holds for a mapping $f : (X, A) \to (Y, B)$. This Ω_0 coincides with the usual Ω_0 if $A = x_0$.

Now, we denote that

$$Q_n^{n+k} = \Omega(\Omega^k(S^{n+k}), S^n).$$

It follows from the homotopy exact sequence associated with the pair $(\Omega^k(S^{n+k}), S^n)$ that the following sequence is exact.

(11.8) $\dots \to \pi_i(S^n) \xrightarrow{E^k} \pi_{i+k}(S^{n+k}) \xrightarrow{I_k} \pi_{i-1}(Q_n^{n+k}) \xrightarrow{p_{k*}} \pi_{i-1}(S^n) \to \dots,$

where $p_k : Q_n^{n+k} \to S^n$ associates each path ℓ of Q_n^{n+k} the starting point $\ell(0) = p_k(\ell)$ and I_k is defined by $I_k = \Omega' \circ j_* \circ \Omega_0^k$.

It follows also from the exact sequence of the triple $(\Omega^{k+h}(S^{n+k+h}), \Omega^k(S^{n+k}), S^n)$, that the following sequence is exact.

$$(11.9).\quad \cdots \to \pi_i(Q_n^{n+k}) \xrightarrow{i_*} \pi_i(Q_n^{n+k+h}) \xrightarrow{I'_k} \pi_{i+k}(Q_{n+k}^{n+k+h}) \xrightarrow{\Delta_k}$$

$$\pi_{i-1}(Q_n^{n+k}) \to \cdots,$$

where I'_k and Δ_k are given by the formulas $I'_k = \Omega' \circ (\Omega_0^k)^{-1} \circ j_* \circ \Omega'^{-1}$ and $\Delta_k = \Omega' \circ \partial \circ \Omega_0^k \circ \Omega'^{-1}$.

Then the commutativity holds in the diagram

$$(11.10)\quad \begin{array}{ccccccc}
\pi_i(S^n) & \xrightarrow{E^k} & \pi_{i+k}(S^{n+k}) & \xrightarrow{I_k} & \pi_{i-1}(Q_n^{n+k}) & \xrightarrow{p_{k*}} & \pi_{i-1}(S^n) \\
\downarrow{\scriptstyle I_h} & & \downarrow{\scriptstyle I_{k+h}} & \nearrow{\scriptstyle I'_h} & & \searrow{\scriptstyle \Delta_h} & \downarrow{\scriptstyle I_h} \\
\pi_{i-h-1}(Q_{n-h}^n) & \xrightarrow{i*} & \pi_{i-h-1}(Q_{n-h}^{n+k}) & & & & \pi_{i-h-2}(Q_{n-h}^n)\ .
\end{array}$$

Next, by use of Theorem 2 of [6], we have that the following homomorphisms are isomorphisms for $i < 4n+k-3$, where p shrinks $E^{n-1}P_n^{n+k-1}$.

$$p_* : \pi_i(E^{n-1}P_n^{n+k+h-1}, E^{n-1}P_n^{n+k-1}) \longrightarrow \pi_i(E^{n-1}P_{n+k}^{n+k+h-1}) \quad ,$$

$$E^k : \pi_i(E^{n-1}P_{n+k}^{n+k+h-1}) \longrightarrow \pi_{i+k}(E^{n+k-1}P_{n+k}^{n+k+h-1}) \ .$$

By use of the composition of these isomorphisms, we have from the homotopy exact sequence of the pair $(E^{n-1}P_n^{n+k+h-1}, E^{n-1}P_n^{n+k-1})$ that the following sequence is exact.

$$(11.11).\quad \cdots \to \pi_i(E^{n-1}P_n^{n+k-1}) \xrightarrow{i*} \pi_i(E^{n-1}P_n^{n+k+h-1}) \xrightarrow{I'_k} \pi_{i+k}(E^{n+k-1}P_{n+k}^{n+k+h-1})$$

$$\xrightarrow{\Delta_k} \pi_{i-1}(E^{n-1}P_n^{n+k-1}) \longrightarrow \cdots$$

for $i < 4n+k-3$, where I'_k and Δ_k are defined by the formulas $I'_k = E^k \circ p_* \circ j_*$ and $\Delta_k = \partial \circ (E^k \circ p_*)^{-1}$.

Let $q : P_n^{n+k-1} \to P^{n+k-1} \cup CP^{n-1}$ be a homotopy inverse of (11.7). Then define a mapping

$$f_n^{n+k} : E^{n-1}P_n^{n+k-1} \to Q_n^{n+k}$$

by the formula $f_n^{n+k} = \Omega(\Omega_0^k F_{n+k}) \circ i \circ E^{n-1}q : E^{n-1}P_n^{n+k-1} \to E^{n-1}(P^{n+k-1} \cup CP^{n-1}) \to \Omega(E^nP^{n+k-1}, E^nP^{n-1}) \to \Omega(\Omega^k(S^{n+k}), S^n) = Q_n^{n+k}$.

<u>Lemma</u> 11.6. <u>The following diagram is commutative.</u>

$$\begin{array}{ccccccc}
\pi_i(E^{n-1}P_n^{n+k-1}) & \xrightarrow{i_*} & \pi_i(E^{n-1}P_n^{n+k+h-1}) & \xrightarrow{I'_k} & \pi_{i+k}(E^{n+k-1}P_{n+k}^{n+k+h-1}) & \xrightarrow{\Delta_k} & \pi_{i-1}(E^{n-1}P_n^{n+k-1}) \\
\downarrow (f_n^{n+k})_* & & \downarrow (f_n^{n+k+h})_* & & \downarrow (f_{n+k}^{n+k+h})_* & & \downarrow (f_n^{n+k})_* \\
\pi_i(Q_n^{n+k}) & \xrightarrow{i_*} & \pi_i(Q_n^{n+k+h}) & \xrightarrow{I'_k} & \pi_{i+k}(Q_{n+k}^{n+k+h}) & \xrightarrow{\Delta_k} & \pi_{i-1}(Q_n^{n+k}) .
\end{array}$$

There is no essential difficulty in the proof of the lemma, so the proof is left to the reader.

Now the main purpose of this Chapter is to prove the following theorem and to apply it to the investigation of E^k.

<u>Theorem</u> 11.7. <u>The homomorphism</u> $(f_n^{n+k})_* : \pi_i(E^{n-1}P_n^{n+k-1}) \to \pi_i(Q_n^{n+k})$ <u>is an isomorphism onto for</u> $i < 3n-2$ <u>and it gives an isomorphism of the 2-primary components for</u> $i < 4n-3$.

<u>Proof</u>. First consider the case that $k = 1$ and $i = 2n-1$. Let $p : P^n \cup CP^{n-1} \to S^n = P_n^n$ be the shrinking map of (11.7). Then the following diagram is commutative.

$$\begin{array}{ccc}
\pi_{2n-1}(E^{n-1}(P^n \cup CP^{n-1})) & \xrightarrow{i_*} & \pi_{2n-1}(\Omega(E^nP^n, E^nP^{n-1})) \\
\downarrow E^{n-1}p_* & & \downarrow (\Omega E^n p)_* \\
\pi_{2n-1}(S^{2n-1}) & \xrightarrow{i_*} & \pi_{2n-1}(\Omega(S^{2n})) .
\end{array}$$

By (11.7), $E^{n-1}p$ is a homotopy equivalence. Thus $E^{n-1}p_*$ is an isomorphism onto. The lower i_* is equivalent to the suspension homomorphism E. Thus it is an isomorphism. The homomorphism $(\Omega E^n p)_*$ is equivalent to the homomorphism $E^np_* : \pi_{2n}(E^nP^n, E^{n-1}P^n) \to \pi_{2n}(S^{2n})$ which is an isomorphism. It follows from the commutativity of the diagram that the upper i_* is an isomorphism onto. The homomorphism

$$(\Omega(\Omega_0 F_{n+1}))_* : \pi_{2n-1}(\Omega(E^nP^n, E^nP^{n-1})) \to \pi_{2n-1}(Q_n^{n+1})$$

is equivalent to the homomorphism $(\Omega_0 F_{n+1})_*$ of Corollary 11.5 which is an isomorphism. Thus the homomorphism $(\Omega(\Omega_0 F_{n+1}))_*$ is an isomorphism onto. Then it follows from the definition of the mapping f_n^{n+1} that

$$(f_n^{n+1})_* = (\Omega(\Omega_0 F_{n+1}))_* \circ i_* \circ (E^{n-1}p)_*^{-1} : \pi_{2n-1}(S^{2n-1}) \to \pi_{2n-1}(Q_n^{n+1})$$

is an isomorphism onto. The injection of the reduced product $(S^n)_\infty$ into

$\Omega(S^{n+1})$ induces the isomorphisms of homotopy groups (Chapter II). Thus it induces isomorphisms

$$i_* : \pi_i((S^n)_\infty, S^n) \to \pi_i(\Omega(S^{n+1}), S^n)$$

and $\Omega i_* : \pi_i(\Omega((S^n)_\infty, S^n)) \to \pi_i(Q_n^{n+1}) = \pi_i(\Omega(\Omega(S^{n+1}), S^n))$.

It follows from Theorem 2.4 that the homomorphism

$$(\Omega\ h_n)_* : \pi_i(\Omega((S^n)_\infty, S^n)) \to \pi_i(\Omega^2(S^{2n+1}))$$

is an isomorphism for $i < 3n-2$ and it gives an isomorphism of the 2-primary components for all i. Consider the diagram

$$\begin{array}{ccccc} & & \pi_i(S^{2n-1}) & & \\ & \swarrow^{(f_n^{n+1})_*} & \downarrow f_* & \searrow^{(\Omega\ h_n \circ f)_*} & \\ \pi_i(Q_n^{n+1}) & \xleftarrow{\Omega i_*} & \pi_i(\Omega((S^n)_\infty, S^n)) & \xrightarrow{(\Omega\ h_n)_*} & \pi_i(\Omega^2(S^{2n+1})) \end{array},$$

where $f : S^{2n-1} \to \Omega((S^n)_\infty, S^n)$ is a representative of $(\Omega i_*)^{-1}(f_n^{n+1})_*(\iota_{2n-1})$. Then the above diagram is commutative. Thus $(\Omega\ h_n \circ f)_*$ is an isomorphism onto for $i = 2n-1$. The group $\pi_{2n-1}(\Omega^2(S^{2n+1}))$ is isomorphic to $\pi_{2n+1}(S^{2n+1}) \approx Z$ and it is generated by the class of the canonical injection $i_0 : S^{2n-1} \to \Omega^2(S^{2n+1})$. Thus $\Omega\ h_n \circ f$ represents $i_{0*}(\iota_{2n-1})$ or $-i_{0*}(\iota_{2n-1})$ and

$$(\Omega\ h_n \circ f)_* = \pm\ i_{0*}\ .$$

Since i_{0*} is equivalent ot $E^2 : \pi_i(S^{2n-1}) \to \pi_{i+2}(S^{2n+1})$, then it follows that $(\Omega\ h_n \circ f)_*$ is an isomorphism onto for $i < 4n-3$. Combining this result with that of $(\Omega\ h_n)_*$, we have that f_* is an isomorphism onto for $i < 3n-2$ and it gives an isomorphism of the 2-primary component for $i < 4n-3$. Since Ωi_* is an isomorphism, then $(f_n^{n+1})_*$ is equivalent to f_*. Consequently the theorem is proved for the case that $k = 1$.

Applying the five lemma to the diagram of Lemma 11.6, where $h = 1$, we have that the assertion of the theorem for $(f_n^{n+k})_*$ implies the assertion for $(f_n^{n+k+1})_*$. Then the theorem is proved by induction on k.

q.e.d.

In the proof, we have a homomorphism

$$H_0 = \Omega_0^{-2} \circ (\Omega\ h_n)_* \circ (\Omega i)_*^{-1} : \pi_i(Q_n^{n+1}) \to \pi_{i+2}(S^{2n+1})$$

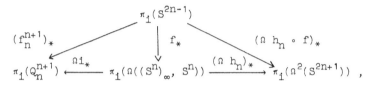

which is an isomorphism for odd n or for $i < 3n-2$ and it gives an isomorphism of the 2-primary components for even n. Then it follows from the definition of H and Δ that the following diagram is commutative.

(11.12).
$$\begin{array}{ccccc} \pi_{i+2}(S^{n+1}) & \xrightarrow{I_1} & \pi_i(Q_n^{n+1}) & \xrightarrow{p_{1*}} & \pi_i(S^n) \\ & \searrow H & \downarrow H_0 & \nearrow \Delta & \\ & & \pi_{i+2}(S^{2n+1}). & & \end{array}$$

We have also in the above proof that

(11.13). $$H_0 \circ (f_n^{n+1})_* = \pm E^2 .$$

In the following, we shall apply the theorem for some special cases which will be needed in the next chapter.

Let $\lambda \in \pi_{2n+2k-2}(E^{n-1}P_n^{n+k-1})$ be the class of the attaching map of the $(2n+2k-1)$-cell $e^{2n+2k-1} = E^{n-1}P_n^{n+k} - E^{n-1}P_n^{n+k-1}$. For the exact sequence

$$\cdots \to \pi_i(E^{n-1}P_n^{n+k-1}) \xrightarrow{i_*} \pi_i(E^{n-1}P_n^{n+k}) \xrightarrow{I'_k} \pi_{i+k}(S^{2n+2k-1}) \xrightarrow{\Delta_k}$$
$$\pi_{i-1}(E^{n-1}P_n^{n+k-1}) \to \cdots$$

of (11.11), $i < 4n+k-3$, we have the following lemma.

Lemma 11.8. *Let* $i < 4n+k-3$.

i). $\Delta_k(E^{k+1}\alpha) = \lambda \circ \alpha$ *for* $\alpha \in \pi_{i-1}(S^{2n+k-2})$.

ii). *Assume that* $I'_k(\alpha) = E^{k+1}\alpha'$ *and* $\alpha' \circ \beta = 0$ *for* $\alpha \in \pi_i(E^{n-1}P_n^{n+k})$, $\alpha' \in \pi_{i-1}(S^{2n+k-2})$ *and* $\beta \in \pi_j(S^{i-1})$, *then*
$$\alpha \circ E\beta \in - i_*\{\lambda, \alpha', \beta\} .$$

Proof. Consider the following diagram.

$$\begin{array}{ccccc} \pi_i(E^{n-1}P_n^{n+k}, E^{n-1}P_n^{n+k-1}) & & \xrightarrow{\partial} & & \pi_{i-1}(E^{n-1}P_n^{n+k-1}) \\ \uparrow f_* & \searrow p_* & & \nearrow \Delta_k \circ E^k & \uparrow f_* \\ & & \pi_i(S^{2n+k-1}) & & \\ & \nearrow p'_* & & \nwarrow E & \\ \pi_i(CS^{2n+k-2}, S^{2n+k-2}) & & \xrightarrow{\partial} & & \pi_{i-1}(S^{2n+k-2}), \end{array}$$

where f is a characteristic mapping of the cell e^{2n+k-1}, $p' = p \circ f$ and $\Delta_k = \partial \circ (E^k \circ p_*)^{-1}$. Except the triangle of the right side, the commutativity holds for the other three triangles and the square. It follows the commutativity of the right triangle. Thus
$$\lambda \circ \alpha = f_*(\alpha) = \Delta_k(E^{k+1}\alpha) ,$$
and then i) is proved.

Since $I'_k = E^k \circ p_* \circ j_*$, then $E^{k+1}\alpha' = I'_k(\alpha) = E^k(p_* j_*(\alpha))$. Since $E^k : \pi_i(S^{2n+k-1}) \to \pi_{i+k}(S^{2n+2k-1})$ is an isomorphism onto for $i < 2(2n+k-1)-1$, then it follows that $p_* j_*(\alpha) = E\alpha'$. Considering that $E^{n-1}P_n^{n+k} = E^{n-1}P_n^{n+k-1} \cup_\lambda CS^{2n+2k-2}$, we have a coextension $\tilde{\alpha} \in \pi_i(E^{n-1}P_n^{n+k})$ of α'. By (1.18), $p_*(\tilde{\alpha}) = E\alpha'$ for $p_* : \pi_i(E^{n-1}P_n^{n+k}) \to \pi_i(S^{2n+2k-1})$. Since $p_* : \pi_i(E^{n-1}P_n^{n+k}, E^{n-1}P_n^{n+k-1}) \to \pi_i(S^{2n+2k-1})$ is an isomorphism onto for $i < 4n+k-3$, then it follows that $j_*(\alpha) = j_*(\tilde{\alpha}) = p_*^{-1}(E\alpha')$. From the exact sequence of the pair $(E^{n-1}P_n^{n+k}, E^{n-1}P_n^{n+k-1})$, we have that $\alpha - \tilde{\alpha}$ is contained in $i_* \pi_i(E^{n-1}P_n^{n+k-1})$. By Proposition 1.8, we have that $\tilde{\alpha} \circ E\beta \in - i_*\{\lambda, \alpha', \beta\}$ and thus

$$\alpha \circ E\beta \in - i_*\{\lambda, \alpha', \beta\} + i_*\pi_i(E^{n-1}P_n^{n+k-1}) \circ E\beta .$$

Since the secondary composition $\{\lambda, \alpha', \beta\}$ is a coset of a subgroup containing $\pi_i(E^{n-1}P_n^{n+k-1}) \circ E\beta$, then $\alpha \circ E\beta \in - i_*\{\lambda, \alpha', \beta\}$, and ii) is proved. q.e.d.

Proposition 11.9. *Let $i < 4n+k-3$ and $k > 0$. Assume that*

$$\Delta_k(\alpha) = i_*(\beta) \quad \text{in} \quad \pi_{i-1}(E^{n-1}P_n^{n+k-1})$$

for $\alpha \in \pi_{i+k}^{2n+2k-1}$ and $\beta \in \pi_{i-1}^{2n-1}$. Then there exists an element γ of π_{i+1}^{n+1} such that

$$\Delta(E^2\alpha) = E^{k-1}\gamma \quad \text{and} \quad H(\gamma) = \pm E^2\beta .$$

Proof. Set $\alpha' = (f_{n+k}^{n+k+1})_* \alpha \in \pi_{i+k}(Q_{n+k}^{n+k+1})$ and $\beta' = (f_n^{n+1})_*\beta \in \pi_{i-1}(Q_n^{n+1})$, then

$$H_0(\alpha') = \pm E^2\alpha, \ H_0(\beta') = \pm E^2\beta \ \text{ and } \ \Delta_k(\alpha') = i_*(\beta'),$$

by (11.13) and Lemma 11.6. By the commutativity of (11.12) and (11.10),

$$I_k(\Delta(\pm E^2\alpha)) = I_k(\Delta(H_0\alpha')) = I_k(p_{1*}\alpha') = \Delta_k\alpha' = i_*\beta'$$

and $I_{k-1}(\Delta(\pm E^2\alpha)) = I'_1 I_k(\Delta(\pm E^2\alpha)) = I'_1(i_*\beta')$.

Then $I_{k-1}(\Delta(E^2\alpha)) = \pm I'_1(i_*\beta') = 0$ by the exactness of the sequence $\pi_{i-1}(Q_n^{n+1}) \to \pi_{i-1}(Q_n^{n+k}) \to \pi_i(Q_{n+1}^{n+k})$ of (11.9). It follows from the exactness of the sequence $\pi_{i+1}(S^{n+1}) \to \pi_{i+k}(S^{n+k}) \to \pi_i(Q_{n+1}^{n+k})$ of (11.8) that there exists an element γ' of $\pi_{i+1}(S^{n+1})$ such that

$$E^{k-1}(\gamma') = \Delta(E^2\alpha) .$$

By the commutativity of the diagram (11.10), we have that

$$i_*\beta' = I_k(\Delta(\pm E^2\alpha)) = \pm I_k(E^{k-1}\gamma') = \pm i_*(I_1(\gamma')) .$$

Then it follows from the exactness of the sequence (11.9) that there exists an element δ of $\pi_i(Q_{n+1}^{n+k})$ such that $\pm\beta' = I_1(\gamma') + \Delta_1(\delta)$. By the commutativity of (11.10), we have that $\Delta_1(\delta) = I_1 p_{k-1*}(\delta)$. Thus $\pm\beta' = I_1(\gamma' + p_{k-1*}(\delta))$. If $k - 1 > 0$, then $E^{k-1}p_{k-1*}(\delta) = 0$ by the exactness of (11.8), and thus we have, by setting $\gamma = \gamma' + p_{k-1*}(\delta)$, that

$$E^{k-1}\gamma = \Delta(E^2\alpha) \quad \text{and} \quad I_1(\gamma) = \pm\beta' .$$

In the case $k = 1$, i_* is the identity. Then we have, by setting $\gamma = \gamma'$ that

$$\gamma = \Delta(E^2\alpha) \quad \text{and} \quad I_1(\gamma) = \pm i_*\beta' = \pm\beta' .$$

We have $H(\gamma) = H_0(I_1(\gamma)) = \pm H_0(\beta')$, by the commutativity of (11.12). Since $H_0(\beta') = \pm E^2\beta$, then we have that $H(\gamma) = \pm E^2\beta$. Finally we remark that the element γ can be chosen from π_{i+1}^{n+1}, then the proof of the proposition is established. q.e.d.

As is well known, in the real projective space $P^\infty = \cup P^n$, the incidence number of e^{2m-1} with e^{2m} is ± 2, and that of e^{2m} with e^{2m+1} is zero. Thus, if n is even then $E^{n-1}P_n^{n+1}$ has the same homotopy type as $S^{2n-1} \vee S^{2n}$ and if n is odd then it has the same homotopy type of $S^{2n-1} \cup_\lambda e^{2n}$ for $\lambda = \pm 2\iota_{2n-1}$. Applying Lemma 11.8 and Proposition 11.9 for $E_n^{n+1}P_n^{n+1}$ of odd n, we have

(11.14). $H(\Delta(E^2\alpha)) = \pm 2\alpha$ <u>for</u> $\alpha \in \pi_{i+1}^{2n+1}$, $i < 4n-2$ <u>and for odd</u> n.

This is, however, a consequence of Proposition 2.5 and Proposition 2.7.

Next consider the squaring operations in $H^*(P^\infty, Z_2)$ which is a polynomial ring over a generator $e \in H^1(P^\infty, Z_2)$ such that $Sq^1 e = e^2$. By use of Cartan's formula, we have the formula

(11.15). $$Sq^t e^r = \binom{r}{t} e^{r+t}.$$

In particular, $Sq^2 e^r = e^{r+2}$ for $r \equiv 2, 3 \pmod 4$ and $Sq^2 e^r = 0$ for $r \equiv 0, 1 \pmod 4$. Consider the complex $E^{n-1}P_n^{n+2} = E^{n-1}P_n^{n+1} \cup e^{2n+1}$. If n is even, then $E^{n-1}P_n^{n+1}$ can be replaced by $S^{2n-1} \vee S^{2n}$. By shrinking S^{2n} we have from $E^{n-1}P_n^{n+2}$ a cell complex $K = S^{2n-1} \cup e^{2n+1}$. $Sq^2 \neq 0$ in $E^{n-1}P_n^{n+2}$ if and only if $Sq^2 \neq 0$ in K, that is, e^{2n+1} is attached to S^{2n-1} by η_{2n-1} (Proposition 8.1). If n is odd, then the attaching map of e^{2n+1} is inessential on e^{2n} and it represents η_{2n-1} if and only if $Sq^2 \neq 0$. Consequently the following result is obtained.

(11.16). $E^{n-1}P_n^{n+2}$ *has the same homotopy type of the following complex* $K(r)$ *if* $n \equiv r \pmod 4$.

$K(0) = S^{2n-1} \vee S^{2n} \cup_\beta e^{2n+1}$, *where* $\beta = \pm 2\iota_{2n}$,

$K(1) = S^{2n-1} \cup_\alpha e^{2n} \vee S^{2n+1}$, *where* $\alpha = \pm 2\iota_{2n-1}$,

$K(2) = S^{2n-1} \vee S^{2n} \cup_\beta e^{2n+1}$, *where* $\beta = \pm 2\iota_{2n} + \eta_{2n-1}$,

$K(3) = S^{2n-1} \cup_\alpha e^{2n} \cup_\beta e^{2n+1}$, *where* $\alpha = \pm 2\iota_{2n-1}$ and $\beta = \eta_{2n-1}$.

Now we have

Proposition 11.10. *Let* $i < 4n-1$.

i). *Let* $n \equiv 2 \pmod 4$. *Assume that* $2\alpha = 0$ for an element α of π_{i-1}^{2n}, *then there exists an element* β *of* π_{i+1}^{n+1} *such that*

$$\Delta(E^5\alpha) = E\beta \quad \text{and} \quad H(\beta) = \eta_{2n+1} \circ E^2\alpha .$$

ii). *Let* $n \equiv 3 \pmod 4$. *For an element* α of π_{i-1}^{2n}, *there exists an element* β of π_{i+1}^{n+1} *such that*

$$\Delta(E^5\alpha) = E\beta \quad \text{and} \quad H(\beta) = \eta_{2n+1} \circ E^2\alpha.$$

Proof. Let $n \equiv 2 \pmod 4$. By (11.16), the characteristic class λ in Lemma 11.8 is the form $\pm 2\iota_{2n} + \eta_{2n-1}$. Since $2\alpha = 0$, then $\lambda \circ \alpha = i_*(\eta_{2n-1} \circ \alpha)$ for the injection i of $S^{2n-1} = E^{n-1}P_n^n$ into $E^{n-1}P_n^{n+1}$. Then, by i) of Lemma 11.8, $\Delta_2(E^3\alpha) = i_*(\eta_{2n-1} \circ \alpha)$. It follows from Proposition 11.9 that there exists an element β of π_{i+1}^{n+1} such that $\Delta(E^5\alpha) = E\beta$ and $H(\beta) = \pm E^2(\eta_{2n-1} \circ \alpha) = \eta_{2n+1} \circ E^2\alpha$. Thus i) is proved. The proof of ii) is similar. q.e.d.

Next we have

Proposition 11.11. *Let* $i < 4n-1$.

i). *Let* $n \equiv 1 \pmod 4$. *Assume that* $2\alpha = 0$ *and* $\eta_{2n-1} \circ \alpha = 0$ *for an element* α of π_{i-2}^{2n}, *then there exists an element* β of π_{i+1}^{n+1} *such that*

$$\Delta(E^7\alpha) = E^2\beta \quad \text{and} \quad H(\beta) \in \{2\iota_{2n+1}, \eta_{2n+1}, E^2\alpha\}_2.$$

ii). *Let* $n \equiv 3 \pmod 4$. *Assume that* $2\alpha = 0$ *for an element* α of π_{i-2}^{2n}, *then there exists an element* β *of* π_{i+1}^{n+1} *such that*

$$\Delta(E^7\alpha) = E^2\beta \quad \text{and} \quad H(\beta) \in \{\eta_{2n+1}, 2\iota_{2n+2}, E^2\alpha\}_2.$$

Proof. i). By (11.16), we may assume that $E^{n-1}P_n^{n+3} = S^{2n-1} \cup e^{2n} \vee S^{2n+1} \cup e^{2n+2}$. By shrinking S^{2n-1} to a point, $E^{n-1}P_n^{n+3}$ becomes $E^{n-1}P_{n+1}^{n+3}$. $E^{n-1}P_{n+1}^{n+3}$ has a similar structure as $K(0)$ of (11.16). Then we see that the characteristic class λ for e^{2n+2} is of the form

$\pm 2\iota_{2n+1} + \gamma$, where $\gamma \in \pi_{2n+1}(E^{n-1}P_n^{n+1})$ is a coextension of η_{2n-1}. By the assumption $2\alpha = 0$, we have that

$$\lambda \circ E\alpha = i'_*(\gamma \circ E\alpha)$$

for the injection i' of $E^{n-1}P_n^{n+1}$ into $E^{n-1}P_n^{n+2}$. By Proposition 1.8,

$$\gamma \circ E(\alpha) = i''_*(\delta)$$

for the injection $i'' : S^{2n-1} \to E^{n-1}P_n^{n+1}$ and for an element δ of $\pm\{2\iota_{2n-1}, \eta_{2n-1}, \alpha\}$. Thus

$$\lambda \circ E(\alpha) = i_*(\delta)$$

for the injection i of S^{2n-1} into $E^{n-1}P_n^{n+2}$. Remark that we may choose δ in π_{i-1}^{2n-1} since $\{2\iota_{2n-1}, \eta_{2n-1}, \alpha\}$ is a coset of a subgroup which contains $2\pi_{i-1}(S^{2n-1})$. Further, we have that $-\{2\iota_{2n-1}, \eta_{2n-1}, \alpha\} = \{2\iota_{2n-1}, \eta_{2n-1}, \alpha\}$. Applying Lemma 11.8, we have that

$$\Delta_3(E^5\alpha) = \lambda \circ E\alpha = i_*(\delta) .$$

Then it follows from Proposition 11.9 that there exists an element β of π_{i+1}^{n+1} such that

$$\Delta(E^7\alpha) = E^2\beta \quad \text{and} \quad H(\beta) = \pm E^2 \delta .$$

The element $\pm E^2\delta$ is contained in $\pm E^2\{2\iota_{2n-1}, \eta_{2n-1}, \alpha\} = E^2\{2\iota_{2n-1}, \eta_{2n-1}, \alpha\} \subset \{2\iota_{2n+1}, \eta_{2n+1}, E^2\alpha\}_2$. Thus i) is proved.

ii). In the case that $n \equiv 3 \pmod 4$, the characteristic class of the cell e^{2n+2} of $E^{n-1}P_n^{n+3}$ is a coextension of $2\iota_{2n}$ in $S^{2n-1} \cup e^{2n+1} \subset E^{n-1}P_n^{n+2}$, where e^{2n+1} is attached to S^{2n-1} by η_{2n-1}. Then the proof of ii) is similar to that of i). q.e.d.

We know the following examples in the previous calculations.

<u>Examples for Proposition</u> 11.10.

i). $(\alpha, \beta) = (\eta_{12}, \sigma' \circ \eta_{14})$, (η_{20}, θ'), $(\bar\nu_{12}, \sigma' \circ \bar\nu_{14})$, $(\varepsilon_{12}, \sigma' \circ \varepsilon_{14})$, etc.

ii). $(\alpha, \beta) = (\iota_6, \nu_4 \circ \eta_7)$, $(\iota_{14}, \sigma_8 \circ \eta_{15} + \bar\nu_8 + \varepsilon_8)$, $(\eta_{14}^2, 4\sigma_8 \circ \nu_{15})$, (ι_{22}, θ), $(\sigma_{14}, \sigma_8 \circ \bar\nu_{15} + \sigma_8 \circ \varepsilon_{15})$, etc.

<u>Examples for ii) of Proposition</u> 11.11: $(\alpha, \beta) = (\eta_{14}, 2\sigma_8 \circ \nu_{15})$, $(\nu_{14}^2, \sigma_8 \circ \varepsilon_{15})$.

<u>Lemma</u> 11.12.

i). <u>Assume that</u> $(i < 4n+k-3)$ $I_{k+1}\alpha = (f_n^{n+k+1})_*\alpha'$ <u>for</u> $\alpha \in \pi_{i+k+2}(S^{n+k+1})$ <u>and</u> $\alpha' \in \pi_i(E^{n-1}P_n^{n+k})$. <u>Then</u>

$H(\alpha) = \pm E^2(I_k' \alpha')$ *and* $(f_n^{n+k+1})_*(\alpha' \circ \beta') = I_{k+1}(\alpha \circ E^{k+2}\beta')$ *for* $\beta' \in \pi_j(S^i)$.

ii). *Assume that* $I_{k+1}(\alpha) = (f_n^{n+k+1})_* i_* \beta$ *for* $\alpha \in \pi_{i+k+1}(S^{n+k+1})$, $\beta \in \pi_{i-1}(S^{2n-1})$ *and the injection homomorphism* $i_* : \pi_{i-1}(S^{2n-1}) \to \pi_{i-1}(E^{n-1}P_n^{n+k})$. *Then there exists an element* γ *of* $\pi_{i+1}(S^{n+1})$ *such that*

$$\alpha = E^k\gamma \quad \textit{and} \quad H(\gamma) = \pm E^2\beta .$$

Proof. i). By the commutativity of (11.12), (11.10), Lemma 11.6 and by (11.13), we have

$$H(\alpha) = H_0 I_1(\alpha) = H_0 I_k' I_{k+1}(\alpha) = H_0 I_k'(f_n^{n+k+1})_*(\alpha')$$
$$= H_0(f_{n+k}^{n+k+1})_* I_k'(\alpha') = \pm E^2(I_k'(\alpha')) .$$

It is easily verified that the following diagram is commutative.

$$\begin{array}{ccc} \pi_{i+1}(X) & \xrightarrow{\ j_*\ } & \pi_{i+1}(X, A) \\ \Big\downarrow \Omega_0 & & \Big\downarrow \Omega' \\ \pi_i(\Omega(X)) & \xrightarrow{\ i_*\ } & \pi_i(\Omega(X, A)) . \end{array}$$

Then, by the definition of I_{k+1},

$$I_{k+1}(\alpha \circ E^{k+1}\beta') = (\Omega' \circ j_* \circ \Omega_0^{k+1})(\alpha \circ E^{k+1}\beta')$$
$$= i_*(\Omega_0^{k+2}(\alpha \circ E^{k+2}\beta')) = i_*(\Omega_0^{k+2}\alpha \circ \beta')$$
$$= I_{k+1}(\alpha) \circ \beta' = (f_n^{n+k+1})_*(\alpha') \circ \beta'$$
$$= (f_n^{n+k+1})_*(\alpha' \circ \beta') .$$

ii). By Lemma 11.6, $I_{k+1}(\alpha) = (f_n^{n+k+1})_*(i_*\beta) = i_*((f_n^{n+1})_*\beta)$.
By the commutativity of (11.10) and by the exactness of (11.9),

$$I_k(\alpha) = I_1' I_{k+1}(\alpha) = I_1' \, i_*((f_n^{n+1})_*\beta) = 0.$$

Then it follows from the exactness of the sequence (11.8) that there exists an element γ' of $\pi_{i+1}(S^{n+1})$ such that $E^k\gamma' = \alpha$. By the commutativity of (11.10),

$$i_*(f_n^{n+1})_*\beta = I_{k+1}(\alpha) = I_{k+1}(E^k\gamma') = i_* I_1 \gamma' .$$

Then it follows from the exactness of the sequence (11.9) that there exists an element δ of $\pi_{i+k+1}(Q_{n+1}^{n+k+1})$ such that $\Delta_1(\delta) = (f_n^{n+1})_*\beta - I_1(\gamma')$. Denote that $\gamma = \gamma' + p_{1*}(\delta)$, then we have that $I_1(\gamma) = I_1(\gamma') + I_1 p_{1*}(\delta) = I_1(\gamma') + \Delta_1(\delta) = (f_n^{n+1})_*\beta$, by the commutativity of (11.10). By the commutativity of (11.12) and by (11.13),

$$H(\gamma) = H_0(I_1(\gamma)) = H_0(f_n^{n+1})_* \beta = \pm E^2\beta .$$

By the exactness of (11.8), we have

$$E^k(\gamma) = E^k(\gamma') + E^{k-1}(E\, p_{1*}(\delta)) = E^k(\gamma') = \alpha .$$

q.e.d.

Proposition 11.13. *Let n be odd and $i < 4n-2$. Assume that $H(\alpha) = E^5\beta$ and $2\beta = 0$ for $\alpha \in \pi^{n+2}_{i+2}$ and $\beta \in \pi^{2n-2}_{i-3}$. Then there exists an element γ of π^{n+1}_{i+1} such that*

$$2\alpha = E\gamma \quad \text{and} \quad H(\gamma) \equiv E^3\beta \circ \eta_i \mod 2E^2\pi^{2n-1}_{i-1}.$$

Proof. By Theorem 11.7, $(f^{n+2}_n)_* : \pi_{i-1}(E^{n-1}P^{n+1}_n) \to \pi_{i-1}(Q^{n+2}_n)$ is an isomorphism of the 2-primary components (an isomorphism for $i+2 = 2n+3$). Then there exists an element α' of $\pi_{i-1}(E^{n-1}P^{n+1}_n)$ such that

$$(f^{n+2}_n)_*\alpha' = I_2(\alpha) .$$

By i) of Lemma 11.12, $E^5\beta = H(\alpha) = \pm E^2(I_1'\alpha')$. Since $i < 4n-2$, then $E^2 : \pi_i(S^{2n+1}) \to \pi_{i+2}(S^{2n+3})$ is an isomorphism onto, and hence $I_1'\alpha' = \pm E^3\beta$. Since n is odd, then $E^{n-1}P^{n+1}_n = S^{2n-1} \cup_\lambda e^{2n}$ for $\lambda = \pm 2\iota_{2n-1}$. By ii) of Lemma 11.8,

$$2\alpha' = \alpha' \circ 2\iota_{i-1} \in \pm i_*\{2\iota_{2n-1}, E\beta, 2\iota_{i-2}\} .$$

By Corollary 3.7, $i_*(E\beta \circ \eta_{i-2}) \in i_*\{2\iota_{2n-1}, E\beta, 2\iota_{i-2}\}$. Thus

$$2\alpha' = i_*(E\beta \circ \eta_{i-2} + 2\delta)$$

for an element δ of $\pi_{i-1}(S^{2n-1})$. It follows then

$$I_2(2\alpha) = (f^{n+2}_n)_*(2\alpha') = (f^{n+2}_n)_*\, i_*(E\beta \circ \eta_{i-2} + 2\delta).$$

Applying ii) of Lemma 11.12, we have that there exists an element γ of π^{n+1}_{i+1} such that $2\alpha = E\gamma$ and $H(\gamma) = \pm E^2(E\beta \circ \eta_{i-2} + 2\delta) \equiv E^3\beta \circ \eta_i \mod 2E^2\pi^{2n-1}_{i-1}$.

q.e.d.

Examples: $(\alpha, \gamma) = (\sigma', \sigma''), (\kappa_7, \bar{\nu}_6 \circ \nu^2_{14}), (\rho'', \rho''')$.

Proposition 11.14. *Let $n \equiv 3 \pmod 4$ and $i < 4n-2$. Assume that $H(\alpha) = E^5\beta$ and $\beta \circ \gamma = 0$ for $\alpha \in \pi^{n+3}_{i+3}$, $\beta \in \pi^{2n}_{i-2}$ and $\gamma \in \pi^{i-2}_j$. Then there exists an element δ of π^{n+1}_{i+1} such that*

$$E^2\delta = \alpha \circ E^5\gamma \quad \text{and} \quad H(\delta) \in \{\eta_{2n+1}, E^2\beta, E^2\gamma\}_2 .$$

Proof. By Theorem 11.7, $(f^{n+3}_n)_* : \pi_{i-1}(E^{n-1}P^{n+2}_n) \to \pi_{i-1}(Q^{n+3}_n)$ is an isomorphism of the 2-primary components (an isomorphism for $i+3 = 2n+5$). Then there exists an element α' of $\pi_{i-1}(E^{n-1}P^{n+2}_n)$ such that

$$(f^{n+3}_n)_*\alpha' = I_3(\alpha) .$$

By i) of Lemma 11.12, $E^5\beta = H(\alpha) = \pm E^2I_2'(\alpha')$. Since $i < 4n-2 < 2(2n+3)-1$, then $E^2 : \pi_{i+1}(S^{2n+3}) \to \pi_{i+3}(S^{2n+5})$ is an isomorphism onto.

Thus $E^3\beta = \pm I_2^!(\alpha')$. By (11.16), the characteristic class of $e^{2n+1} = E^{n-1}P_n^{n+2} - E^{n-1}P_n^{n+1}$ is $\lambda = i_*^! \eta_{2n-1}$ for the injection i' of S^{2n-1} into $E^{n-1}P_n^{n+1}$. By ii) of Lemma 11.8, we have

$$\alpha' \circ E\gamma \in -i_*''\{i_*^! \eta_{2n-1}, \beta, \gamma\} \subset i_*\{\eta_{2n-1}, \beta, \gamma\}$$

for the injections $i'' : E^{n-1}P_n^{n+1} \subset E^{n-1}P_n^{n+2}$ and $i = i'' \circ i'$. By i) of Lemma 11.12,

$$I_3(\alpha \circ E^5\gamma) = (f_n^{n+3})_*(\alpha' \circ E\gamma) \in (f_n^{n+3})_* i_*\{\eta_{2n-1}, \beta, \gamma\}.$$

Then it follows from ii) of Lemma 11.12 that there exists an element δ of $\pi_{i+1}(S^{n+1})$ such that

$$\alpha \circ E^5\gamma = E^2\delta$$

and $H(\delta) \in \pm E^2\{\eta_{2n-1}, \beta, \gamma\} \subset \{\eta_{2n+1}, E^2\beta, E^2\gamma\}_2$. The proof of the fact that we can choose δ in π_{i+1}^{n+1} will be left to the reader. q.e.d.

<u>Proposition</u> 11.15. <u>Let</u> $n \equiv 2 \pmod 4$ <u>and</u> $i < 4n-2$. <u>Assume that</u> $H(\alpha) = E^5\beta$ <u>for</u> $\alpha \in \pi_{i+3}^{n+3}$ <u>and</u> $\beta \in \pi_{i-2}^{2n}$. <u>Then there exists an element</u> γ <u>of</u> π_{i+1}^{n+1} <u>such that</u>

$$2\alpha = E^2\gamma \quad \underline{\text{and}} \quad H(\gamma) \in \{\eta_{2n+1}, E^2\beta, 2\iota_i\}_2 .$$

<u>Proof</u>. By a similar discussion to the previous two Propositions, we have an element $\alpha' \in \pi_{i-1}(E^{n-1}P_n^{n+2})$ such that

$$(f_n^{n+3})_*\alpha' = I_3(\alpha) \quad \text{and} \quad E^3\beta = \pm I_2^! \alpha'.$$

By (11.16), the characteristic class of $e^{2n+1} = E^{n-1}P_n^{n+2} - E^{n-1}P_n^{n+1}$ is $\lambda = i_{1*}\eta_{2n-1} + i_{2*}(2\iota_{2n})$ for the injections i_1 of S^{2n-1} and i_2 of S^{2n} into $E^{n-1}P_n^{n+1}$. By i) of Lemma 11.8 and by the exactness of (11.11),

$$\lambda \circ \beta = \Delta_2(E^3\beta) = \Delta_2(\pm I_2^!\alpha') = 0 .$$

Since $i - 4 < 4n-6 < 2(2n-2)-1$, $E^2 : \pi_{i-4}(S^{2n-2}) \to \pi_{i-2}(S^{2n})$ is an isomorphism onto. Let $\beta' = E^{-2}\beta$. By (1.7),

$$0 = \lambda \circ \beta = (i_{1*}\eta_{2n-1} \pm i_{2*}(2\iota_{2n})) \circ E^2\beta'$$
$$= i_{1*}(\eta_{2n-1} \circ E^2\beta') \pm i_{2*}(2E^2\beta') .$$

By Theorem 4.8 of [26], i_{1*} and i_{2*} are isomorphism into disjoint direct factors of $\pi_{i-2}(S^{2n-1} \vee S^{2n})$. It follows that

$$\eta_{2n-1} \circ E^2\beta' = 0 \quad \text{and} \quad 2E^2\beta' = \beta \circ 2\iota_{i-2} = 0.$$

By Proposition 3.1,

$$\beta \circ \eta_{i-2} = E^2\beta' \circ \eta_{i-2} = \eta_2 \# \beta' = \eta_{2n} \circ E^3\beta' = 0.$$

Apply ii) of Lemma 11.8 to the relation $I_2^! \alpha' = \pm E^3\beta$. Then

$$2\alpha' \in - i'_*\{i_{1*}\,\eta_{2n-1} \pm i_{2*}\,2\iota_{2n},\ \beta,\ 2\iota_{i-2}\}$$
$$\subset i'_*\,i_{1*}\{\eta_{2n-1},\ \beta,\ 2\iota_{i-2}\} \pm i'_*\,i_{2*}\{2\iota_{2n},\ E^2\beta',\ 2\iota_{i-2}\}\ ,$$

for the injection i' of $E^{n-1}P_n^{n+1}$ into $E^{n-1}P_n^{n+2}$. Since $\beta \circ \eta_{i-2} = 0$, it follows from Corollary 3.7 that

$$\{2\iota_{2n},\ E^2\beta',\ 2\iota_{i-2}\} = 2\iota_{2n} \circ \pi_{i-1}(S^{2n}) + \pi_{i-1}(S^{2n}) \circ 2\iota_{i-1}.$$

Since $i-1 < 4n-1$, then $\pi_{i-1}(S^{2n})$ is stable and $2\iota_{2n} \circ \pi_{i-1}(S^{2n}) = \pi_{i-1}(S^{2n}) \circ 2\iota_{i-1}$. Since $i'_*\,\lambda = 0$, then $i'_*\,i_{1*}\,\eta_{2n-1} = \pm\, i'_*\,i_{2*}\,2\iota_{2n}$. Therefore we have that

$$i'_*\,i_{2*}\{2\iota_{2n},\ \beta,\ 2\iota_{i-2}\} = i'_*\,i_{2*}(2\iota_{2n} \circ \pi_{i-1}(S^{2n}))$$
$$= i'_*\,i_{1*}(\eta_{2n-1} \circ \pi_{i-1}(S^{2n}))\ .$$

The secondary composition $\{\eta_{2n-1},\ \beta,\ 2\iota_{i-2}\}$ is a coset of a subgroup containing $\eta_{2n-1} \circ \pi_{i-1}(S^{2n})$. Thus

$$2\alpha' \in i_*\{\eta_{2n-1},\ \beta,\ 2\iota_{i-2}\}$$

for the injection i of S^{2n-1} into $E^{n-1}P_n^{n+2}$.

Now, by a similar way to the proof of Proposition 11.14, we have the existence of an element γ of π_{i+1}^{n+1} such that $2\alpha = E^2\gamma$ and

$$H(\gamma) \in E^2\{\eta_{2n-1},\ \beta,\ 2\iota_{i-2}\} \subset \{\eta_{2n+1},\ E^2\beta,\ 2\iota_i\}_2.$$

q.e.d.

Proposition 11.16. *Let* $n \equiv 2 \pmod 4$ *and* $i < 4n-2$. *Assume that* $H(\alpha) = E^3\beta$ *for* $\alpha \in \pi_{i+2}^{n+2}$ *and* $\beta \in \pi_{i-1}^{2n}$. *Then there exitst an element* γ *of* π_{i+1}^{n+1} *such that*

$$\Delta(E^5\beta) \pm 2\alpha = E\gamma \quad \text{and} \quad H(\gamma) \equiv \eta_{2n+1} \circ E^2\beta \mod 2\pi_{i+1}^{2n+1}\ .$$

Proof. By Propositions 2.5 and 2.7, $H(\Delta(E^5\beta)) = \pm\, 2E^3\beta = \pm\, H(2\alpha)$. It follows from the exactness of (4.4) that there exists an element γ of π_{i+1}^{n+1} such that

$$E\gamma = \Delta(E^5\beta) \pm 2\alpha\ .$$

By Theorem 11.7, there are elements $\alpha' \in \pi_{i-1}(E^{n-1}P_n^{n+1})$ and $\gamma' \in \pi_{i-1}^{2n-1}$ such that $(f_n^{n+2})_*\alpha' = I_2\alpha$ and $(f_n^{n+1})_*\gamma' = I_1\gamma$. We have

$$\begin{aligned}
(f_n^{n+2})_* i_*\gamma' &= i_*(f_n^{n+1})_*\gamma' = i_* I_1\gamma && \text{by Lemma 11.6}\\
&= I_2E\gamma = I_2(\Delta(E^5\beta) \pm 2\alpha) && \text{by (11.10)}\\
&= I_2\Delta H_0(f_n^{n+1})_*E^3\beta \pm 2(f_n^{n+2})_*\,\alpha' && \text{by (11.13)}\\
&= I_2\,p_{1*}(f_n^{n+1})_*E^3\beta \pm 2(f_n^{n+2})_*\,\alpha' && \text{by (11.12)}
\end{aligned}$$

$$= \Delta_2(f_n^{n+1})_* E^3\beta \pm 2(f_n^{n+2})_* \alpha' \qquad \text{by (11.10)}$$

$$= (f_n^{n+2})_*(\Delta_2(E^3\beta) \pm 2\alpha') \qquad \text{by Lemma 11.6.}$$

Then it follows from Theorem 11.7 that

$$i_*\gamma' = \Delta_2(E^3\beta) \pm 2\alpha' .$$

By (11.16), $E^{n-1}P_n^{n+2} = S^{2n-1} \vee S^{2n} \cup_\lambda e^{2n+1}$ for $\lambda = i_* \eta_{2n-1} \pm i'_* (2\iota_{2n})$ for the injections i of S^{2n-1} and i' of S^{2n} into $S^{2n-1} \vee S^{2n}$. By i) of Lemma 11.8,

$$\Delta_2(E^3\beta) = \lambda \circ \beta = (i_* \eta_{2n-1} \pm i'_* (2\iota_{2n})) \circ \beta$$

$$= i_*(\eta_{2n-1} \circ \beta) \pm i'_*(2\beta) ,$$

where we remark that β is a suspension element. We have a direct sum decomposition $\pi_{i-1}(S^{2n-1} \vee S^{2n}) = i_* \pi_{i-1}(S^{2n-1}) \oplus i'_* \pi_{i-1}(S^{2n})$, where i_* and i'_* are isomorphisms into. Thus

$$\alpha' = i_*\alpha'_1 + i'_*\alpha'_2 \quad \text{for } \alpha'_1 \in \pi_{i-1}^{2n-1} \text{ and } \alpha'_2 \in \pi_{i-1}^{2n} .$$

It is obvious from the definition of I'_1 that $I'_1\alpha' = E\alpha'_2$. By i) of Lemma 11.12, $E^3\beta = H(\alpha) = \pm E^2(I'_1\alpha') = \pm E^3\alpha'_2$. Since $i-1 < 4n-3 < 4n-1$, then $E^2 : \pi_{i-1}(S^{2n}) \to \pi_{i+1}(S^{2n+3})$ is an isomorphism onto. Thus $\alpha'_2 = \pm \beta$ and $2\alpha' = i_*(2\alpha'_1) \pm i'_*(2\beta)$. Then we have that

$$i_*\gamma' \equiv i_*(\eta_{2n-1} \circ \beta) \bmod 2i_*\pi_{i-1}^{2n-1} .$$

Since i_* is an isomorphism into, it follows that

$$\gamma' \equiv \eta_{2n-1} \circ \beta \bmod 2\, \pi_{i-1}^{2n-1} .$$

We have finally,

$$H(\gamma) = H_0 I_1(\gamma) = H_0(f_n^{n+1})_*(\gamma') = \pm E^2\gamma' \qquad \text{by (11.12), (11.13)}$$

$$\equiv E^2(\eta_{2n-1} \circ \beta) \quad \bmod \quad 2E^2\pi_{i-1}^{2n-1} = 2\pi_{i+1}^{2n+1}$$

$$= \eta_{2n+1} \circ E^2\beta .$$

q.e.d.

We may continue our discussion for $E^{n-1}P_n^{n+4}$ and so on, but it will be more complicated in general. Here, we consider the following special case, which will be applied in the next chapter.

<u>Lemma</u> 11.17. <u>Let</u> $n \equiv 4 \pmod 8$ <u>and</u> $n \geq 12$. <u>Assume that</u> $H(\alpha) = \nu_{2n+7}$ <u>for an element</u> $\alpha \in \pi_{2n+10}^{n+4}$, <u>then there exists an element</u> β <u>of</u> π_{2n+7}^{n+1} <u>such that</u>

$$\pm \Delta(\nu_{2n+9}) = 2\alpha - E^3\beta \quad \underline{\text{and}} \quad H(\beta) = \nu_{2n+1}^2 .$$

Proof. By Propositions 2.5 and 2.7, $H(\Delta(\nu_{2n+9})) = \pm H(2\alpha)$. It follows from the exactness of the sequence (4.4) that there exists an element β' of π^{n+3}_{2n+9} such that $2\alpha - E\beta' = \pm \Delta(\nu_{2n+9})$. $\pi^{2n+5}_{2n+9} = \pi^{2n+3}_{2n+8} = 0$ by Propositions 5.8 and 5.9. Then it follows from the exactness of (4.4) that $E : \pi^{2n+1}_{2n+7} \to \pi^{2n+2}_{2n+8}$ and $E : \pi^{2n+2}_{2n+8} \to \pi^{n+3}_{2n+9}$ are homomorphisms onto. Thus there exists an element β of π^{n+1}_{2n+7} such that

$$\pm \Delta(\nu_{2n+9}) = 2\alpha - E^3\beta.$$

Now we shall show that the assumption $H(\beta) = 0$ implies a contradiction. Assume that $H(\beta) = 0$, then $\beta = E\beta''$ for some $\beta'' \in \pi^{n}_{2n+6}$. By the exactness of (11.8),

$$I_4(2\alpha - \Delta(\pm \nu_{2n+9})) = I_4E^3\beta = I_4E^4\beta'' = 0.$$

we have also,

$$\begin{aligned} I_4\, \Delta(\nu_{2n+9}) &= I_4\, \Delta(E^2\nu_{2n+7}) = \pm I_4\, \Delta\, H_0(f^{n+5}_{n+4})_*\nu_{2n+7} && \text{by (11.13)} \\ &= \pm I_4\, p_{1*}(f^{n+5}_{n+4})_*\, \nu_{2n+7} && \text{by (11.12)} \\ &= \pm \Delta_4(f^{n+5}_{n+4})_*\, \nu_{2n+7} && \text{by (11.10)} \\ &= \pm (f^{n+4}_{n})_*\, \Delta_4(\nu_{2n+7}) && \text{by Lemma 11.6} \end{aligned}$$

Thus $(f^{n+4}_n)_*\, \Delta_4(\nu_{2n+7}) = \pm 2I_4(\alpha)$. Since this $(f^{n+4}_n)_*$ is an isomorphism onto, by Theorem 11.7, then it follows that there exists an element α' of $\pi_{2n+5}(E^{n-1}P^{n+3}_n)$ such that

$$\Delta_4(\nu_{2n+7}) = 2\alpha'.$$

By i) of Lemma 11.8,

$$2\alpha' = \Delta_4(\nu_{2n+7}) = \lambda \circ \nu_{2n+2}$$

for the class λ of the attaching map of $e^{2n+3} = E^{n-1}P^{n+4}_n - E^{n-1}P^{n+3}_n$. By attaching $(2n+6)$-cell e^{2n+6} to $E^{n-1}P^{n+3}_n$ by a representative of α', we have a cell complex

$$K = E^{n-1}P^{n+4}_n \cup_{\alpha'} e^{2n+6} = E^{n-1}P^{n+3}_n \cup_\lambda e^{2n+3} \cup_{\alpha'} e^{2n+6}.$$

We consider that K is of the form

$$K = E^{n-1}P^{n+3}_n \cup_\gamma C(S^{2n+2} \vee S^{2n+5}),$$

where γ is given by λ and α'. Then we have a relation

$$\gamma \circ (i_{1*}\,\nu_{2n+2} - i_{2*}(2\iota_{2n+5})) = 0,$$

for the injections i_1 and i_2 of S^{2n+2} and S^{2n+5} into $S^{2n+2} \vee S^{2n+5}$. Let $\delta \in \pi_{2n+6}(K)$ be a coextension of $(i_{1*}\,\nu_{2n+2} - i_{2*}(2\iota_{2n+5}))$

and construct a cell complex

$$M = K \cup_{\delta} e^{2n+7} .$$

Now it is verified easily that $H^i(M, Z_2) = \{u_i\} \approx Z_2$ for $i = 0$, $2n-1$, $2n$, $2n+1$, $2n+2$, $2n+3$, $2n+6$, $2n+7$ and $H^i(M, Z_2) = 0$ otherwise. By the definition of M, we see that

$$Sq^1 u_{2n+6} = Sq^4 u_{2n+3} = u_{2n+7} .$$

Since $n \equiv 4 \pmod 8$, we have from (11.15), $Sq^4 e^n = e^{n+4}$ and $Sq^1 e^n = Sq^2 e^n = 0$. Thus we have

$$Sq^4 u_{2n-1} = u_{2n+3} \quad \text{and} \quad Sq^1 u_{2n-1} = Sq^2 u_{2n-1} = 0 .$$

Consider the relation $Sq^4 Sq^4 = Sq^6 Sq^2 + Sq^7 Sq^1$ of Adem. Then we have

$$u_{2n+7} = Sq^4 Sq^4 u_{2n-1} = Sq^6 Sq^2 u_{2n-1} + Sq^7 Sq^1 u_{2n-1} = 0 ,$$

but this contradicts $u_{2n+7} \neq 0$. Therefore $H(\beta) \neq 0$. Since ν_{2n+1}^2 generates $\pi_{2n+7}^{2n+1} \approx Z_2$, then we have that $H(\beta) = \nu_{2n+1}^2$.

q.e.d.

Remark that Lemma 11.17 still holds for $n = 4$ (cf. Theorem 7.3).

CHAPTER XII.

2-Primary Components of $\pi_{n+k}(S^n)$ for $16 \leq k \leq 19$.

i). Some new elements.

First we have

Lemma 12.1. There exists an element ζ' of π_{22}^6 such that $H(\zeta') \equiv \zeta_{11} \bmod 2\zeta_{11}$ and $E\zeta' \equiv \sigma' \circ \eta_{14} \circ \varepsilon_{15} \bmod 2\pi_{23}^7$.

Proof. Consider the composition $\sigma'' \circ \varepsilon_{13} \in \pi_{21}^6$. By Lemma 5.14, $E(\sigma'' \circ \varepsilon_{13}) = E\sigma'' \circ \varepsilon_{14} = 2\sigma' \circ \varepsilon_{14} = \sigma' \circ 2\varepsilon_{14} = 0$. Since $E : \pi_{21}^6 \to \pi_{22}^7$ is an isomorphism into (Theorem 10.5), it follows that $\sigma'' \circ \varepsilon_{13} = 0$. Then the following secondary composition is defined:

$$\{\sigma'', \varepsilon_{13}, 2\iota_{21}\}_1 \in \pi_{22}(S^6)/(\sigma'' \circ \pi_{22}^{13} + 2\pi_{22}(S^6)).$$

Since $2\pi_{22}(S^6)$ contains the odd component of $\pi_{22}(S^6)$, then we may choose an element ζ' of $\pi_{22}^6 \cap \{\sigma'', \varepsilon_{13}, 2\iota_{21}\}_1$. By Proposition 2.3 and Lemma 5.4,

$$H(\zeta') \in H\{\sigma'', \varepsilon_{13}, 2\iota_{21}\}_1 \subset \{H\sigma'', \varepsilon_{13}, 2\iota_{21}\}_1$$
$$= \{\eta_{11}^2, \varepsilon_{13}, 2\iota_{21}\}_1.$$

By Lemma 9.1, $\zeta_{11} + 2\pi_{22}(S^{11}) = \{\eta_{11}^2, \varepsilon_{13}, 2\iota_{21}\}_1$. It follows that $H(\zeta') \equiv \zeta_{11} \bmod 2\pi_{22}(S^{11})$. Since both of $H(\zeta')$ and ζ_{11} belong to π_{22}^{11}, which is generated by ζ_{11}, then we have that $H(\zeta') \equiv \zeta_{11} \bmod 2\zeta_{11}$. Next we have

$$E\zeta' \in E\{\sigma'', \varepsilon_{13}, 2\iota_{21}\}_1 \subset -\{E\sigma'', \varepsilon_{14}, 2\iota_{22}\}_1$$
$$= -\{2\sigma', \varepsilon_{14}, 2\iota_{22}\}_1 \qquad \text{by Lemma 5.14}$$

and

$$\sigma' \circ \eta_{14} \circ \varepsilon_{15} = -\sigma' \circ \varepsilon_{14} \circ \eta_{22}$$
$$\in -\sigma' \circ \{2\iota_{14}, \varepsilon_{14}, 2\iota_{22}\}_1 \qquad \text{by Corollary 3.7}$$
$$\subset -\{2\sigma', \varepsilon_{14}, 2\iota_{22}\}_1.$$

The secondary composition $-\{2\sigma', \varepsilon_{14}, 2\iota_{22}\}_1$ is a coset of $2\sigma' \circ E\pi_{23}^{14}$ $+ \pi_{23}(S^7) \circ 2\iota_{23} = \sigma' \circ E2\pi_{23}^{14} + 2\pi_{23}(S^7) = 2\pi_{23}(S^7)$.
Thus $E\zeta' \equiv \sigma' \circ \eta_{14} \circ \varepsilon_{15} \mod 2\pi_{23}(S^7)$. Since $E\zeta'$ and $\sigma' \circ \eta_{14} \circ \varepsilon_{15}$ belongs to π_{23}^7, then $E\zeta' \equiv \sigma' \circ \eta_{14} \circ \varepsilon_{15} \mod 2\pi_{23}^7$. q.e.d.

Let $\bar{\mu}_3$ be an element of the secondary composition

$$\{\mu_3, 2\iota_{12}, 8\sigma_{12}\}_1$$

and denote that

$$\bar{\mu}_n = E^{n-3}\bar{\mu}_3 \quad \text{for } n \geq 3 \text{ and } \bar{\mu} = E^{\infty}\bar{\mu}_3 .$$

<u>Lemma</u> 12.2. <u>$H(\bar{\mu}_3) = \rho^{IV}$ and $2\bar{\mu}_n = 2\bar{\mu} = 0$ for $n \geq 3$.</u>

<u>Proof</u>.
$$\begin{aligned} H(\bar{\mu}_3) &\in H\{\mu_3, 2\iota_{12}, 8\sigma_{12}\}_1 \\ &\subset \{H\mu_3, 2\iota_{12}, 8\sigma_{12}\}_1 && \text{by Proposition 2.3} \\ &= \{\sigma''', 2\iota_{12}, 8\sigma_{12}\}_1 && \text{by Lemma 6.5.} \end{aligned}$$

ρ^{IV} is, by its definition, contained in $\{\sigma''', 2\iota_{12}, 8\sigma_{12}\}_1$ which is a coset of

$$\sigma''' \circ E\pi_{19}^{11} + \pi_{13}^5 \circ 8\sigma_{13} = \{\sigma''' \circ \bar{\nu}_{12}\} + \{\sigma''' \circ \varepsilon_{12}\} + 8\pi_{13}^5 \circ \sigma_{13} .$$

$E(\sigma''' \circ \bar{\nu}_{12}) = 2\sigma'' \circ \bar{\nu}_{12} = 0$ by Lemma 5.14. Since $E : \pi_{20}^5 \to \pi_{21}^6$ is an isomorphism into, it follows that $\sigma''' \circ \bar{\nu}_{12} = 0$. Similarly, $\sigma''' \circ \varepsilon_{12} = 0$. By Theorem 7.1, $8\pi_{13}^5 = 0$. Thus $\{\sigma''', 2\iota_{13}, 8\sigma_{13}\}_1$ is a coset of the trivial group 0. We have then $H(\bar{\mu}_3) = \rho^{IV}$. Next,

$$\begin{aligned} 2\bar{\mu}_3 &= 2\iota_3 \circ \bar{\mu}_3 && \text{by Lemma 4.5} \\ &\in 2\iota_3 \circ \{\mu_3, 2\iota_{11}, 8\sigma_{11}\} \\ &= \{2\iota_3, \mu_3, 2\iota_{11}\} \circ -8\sigma_{12} && \text{by Proposition 1.4} \end{aligned}$$

and
$$\begin{aligned} 0 &= \mu_3 \circ \eta_{11} \circ (-8\sigma_{12}) \\ &\in \{2\iota_3, \mu_{11}, 2\iota_{11}\} \circ -8\sigma_{12} && \text{by Corollary 3.7.} \end{aligned}$$

$\{2\iota_3, \mu_3, 2\iota_{11}\} \circ 8\sigma_{12}$ is a coset of

$$2\iota_3 \circ \pi_{13}^3 \circ 8\sigma_{13} = 2\iota_3 \circ 8\pi_{13}^3 \circ \sigma_{13} = 0 .$$

Thus $2\bar{\mu}_3 = 0$ and $2\bar{\mu}_n = 2\mu = 0$ for $n \geq 3$. q.e.d.

<u>Lemma</u> 12.3. <u>There exists an element $\bar{\varepsilon}'$ of π_{20}^3 such that $2\bar{\varepsilon}' = \eta_3^2 \circ \bar{\varepsilon}_5$ and $E\bar{\varepsilon}' = E\nu' \circ \kappa_7$.</u>

<u>Proof</u>.
$$\begin{aligned} H(E\nu' \circ \kappa_7) &= E(\nu' \# \nu') \circ H(\kappa_7) && \text{by Proposition 2.2} \\ &= E^4\nu' \circ E^7\nu' \circ H(\kappa_7) \\ &= 4\nu_7^2 \circ H(\kappa_7) = 0 . \end{aligned}$$

It follows from the exactness of the sequence (4.4), that there exists an element $\bar{\varepsilon}'$ of π_{20}^3 such that $E\bar{\varepsilon}' = E\nu' \circ \kappa_7$. Next,

$$E2\bar{\varepsilon}' = 2\iota_4 \circ E\bar{\varepsilon}' = 2\iota_4 \circ E\nu' \circ \kappa_7 \qquad \text{by (2.1)}$$

$$= E2\nu' \circ \kappa_7 = \eta_4^3 \circ \kappa_7 \qquad \text{by (5.3)}$$

$$= \eta_4^2 \circ \bar{\varepsilon}_6 = E(\eta_3^2 \circ \bar{\varepsilon}_5) \qquad \text{by (10.23)}.$$

By Lemma 4.5, it follows that $2\bar{\varepsilon}' = \eta_3^2 \circ \bar{\varepsilon}_5$. q.e.d.

Choose elements $\bar{\mu}' \in \pi_{22}^3$ and $\bar{\zeta}_5 \in \pi_{24}^5$ as follows.

$$\bar{\mu}' \in \{\mu', 4\iota_{14}, 4\sigma_{14}\}_1$$

and

$$\bar{\zeta}_5 \in \{\zeta_5, 8\iota_{16}, 2\sigma_{16}\}_1 .$$

Denote that

$$\bar{\zeta}_n = E^{n-5}\bar{\zeta}_5 \quad \text{for} \quad n \ge 5 \quad \text{and} \quad \bar{\zeta} = E^\infty\bar{\zeta}_5 .$$

<u>Lemma</u> 12.4. $H(\bar{\mu}') \equiv \bar{\mu}_5 \bmod E^3\pi_{19}^2$, $H(\bar{\zeta}_5) = 8\rho'$, $2\bar{\mu}' = \eta_3^2 \circ \bar{\mu}_5$ <u>and</u> $2\bar{\zeta}_5 = E^2\bar{\mu}'$.

<u>Proof</u>. First, we have

$$H(\bar{\mu}') \in H\{\mu', 4\iota_{14}, 4\sigma_{14}\}_1$$

$$\subset \{H\mu', 4\iota_{14}, 4\sigma_{14}\}_1 \qquad \text{by Proposition 2.3}$$

$$= \{\mu_5, 4\iota_{14}, 4\sigma_{14}\}_1 \qquad \text{by (7.7)}$$

and

$$\bar{\mu}_5 = E^2\bar{\mu}_3 \in E^2\{\mu_3, 2\iota_{12}, 8\sigma_{12}\}_1$$

$$\subset \{\mu_5, 4\iota_{14}, 4\sigma_{14}\}_1 \qquad \text{by Propositions 1.3 and 1.2.}$$

The secondary composition $\{\mu_5, 4\iota_{14}, 4\sigma_{14}\}_1$ is a coset of

$$\mu_5 \circ E\pi_{21}^{13} + \pi_{15}^5 \circ 4\sigma_{15},$$

which is generated by $\mu_5 \circ \bar{\nu}_{14}$, $\mu_5 \circ \varepsilon_{14}$, $\eta_5 \circ \mu_6 \circ 4\sigma_{15}$ and $\nu_5 \circ \sigma_8 \circ 4\sigma_{15}$ (Theorem 7.1 and Theorem 7.3). Obviously, $\eta_5 \circ \mu_6 \circ 4\sigma_{15} = 4\eta_5 \circ \mu_6 \circ \sigma_{15} = 0$. By (7.10) and Lemma 10.7, $\nu_5 \circ \sigma_8 \circ 4\sigma_{15} = 4(\nu_5 \circ \sigma_8) \circ \sigma_{15} = \eta_5^2 \circ \varepsilon_7 \circ \sigma_{15} = 0$.

Next we have

(12.1). $$H(\mu_3 \circ \bar{\nu}_{12}) = H(\mu_3 \circ \varepsilon_{12}) = 0 .$$

By Proposition 2.2 and Lemma 6.5, $H(\mu_3 \circ \overline{\nu_{12}}) = H(\mu_3) \circ \bar{\nu}_{12} = \sigma''' \circ \bar{\nu}_{12}$. By Lemma 5.14, $E(\sigma''' \circ \bar{\nu}_{12}) = 2\sigma'' \circ \bar{\nu}_{13} = \sigma'' \circ 2\bar{\nu}_{13} = 0$. Since $E : \pi_{20}^5 \to \pi_{21}^6$ is an isomorphism into (Theorem 10.5), then it follows that $\sigma''' \circ \bar{\nu}_{12} = 0$. Similarly $H(\mu_3 \circ \varepsilon_{12}) = 0$.

By the exactness of the sequence (4.4), we have that $\mu_3 \circ \bar{\nu}_{12}$ and $\mu_3 \circ \varepsilon_{12}$ are in $E\pi_{19}^2$. Then we have obtained that

$$\mu_5 \circ E\pi_{21}^{13} + \pi_{15}^5 \circ 4\sigma_{15} \subset E^3\pi_{19}^2 ,$$

and thus the relation $H(\bar{\mu}') \equiv \mu_5 \bmod E^3\pi_{19}^2$ has proved.

Next, we have

$$2\bar{\mu}' = 2\iota_3 \circ \bar{\mu}' \in 2\iota_3 \circ \{\mu', 4\iota_{14}, 4\sigma_{14}\}_1$$

$$\subset \{2\mu', 4\iota_{14}, 4\sigma_{14}\}_1 \qquad \text{by Lemma 4.5}$$

and

$$\eta_3^2 \circ \bar{\mu}_5 \in \eta_3^2 \circ E^2\{\mu_3, 2\iota_{12}, 8\sigma_{12}\}_1$$

$$\subset \eta_3^2 \circ \{\mu_5, 2\iota_{14}, 8\sigma_{14}\}_1 \qquad \text{by Proposition 1.3}$$

$$\subset \{\eta_3^2 \circ \mu_5, 4\iota_{14}, 4\sigma_{14}\}_1 \qquad \text{by Proposition 1.2}$$

$$= \{2\mu', 4\iota_{14}, 4\sigma_{14}\}_1 \qquad \text{by (7.7).}$$

The secondary composition $\{2\mu', 4\iota_{14}, 4\sigma_{14}\}_1$ is a coset of

$$2\mu' \circ E\pi_{21}^{13} + \pi_{15}^5 \circ 4\sigma_{15} .$$

As is seen in the above discussion, we have $\pi_{15}^5 \circ 4\sigma_{15} = 0$. Since $2\pi_{21}^{13} = 0$ (Theorem 7.1), $2\mu' \circ E\pi_{21}^{13} = \mu' \circ E2\pi_{21}^{13} = 0$. Therefore we conclude that $2\bar{\mu}' = \eta_3^2 \circ \bar{\mu}_5$.

Similarly, we can prove that

$$H(\bar{\zeta}_5),\ 8\rho' \in \{8\sigma_9, 8\iota_{16}, 2\sigma_{16}\}_1$$

and

$$2\bar{\zeta}_5,\ E^2\bar{\mu}' \in \{2\zeta_5, 4\iota_{16}, 4\sigma_{16}\}_1$$

and that these secondary compositions are cosets of the trivial group 0. Then we have that $H(\bar{\zeta}_5) = 8\rho'$ and $2\bar{\zeta}_5 = E^2\bar{\mu}'$. The details are left to the reader. q.e.d.

By Lemma 10.7 and (7.17), we have the relations

$$\nu_6 \circ (\varepsilon_9 + \bar{\nu}_9) = 0 \quad \text{and} \quad (\varepsilon_8 + \bar{\nu}_8) \circ \sigma_{16} = 0 .$$

Then the secondary composition

$$\{\nu_6, \varepsilon_9 + \bar{\nu}_9, \sigma_{17}\}_1$$

is defined. Let $\bar{\sigma}_6$ be an element of this secondary composition and denote that

$$\bar{\sigma}_n = E^{n-6}\bar{\sigma}_6 \quad \text{for } n \geq 6 \text{ and } \bar{\sigma} = E^{\infty}\bar{\sigma}_6 .$$

<u>Lemma</u> 12.5. $H(\bar{\sigma}_6) \equiv \sigma_{11}^2 \bmod 2\sigma_{11}^2$.

<u>Proof</u>. $H(\bar{\sigma}_6) \in H\{\nu_6, \varepsilon_9 + \bar{\nu}_9, \sigma_{17}\}_1$

$$= \Delta^{-1}(\nu_5 \circ (\varepsilon_8 + \bar{\nu}_8)) \circ \sigma_{18} \quad \text{by Proposition 2.4.}$$

By (7.17), we have that $\Delta^{-1}(\nu_5 \circ (\varepsilon_8 + \bar{\nu}_8))$ is a coset of $2\pi_{18}^{11} = \{2\sigma_{11}\}$ containing σ_{11}. Then the lemma is proved q.e.d.

ii). <u>The groups</u> $\pi_{n+k}(S^n)$ <u>for</u> $16 \leq k \leq 19$ <u>and</u> $2 \leq n \leq 9$.

By use of the elements introduced in the previous i), our results are stated as follows.

Theorem 12.6. $\pi^2_{18} = \{\eta_2 \circ \bar{\varepsilon}_3\} \approx Z_2$,

$\pi^n_{n+16} = \{\mu_n \circ \sigma_{n+9}\} \oplus \{\eta_n \circ \bar{\varepsilon}_{n+1}\} \approx Z_2 \oplus Z_2$ for $n = 3$ and $n = 5$,

$\pi^4_{20} = \{\nu_4^2 \circ \sigma_{10} \circ \nu_{17}\} \oplus \{\mu_4 \circ \sigma_{13}\} \oplus \{\eta_4 \circ \bar{\varepsilon}_5\} \approx Z_2 \oplus Z_2 \oplus Z_2$,

$\pi^6_{22} = \{\zeta'\} \oplus \{\mu_6 \circ \sigma_{15}\} \oplus \{\eta_6 \circ \bar{\varepsilon}_7\} \approx Z_8 \oplus Z_2 \oplus Z_2$,

$\pi^7_{23} = \{\sigma' \circ \mu_{14}\} \oplus \{E\zeta'\} \oplus \{\mu_7 \circ \sigma_{16}\} \oplus \{\eta_7 \circ \bar{\varepsilon}_8\} \approx Z_2 \oplus Z_2 \oplus Z_2 \oplus Z_2$,

$\pi^8_{24} = \{\sigma_8 \circ \nu^3_{15}\} \oplus \{\sigma_8 \circ \mu_{15}\} \oplus \{\sigma_8 \circ \eta_{15} \circ \varepsilon_{16}\} \oplus \{E\sigma' \circ \mu_{15}\}$
$\oplus \{E^2\zeta'\} \oplus \{\mu_8 \circ \sigma_{17}\} \oplus \{\eta_8 \circ \bar{\varepsilon}_9\} \approx Z_2 \oplus Z_2 \oplus Z_2 \oplus Z_2 \oplus Z_2 \oplus Z_2 \oplus Z_2$,

$\pi^9_{25} = \{\sigma_9 \circ \nu^3_{16}\} \oplus \{\sigma_9 \circ \mu_{16}\} \oplus \{\sigma_9 \circ \eta_{16} \circ \varepsilon_{17}\} \oplus \{\mu_9 \circ \sigma_{18}\}$
$\approx Z_2 \oplus Z_2 \oplus Z_2 \oplus Z_2$.

Theorem 12.7. $\pi^2_{19} = \{\eta_2 \circ \mu_3 \circ \sigma_{12}\} \oplus \{\eta_2^2 \circ \bar{\varepsilon}_4\} \approx Z_2 \oplus Z_2$,

$\pi^3_{20} = \{\bar{\varepsilon}'\} \oplus \{\bar{\mu}_3\} \oplus \{\eta_3 \circ \mu_4 \circ \sigma_{13}\} \approx Z_4 \oplus Z_2 \oplus Z_2$,

$\pi^4_{21} = \{\nu_4 \circ \sigma' \circ \sigma_{14}\} \oplus \{\nu_4 \circ \kappa_7\} \oplus \{E\bar{\varepsilon}'\} \oplus \{\bar{\mu}_4\} \oplus \{\eta_4 \circ \mu_5 \circ \sigma_{14}\}$
$\approx Z_8 \oplus Z_4 \oplus Z_4 \oplus Z_2 + Z_2$,

$\pi^5_{22} = \{\nu_5 \circ \kappa_8\} \oplus \{\bar{\mu}_5\} \oplus \{\eta_5 \circ \mu_6 \circ \sigma_{15}\} \approx Z_4 \oplus Z_2 \oplus Z_2$,

$\pi^6_{23} = \{\Delta(E\theta)\} \oplus \{\nu_6 \circ \kappa_9\} \oplus \{\bar{\mu}_6\} \oplus \{\eta_6 \circ \mu_7 \circ \sigma_{16}\} \approx Z_2 \oplus Z_2 \oplus Z_2 \oplus Z_2$,

$\pi^7_{24} = \{\sigma' \circ \eta_{14} \circ \mu_{15}\} \oplus \{\nu_7 \circ \kappa_{10}\} \oplus \{\bar{\mu}_7\} \oplus \{\eta_7 \circ \mu_8 \circ \sigma_{17}\}$
$\approx Z_2 \oplus Z_2 \oplus Z_2 \oplus Z_2$,

$\pi^8_{25} = \{\sigma_8 \circ \eta_{15} \circ \mu_{16}\} \oplus \{E\sigma' \circ \eta_{15} \circ \mu_{16}\} \oplus \{\nu_8 \circ \kappa_{11}\} \oplus \{\bar{\mu}_8\}$
$\oplus \{\eta_8 \circ \mu_9 \circ \sigma_{18}\} \approx Z_2 \oplus Z_2 \oplus Z_2 \oplus Z_2 \oplus Z_2$,

$\pi^9_{26} = \{\sigma_9 \circ \eta_{16} \circ \mu_{17}\} \oplus \{\nu_9 \circ \kappa_{12}\} \oplus \{\bar{\mu}_9\} \oplus \{\eta_9 \circ \mu_{10} \circ \sigma_{19}\}$
$\approx Z_2 \oplus Z_2 \oplus Z_2 \oplus Z_2$.

Theorem 12.8. $\pi^2_{20} = \{\eta_2 \circ \bar{\varepsilon}'\} \oplus \{\eta_2 \circ \bar{\mu}_3\} \oplus \{\eta_2^2 \circ \mu_4 \circ \sigma_{13}\}$
$\approx Z_4 \oplus Z_2 \oplus Z_2$,

$\pi^3_{21} = \{\mu' \circ \sigma_{14}\} \oplus \{\nu' \circ \bar{\varepsilon}_6\} \oplus \{\eta_3 \circ \bar{\mu}_4\} \approx Z_4 \oplus Z_2 \oplus Z_2$,

$\pi^4_{22} = \{\nu_4 \circ \rho''\} \oplus \{\nu_4 \circ \sigma' \circ \bar{\nu}_{14}\} \oplus \{\nu_4 \circ \sigma' \circ \varepsilon_{14}\} \oplus \{\nu_4 \circ \bar{\varepsilon}_7\}$
$\oplus \{E\mu' \circ \sigma_{15}\} \oplus \{E\nu' \circ \bar{\varepsilon}_7\} \oplus \{\eta_4 \circ \bar{\mu}_5\} \approx Z_8 \oplus Z_2 \oplus Z_2 \oplus Z_2 \oplus Z_4 \oplus Z_2 \oplus Z_2$,

$\pi^5_{23} = \{\zeta_5 \circ \sigma_{16}\} \oplus \{\nu_5 \circ \bar{\varepsilon}_8\} \oplus \{\eta_5 \circ \bar{\mu}_6\} \approx Z_8 \oplus Z_2 \oplus Z_2$,

$\pi^6_{24} = \{\Delta(E\theta) \circ \eta_{23}\} \oplus \{\zeta_6 \circ \sigma_{17}\} \oplus \{\eta_6 \circ \bar{\mu}_7\} \approx Z_2 \oplus Z_8 \oplus Z_2$,

$\pi^7_{25} = \{\zeta_7 \circ \sigma_{18}\} \oplus \{\eta_7 \circ \bar{\mu}_8\} \approx Z_8 \oplus Z_2$,

$\pi^8_{26} = \{\sigma_8 \circ \zeta_{15}\} \oplus \{\zeta_8 \circ \sigma_{19}\} \oplus \{\eta_8 \circ \bar{\mu}_9\} \approx Z_8 \oplus Z_8 \oplus Z_2,$

$\pi^9_{27} = \{\sigma_9 \circ \zeta_{16}\} \oplus \{\eta_9 \circ \bar{\mu}_{10}\} \approx Z_8 \oplus Z_2.$

Theorem 12.9. $\pi^2_{21} = \{\eta_2 \circ \mu' \circ \sigma_{14}\} \oplus \{\eta_2 \circ \nu' \circ \bar{\varepsilon}_6\} \oplus \{\eta^2_2 \circ \bar{\mu}_4\} \approx Z_4 \oplus Z_2 \oplus Z_2,$

$\pi^3_{22} = \{\bar{\mu}'\} \oplus \{\nu' \circ \mu_6 \circ \sigma_{15}\} \approx Z_4 \oplus Z_2,$

$\pi^4_{23} = \{\nu_4 \circ \sigma' \circ \mu_{14}\} \oplus \{\nu_4 \circ E\zeta'\} \oplus \{\nu_4 \circ \mu_7 \circ \sigma_{16}\} \oplus \{\nu_4 \circ \eta_7 \circ \bar{\varepsilon}_8\}$
$\oplus \{E\bar{\mu}'\} \oplus \{E\nu' \circ \mu_7 \circ \sigma_{16}\} \approx Z_2 \oplus Z_2 \oplus Z_2 \oplus Z_2 \oplus Z_4 \oplus Z_2,$

$\pi^5_{24} = \{\bar{\zeta}_5\} \oplus \{\nu_5 \circ \mu_8 \circ \sigma_{17}\} \approx Z_8 \oplus Z_2,$

$\pi^6_{25} = \{\bar{\zeta}_6\} \oplus \{\bar{\sigma}_6\} \approx Z_8 \oplus Z_{32},$

$\pi^n_{n+19} = \{\bar{\zeta}_n\} \oplus \{\bar{\sigma}_n\} \approx Z_8 \oplus Z_2$ for $n = 7, 8$ and 9.

Proof of Theorem 12.6.

The result $\pi^2_{18} = \{\eta_2 \circ \bar{\varepsilon}_3\} \approx Z_2$ is a direct consequence of (5.2) and Theorem 10.5. By Lemma 12.2, $\Delta(\rho^{IV}) = \Delta H(\bar{\mu}_3) = 0$. By Lemma 5.2, $\Delta(\bar{\varepsilon}_5) = \Delta(E^2\bar{\varepsilon}_3) = 0$. Since ρ^{IV} and $\bar{\varepsilon}_5$ generate π^5_{20}, By Theorem 10.5, then $\Delta(\pi^5_{20}) = 0$. It follows from (4.4), that the following sequence is exact:

$$0 \longrightarrow \pi^2_{18} \xrightarrow{E} \pi^3_{19} \xrightarrow{H} \pi^5_{19} \xrightarrow{\Delta} \pi^2_{17},$$

where the last homomorphism Δ is onto by (10.13). Since $\pi^5_{19} \approx Z_2 \oplus Z_2$ and $\pi^2_{17} \approx Z_2$ (Theorems 10.3 and 10.5), then we have that $H\pi^3_{19} \approx Z_2$. By Proposition 2.2, Lemma 6.5, Lemma 5.14, and by Theorem 10.3,

$$EH(\mu_3 \circ \sigma_{12}) = EH(\mu_3) \circ E\sigma_{12} = E\sigma''' \circ \sigma_{13} = 2\sigma'' \circ \sigma_{13} \neq 0.$$

Thus $H(\mu_3 \circ \sigma_{12}) \neq 0$. Obviously $2(\mu_3 \circ \sigma_{12}) = 2\mu_3 \circ \sigma_{12} = 0$. Consequently, we have from the above exact sequence that

$$\pi^3_{19} = \{\mu_3 \circ \sigma_{12}\} \oplus \{\eta_3 \circ \bar{\varepsilon}_4\} \approx Z_2 \oplus Z_2.$$

By (5.6) and Theorem 7.7,

$$\pi^4_{20} = \{\nu^2_4 \circ \sigma_{10} \circ \nu_{17}\} \oplus \{\mu_4 \circ \sigma_{13}\} \oplus \{\eta_4 \circ \bar{\varepsilon}_5\} \approx Z_2 \oplus Z_2 \oplus Z_2.$$

Consider the exact sequence

$$\pi^9_{22} \xrightarrow{\Delta} \pi^4_{20} \xrightarrow{E} \pi^5_{21} \xrightarrow{H} \pi^9_{21}$$

of (4.4). $\pi^9_{21} = 0$ by Theorem 7.6. Then E is onto. $\pi^9_{22} = \{\sigma_9 \circ \nu^2_{16}\} \approx Z_2$, by Theorem 7.7. We have

$$\Delta(\sigma_9 \circ \nu^2_{16}) = \Delta(\sigma_9) \circ \nu^2_{14} \qquad \text{by Proposition 2.5}$$

$$= \nu_4 \circ \sigma' \circ \nu_{14}^2 + E\varepsilon' \circ \nu_{14}^2 \qquad \text{by (7.16)}$$
$$= \nu_4^2 \circ \sigma_{10} \circ \nu_{17} + E\varepsilon' \circ \nu_{14}^2 \qquad \text{by (7.19).}$$

The element $E\varepsilon' \circ \nu_{14}^2$ is in the subgroup

$$E\pi_{16}^3 \circ \nu_{17} = \{E\nu' \circ \eta_7 \circ \mu_8 \circ \nu_{17}\} \qquad \text{by Theorem 7.6}$$
$$= \{E\nu' \circ \mu_7 \circ \eta_{16} \circ \nu_{17}\} = 0 \text{ by (5.9).}$$

Thus $\Delta(\sigma_9 \circ \nu_{16}^2) = \nu_4^2 \circ \sigma_{10} \circ \nu_{17}$, and from the above exact sequence we have that

$$\pi_{21}^5 = \{\mu_5 \circ \sigma_{14}\} \oplus \{\eta_5 \circ \bar{\varepsilon}_6\} \approx Z_2 \oplus Z_2$$

and that $\Delta : \pi_{22}^9 \to \pi_{20}^4$ is an isomorphism into. It follows from the exactness of the sequence (4.4) that

(12.2). $E : \pi_{21}^4 \to \pi_{22}^5$ <u>is onto</u>.

By Proposition 2.5 and (7.20),

$$\Delta(\sigma_{13}) \circ \nu_{18} = \Delta(\sigma_{13} \circ \nu_{20}) = 0.$$

Then the secondary composition

$$\{\Delta(\sigma_{13}), \nu_{18}, \eta_{21}\}_1 \in \pi_{23}^6/(\pi_{22}^6 \circ \eta_{22})$$

is defined. We have

(12.3). $H\{\Delta(\sigma_{13}), \nu_{18}, \eta_{21}\}_1 = \theta'$ <u>and</u> $2\{\Delta(\sigma_{13}), \nu_{18}, \eta_{21}\}_1 = 0$.

For, we have

$$2\{\Delta(\sigma_{13}), \nu_{18}, \eta_{21}\}_1 = \{\Delta(\sigma_{31}), \nu_{18}, \eta_{21}\}_1 \circ 2\iota_{23}$$
$$= \Delta(\sigma_{31}) \circ E\{\nu_{17}, \eta_{20}, 2\iota_{21}\} \qquad \text{by Proposition 1.4}$$
$$\subset \Delta(\sigma_{31}) \circ E\pi_{22}^{17} = 0 \qquad \text{by Proposition 5.9}$$

and $H\{\Delta(\sigma_{13}), \nu_{18}, \eta_{21}\}_1 \subset \{H\Delta(\sigma_{13}), \nu_{18}, \eta_{21}\}_1$ by Proposition 2.3

$$= \{\pm 2\sigma_{11}, \nu_{18}, \eta_{21}\}_1 \qquad \text{by Propositions 2.2 and 2.7}$$
$$\subset \pm \{\sigma_{11}, 2\nu_{18}, \eta_{21}\}_1 \qquad \text{by Proposition 1.2.}$$

It is easy to see that $\{\sigma_{11}, 2\nu_{18}, \eta_{21}\}_1$ consists of the element θ' only. Then (12.3) is proved.

Since θ' generates π_{23}^{11}, then it follows from (12.3) that $H : \pi_{23}^6 \to \pi_{23}^{11}$ is onto. By the exactness of the sequence (4.4), we have an exact sequence

$$0 \to \pi_{21}^5 \xrightarrow{E} \pi_{22}^6 \xrightarrow{H} \pi_{22}^{11}.$$

$\pi_{22}^{11} = \{\zeta_{11}\} \approx Z_8$ by Theorem 7.4. Then it follows from Lemma 12.1 that H is onto and π_{22}^6 is generated by $E\pi_{21}^5$ and ζ'. By Proposition 2.2 and Proposition 2.7, we have that $H(\Delta(\zeta_{13})) = \pm 2\zeta_{11}$. Then there exists an odd integer x such that $H(\Delta(\zeta_{13})) = 2xH\zeta'$. By the exactness of (4.4), it follows that

$$\Delta(\zeta_{13}) \equiv 2x\zeta' \bmod E\pi^5_{21}.$$

Since $2\pi^5_{21} = 0$, then $\Delta(2\zeta_{13}) = 4x\zeta'$. Further, we have

$$\begin{aligned} 8x\zeta' = \Delta(4\zeta_{13}) &= \Delta(\eta^2_{13} \circ \mu_{15}) && \text{by (7.14)} \\ &= \Delta(\eta_{13}) \circ \eta_{12} \circ \mu_{13} && \text{by Proposition 2.5} \\ &= \Delta H\sigma' \circ \eta_{12} \circ \mu_{13} = 0 && \text{by Lemma 5.14.} \end{aligned}$$

Since x is odd, then this implies that $8\zeta' = 0$. This means that the above exact sequence splits and thus

$$\pi^6_{22} = \{\zeta'\} \oplus \{\mu_6 \circ \sigma_{15}\} \oplus \{\eta_6 \circ \bar{\varepsilon}_7\} \approx Z_8 \oplus Z_2 \oplus Z_2.$$

Consider the exact sequence

$$\pi^{13}_{24} \xrightarrow{\Delta} \pi^6_{22} \xrightarrow{E} \pi^7_{23} \xrightarrow{H} \pi^{13}_{23}$$

of (4.4), where $\pi^{13}_{24} = \{\zeta_{13}\} \approx Z_8$ and $\pi^{13}_{23} = \{\eta_{13} \circ \mu_{14}\} \approx Z_2$, by Theorem 7.3 and Theorem 7.4. By Proposition 2.2 and Lemma 5.14,

$$H(\sigma' \circ \mu_{14}) = H(\sigma') \circ \mu_{14} = \eta_{13} \circ \mu_{14}.$$

Obviously $2(\sigma' \circ \mu_{14}) = 0$. Then it follows from the exactness of the above sequence that

$$\pi^7_{23} = E\pi^6_{22} \oplus \{\sigma' \circ \mu_{14}\} \approx E\pi^6_{22} \oplus Z_2 .$$

In the second relation of Lemma 12.1, we have that $2\pi^7_{23} = 2E\pi^6_{22} = E2\pi^6_{22} = \{2E\zeta'\}$. Thus $E\zeta' \equiv \sigma' \circ \eta_{14} \circ \varepsilon_{15} \bmod 2E\zeta'$. Since $2(\sigma' \circ \eta_{14} \circ \varepsilon_{15}) = 0$ and since $E\zeta'$ is in the 2-primary component, then we have that $2E\zeta' = 0$ and

(12.4) $E\zeta' = \sigma' \circ \eta_{14} \circ \varepsilon_{15}$ <u>and</u> $\Delta(\zeta_{13}) = \pm 2\zeta'$.

Consequently, we have the result

$$\pi^7_{23} = \{\sigma' \circ \mu_{14}\} \oplus \{E\zeta'\} \oplus \{\mu_7 \circ \sigma_{16}\} \oplus \{\eta_7 \circ \bar{\varepsilon}_8\} \approx Z_2 \oplus Z_2 \oplus Z_2 \oplus Z_2$$

By (5.15) and Theorem 7.3, we obtain the result on π^8_{24} in Theorem 12.6.

By (10.17), we have the exact sequence

$$\pi^{17}_{26} \xrightarrow{\Delta} \pi^8_{24} \xrightarrow{E} \pi^9_{25} \longrightarrow 0$$

of (4.4). $\pi^{17}_{26} = \{\nu^3_{17}\} \oplus \{\mu_{17}\} \oplus \{\eta_{17} \circ \varepsilon_{18}\} \approx Z_2 \oplus Z_2 \oplus Z_2$ by Theorem 7.2. We have, by use of Proposition 2.5 and (5.16),

$$\Delta(\nu^3_{17}) = \Delta(\nu_{17}) \circ \nu^2_{18} = \nu_8 \circ \sigma_{11} \circ \nu^2_{18} \qquad \text{by (7.19),}$$

$$\Delta(\mu_{17}) = \Delta(\iota_{17}) \circ \mu_{15} = E\sigma' \circ \mu_{15}$$

and $\Delta(\eta_{17} \circ \varepsilon_{18}) = \Delta(\iota_{17}) \circ \eta_{15} \circ \varepsilon_{16} = E\sigma' \circ \eta_{15} \circ \varepsilon_{16} = E^2\zeta'$ by (12.4).

Then the result

$$\pi^9_{25} = \{\sigma_9 \circ \nu^3_{16}\} \oplus \{\sigma_9 \circ \mu_{16}\} \oplus \{\sigma_9 \circ \eta_{16} \circ \varepsilon_{17}\} \oplus \{\mu_9 \circ \sigma_{18}\}$$
$$\approx Z_2 \oplus Z_2 \oplus Z_2 \oplus Z_2$$

is obtained by virtue of the following lemma.

Lemma 12.10. $\nu_n \circ \sigma_{n+3} \circ \nu^2_{n+10} = \eta_n \circ \bar{\varepsilon}_{n+1}$ for $n \geq 5$ and $\varepsilon_n \circ \varepsilon_{n+3} = \varepsilon_n \circ \bar{\nu}_{n+8} = \eta_n \circ \bar{\varepsilon}_{n+1} = \bar{\varepsilon}_n \circ \eta_{n+15}$ for $n \geq 3$.

Proof. By the definition of $\bar{\varepsilon}_3$, $\bar{\varepsilon}_3 \circ \eta_{18} \in \{\varepsilon_3, 2\iota_{11}, \nu^2_{11}\}_1 \circ \eta_{18}$. By Proposition 1.4, $\bar{\varepsilon}_3 \circ \eta_{18} \in \varepsilon_3 \circ E\{2\iota_{10}, \nu^2_{10}, \eta_{16}\}$. By (7.6), $\varepsilon_3 \circ \varepsilon_{11} \in \varepsilon_3 \circ E\{2\iota_{10}, \nu^2_{10}, \eta_{16}\}$. The composition $\varepsilon_3 \circ E\{2\iota_{10}, \nu^2_{10}, \eta_{16}\}$ is a coset of the subgroup

$$\varepsilon_3 \circ E\pi^{10}_{17} \circ \eta_{17} = \{\varepsilon_3 \circ \sigma_{11} \circ \eta_{17}\} = 0 \quad \text{by Lemma 10.7}$$

Thus $\bar{\varepsilon}_3 \circ \eta_{18} = \varepsilon_3 \circ \varepsilon_{11}$. By Proposition 3.1, $\bar{\varepsilon}_5 \circ \eta_{20} = \bar{\varepsilon}_3 \# \eta_2 = \eta_5 \circ \bar{\varepsilon}_6$. Since $E^2 : \pi^3_{19} \to \pi^5_{21}$ is an isomorphism, then it follows that $\bar{\varepsilon}_3 \circ \eta_{18} = \eta_3 \circ \bar{\varepsilon}_4$. By Lemma 6.4 and Lemma 10.7,

$$\varepsilon_3 \circ \bar{\nu}_{11} = \varepsilon_3 \circ (\varepsilon_{11} + \sigma_{11} \circ \eta_{18})$$
$$= \varepsilon_3 \circ \varepsilon_{11} + \varepsilon_3 \circ \sigma_{11} \circ \eta_{18} = \varepsilon_3 \circ \varepsilon_{11}.$$

Consequently we have $\varepsilon_3 \circ \bar{\nu}_{11} = \varepsilon_3 \circ \varepsilon_{11} = \bar{\varepsilon}_3 \circ \eta_{18} = \eta_3 \circ \bar{\varepsilon}_4$ and thus $\varepsilon_n \circ \bar{\nu}_{n+8} = \varepsilon_n \circ \varepsilon_{n+8} = \bar{\varepsilon}_n \circ \eta_{n+15} = \eta_n \circ \bar{\varepsilon}_{n+1}$ for $n \geq 3$. Next consider the secondary composition $\{\nu^2_5, 2\iota_{11}, \nu^2_{11}\}_1$. We have, by Proposition 2.6 and by (5.13), that $\nu^3_9 = \pm \nu_9 \circ \nu^2_{12}$ is contained in $H\{\nu^2_5, 2\iota_{11}, \nu^2_{11}\}_1 = \Delta^{-1}(\nu^2_4 \circ 2\iota_{10}) \circ \nu^2_{12}$. In Theorem 7.7 and its proof, we see that $\pi^5_{18} = \{\nu_5 \circ \sigma_8 \circ \nu_{15}\} \oplus \{\nu_5 \circ \eta_8 \circ \mu_9\} \approx Z_2 \oplus Z_2$, $H(\nu_5 \circ \sigma_8 \circ \nu_{15}) = \nu^3_9$ and $H(\nu_5 \circ \eta_8 \circ \mu_9) = 0$. Thus the secondary composition $\{\nu^2_5, 2\iota_{11}, \nu^2_{11}\}_1$ contains $\nu_5 \circ \sigma_8 \circ \nu_{15}$ or $\nu_5 \circ \sigma_8 \circ \nu_{15} + \nu_5 \circ \eta_8 \circ \mu_9$. Since $\nu_5 \circ \eta_8 \circ \mu_9 \circ \nu_{18} = \nu_5 \circ \mu_8 \circ \eta_{17} \circ \nu_{18} = 0$ by (5.9), then we have that

$$\nu_5 \circ \sigma_8 \circ \nu^2_{15} \in \{\nu^2_5, 2\iota_{11}, \nu^2_{11}\}_1 \circ \nu_{18}$$
$$\subset \{\nu^2_5, 2\iota_{11}, \nu^3_{11}\}_1 \qquad \text{by Proposition 1.2}$$
$$= \{\nu^2_5, 2\iota_{11}, \eta_{11} \circ \bar{\nu}_{12}\}_1 \qquad \text{by Lemma 6.3.}$$

By (7.6) and Proposition 1.2,

$$\varepsilon_5 \circ \bar{\nu}_{13} \in \{\nu^2_5, 2\iota_{11}, \eta_{11}\}_1 \circ \bar{\nu}_{13}$$
$$\subset \{\nu^2_5, 2\iota_{11}, \eta_{11} \circ \bar{\nu}_{12}\}_1 .$$

The secondary composition $\{\nu^2_5, 2\iota_{11}, \eta_{11} \circ \bar{\nu}_{12}\}_1$ is a coset of the subgroup $\nu^2_5 \circ E\pi^{10}_{20} + \pi^5_{12} \circ \eta_{12} \circ \bar{\nu}_{13}$, by (4.7) $E(\nu^2_5 \circ E\pi^{10}_{20}) = \nu^2_6 \circ \eta_{12} \circ \mu_{13} = 0$.

by Theorem 7.3 and (5.9). $E(\pi_{12}^5 \circ \eta_{12} \circ \bar{\nu}_{13}) = E\sigma''' \circ \eta_{12} \circ \bar{\nu}_{13} = 0$.
Since $E : \pi_{21}^5 \to \pi_{22}^6$ is an isomorphism into, then it follows that $\nu_5^2 \circ$
$E\pi_{20}^{10} + \pi_{12}^5 \circ \eta_{12} \circ \bar{\nu}_{13} = 0$. Therefore we have that $\varepsilon_5 \circ \bar{\nu}_{13} = \nu_5 \circ \sigma_8 \circ$
ν_{15}^2 and $\nu_n \circ \sigma_{n+3} \circ \nu_{n+10}^2 = \varepsilon_n \circ \bar{\nu}_{n+8} = \eta_n \circ \bar{\varepsilon}_{n+1}$ for $n \geq 5$.

q.e.d.

In the above computation of $\Delta : \pi_{26}^{17} \to \pi_{24}^8$, we see that this Δ
is an isomorphism into. It follows from the exactness of the sequence (4.4)
that

(12.5). $E : \pi_{25}^8 \to \pi_{26}^9$ is onto.

<u>Proof of Theorem</u> 12.7.

The result of π_{19}^2 is a direct consequence of (5.2) and Theorem
12.6.

Consider the exact sequence
$$\pi_{21}^5 \xrightarrow{\Delta} \pi_{19}^2 \xrightarrow{E} \pi_{20}^3 \xrightarrow{H} \pi_{20}^5$$
of (4.4). By Lemma 5.2, we have that $\Delta(\mu_5 \circ \sigma_{14}) = \Delta(\eta_5 \circ \bar{\varepsilon}_6) = 0$. Since
$\mu_5 \circ \sigma_{14}$ and $\eta_5 \circ \bar{\varepsilon}_6$ generates π_{21}^5 (Theorem 12.6), then Δ is trivial
and E is an isomorphism into. By Theorem 10.5, $\pi_{20}^5 \approx Z_2 \oplus Z_2$. Thus
π_{20}^3 has at most 16 elements. By Lemma 12.2, $\bar{\mu}_3$ is an element of order
2 which is not contained in $E\pi_{19}^2$, since $H(\bar{\mu}_3) \neq 0$. By Lemma 12.3, $\bar{\varepsilon}'$
is an element of order 4 which is not contained in $E\pi_{19}^2 \approx Z_2 \oplus Z_2$. Then
it is an algebraic consequence that
$$\pi_{20}^3 = \{\bar{\varepsilon}'\} \oplus \{\bar{\mu}_3\} \oplus \{\eta_3 \circ \mu_4 \circ \sigma_{13}\} \approx Z_4 \oplus Z_2 \oplus Z_2.$$
From this, (5.6) and from Theorem 10.3, the result on π_{21}^4 follows.

By (12.2) and by the exactness of (4.4), we have an isomorphism
$$E : \pi_{21}^4/\Delta\pi_{23}^9 \approx \pi_{22}^5 .$$
The group π_{23}^9 is generated by σ_9^2 and κ_9, by Theorem 10.5. We have
$$\Delta(\sigma_9^2) = \Delta(\sigma_9) \circ \sigma_{14} \qquad \text{by Proposition 2.5}$$
$$= x(\nu_4 \circ \sigma' \circ \sigma_{14}) \pm E\varepsilon' \circ \sigma_{14} \quad \text{by (7.16)}$$
and $\quad \Delta(\kappa_9) = \Delta(\iota_9) \circ \kappa_7 \qquad \text{by Proposition 2.5}$
$$= \pm (2\nu_4 - E\nu') \circ \kappa_7 \qquad \text{by (5.8)}$$
$$= \pm (\nu_4 \circ (2\iota_7 \circ \kappa_7) - \bar{\varepsilon}') \quad \text{by Lemma 12.3}$$
$$= \pm (2(\nu_4 \circ \kappa_7) - \bar{\varepsilon}') \qquad \text{by Lemma 4.5,}$$
where x is an odd integer. It follows that
$$\pi_{22}^5 = \{\nu_5 \circ \kappa_8\} \oplus \{\bar{\mu}_5\} \oplus \{\eta_5 \circ \mu_6 \circ \sigma_{15}\} \approx Z_4 \oplus Z_2 \oplus Z_2,$$

and that

(12.6). <u>the kernel of</u> $\Delta : \pi_{23}^9 \to \pi_{21}^4$ <u>is</u> $\{8\sigma_9^2\} \approx Z_2$.

Next we prove

<u>Lemma</u> 12.11. $H(\Delta(E\theta)) = \theta'$.

<u>Proof</u>. Let α be an element of $\{\Delta(\sigma_{13}), \nu_{18}, \eta_{21}\}_1$. By Propositions 1.2 and 1.3,

$$E\alpha \in E\{\Delta(\sigma_{13}), \nu_{18}, \eta_{21}\}_1 \subset \{E\Delta(\sigma_{13}), \nu_{19}, \eta_{22}\}_1$$
$$= \{0, \nu_{19}, \eta_{22}\}_1 = \pi_{23}^7 \circ \eta_{23}.$$

Thus $E\alpha = E\beta \circ \eta_{23} + x\sigma' \circ \mu_{14} \circ \eta_{23}$ for an element $\beta \in \pi_{22}^6$ and an integer x, by Theorem 12.6. Since $HE = 0$, it follows that $xH(\sigma' \circ \mu_{14} \circ \eta_{23}) = 0$. By Proposition 2.2, Lemma 5.14 and by Theorem 7.4, $H(\sigma' \circ \mu_{14} \circ \eta_{23}) = H(\sigma') \circ \mu_{14} \circ \eta_{23} = \eta_{13} \circ \mu_{14} \circ \eta_{23} = \eta_{13}^2 \circ \mu_{15} = 4\zeta_{13} \neq 0$. Thus $x \equiv 0$ (mod 2) and $E\alpha = E\beta \circ \eta_{23}$. Since $\{\Delta(\sigma_{13}), \nu_{18}, \eta_{21}\}_1$ is a coset of the subgroup $\pi_{22}^6 \circ \eta_{22}$, then $\alpha - \beta \circ \eta_{22}$ is contained in this secondary composition. By (12.3), $H(\alpha - \beta \circ \eta_{22}) = \theta'$ and $2(\alpha - \beta \circ \eta_{22}) = 0$. Thus $\alpha - \beta \circ \eta_{22}$ is an element of order 2 and it vanishes under the suspension : $E(\alpha - \beta \circ \eta_{22}) = 0$. By the exactness of (4.4) and by Theorem 7.6, we have that the kernel of $E : \pi_{23}^6 \to \pi_{24}^7$ is $\Delta\pi_{25}^{13} = \{\Delta(E\ \theta)\}$ and it has at most two elements. Therefore, we obtain that $\Delta(E\theta) = \alpha - \beta \circ \eta_{22}$ and $H(\Delta(E\theta)) = \theta'$. q.e.d.

Consider the exact sequence

$$\pi_{24}^{11} \xrightarrow{\Delta} \pi_{22}^5 \xrightarrow{E} \pi_{23}^6 \xrightarrow{H} \pi_{23}^{11}$$

of (4.4). By use of Proposition 2.5, we have

$$\Delta(\theta' \circ \eta_{23}) = \Delta(\theta') \circ \eta_{21} = H\Delta(E\theta) \circ \eta_{21} = 0$$

and
$$\Delta(\sigma_{11} \circ \nu_{18}^2) = \Delta(\sigma_{11}) \circ \nu_{16}^2$$
$$= \nu_5 \circ \varepsilon_8 \circ \nu_{16}^2 + \nu_5 \circ \bar{\nu}_8 \circ \nu_{16}^2 \qquad \text{by (7.17)}$$
$$= \nu_5 \circ E^5\nu' \circ \bar{\nu}_{11} \circ \nu_{19} + \nu_5 \circ \bar{\nu}_8 \circ \nu_{16}^2 \qquad \text{by (7.12)}$$
$$= 2\nu_5^2 \circ \bar{\nu}_{11} \circ \nu_{19} + \nu_5 \circ \bar{\nu}_8 \circ \nu_{16}^2$$
$$= \nu_5 \circ \bar{\nu}_8 \circ \nu_{16}^2 .$$

By Lemma 10.1, $\nu_5 \circ \bar{\nu}_8 \circ \nu_{16}^2 \equiv 2(\nu_5 \circ \kappa_8)$ mod $\nu_5^2 \circ \zeta_{11}$. By (10.7), Lemma 5.14, (7.16), and Theorem 7.3,

$$\nu_5^2 \circ \zeta_{11} = \nu_5 \circ E^2(\nu_6 \circ \zeta_9) = \nu_5 \circ 2E^2(\sigma'' \circ \sigma_{13})$$
$$= \nu_5 \circ 4E\sigma' \circ \sigma_{15} = 8(\nu_5 \circ \sigma_8) \circ \sigma_{15} = 0 .$$

Thus we have the relation $\Delta(\sigma_{11} \circ \nu_{18}^2) = 2\nu_5 \circ \kappa_8$, and

$$E\pi_{22}^5 = \{\nu_6 \circ \kappa_9\} \oplus \{\bar{\mu}_6\} \oplus \{\eta_6 \circ \mu_7 \circ \sigma_{16}\} \approx Z_2 \oplus Z_2 \oplus Z_2,$$

since $\theta' \circ \eta_{23}$ and $\sigma_{11} \circ \nu_{18}^2$ generate π_{24}^{11}, by Theorem 7.7. We have also

(12.7). <u>the kernel of</u> $\Delta : \pi_{24}^{11} \to \pi_{22}^5$ <u>is</u> $\{\theta' \circ \eta_{23}\} = \{H(\Delta(E\theta) \circ \eta_{23})\} \approx Z_2$.

$\pi_{23}^{11} = \{\theta'\} \approx Z_2$, by Theorem 7.6. It follows from Lemma 12.11 and from the above exact sequence that

$$\pi_{23}^6 = \{\Delta(E\theta)\} \oplus E\pi_{22}^5 \approx Z_2 \oplus Z_2 \oplus Z_2 \oplus Z_2.$$

Since $\pi_{25}^{13} = \{E\theta\} \approx Z_2$, then it follows from the exactness of (4.4) that

$$E\pi_{23}^6 = E^2\pi_{22}^5 \approx Z_2 \oplus Z_2 \oplus Z_2,$$

(12.8) $E : \pi_{24}^6 \to \pi_{25}^7$ <u>is onto</u>

and that the following sequence is exact:

$$0 \longrightarrow E\pi_{23}^6 \longrightarrow \pi_{24}^7 \xrightarrow{H} \pi_{24}^{13} \xrightarrow{\Delta} \pi_{22}^6$$

It follows from (12.4) and Theorem 7.4 that the last homomorphism Δ has the kernel $\{4\zeta_{13}\} = \{\eta_{13}^2 \circ \mu_{15}\} \approx Z_2$. As is seen in the proof of Lemma 12.11, $H(\sigma' \circ \eta_{14} \circ \mu_{15}) = \eta_{13}^2 \circ \mu_{15}$. Since $2(\sigma' \circ \eta_{14} \circ \mu_{15}) = 0$, then it follows from the above sequence that

$$\begin{aligned} \pi_{24}^7 &= \{\sigma' \circ \eta_{14} \circ \mu_{15}\} \oplus E^2\pi_{22}^5 \\ &= \{\sigma' \circ \eta_{14} \circ \mu_{15}\} \oplus \{\nu_7 \circ \kappa_{10}\} \oplus \{\bar{\mu}_7\} \oplus \{\eta_7 \circ \mu_8 \circ \sigma_{17}\} \\ &\approx Z_2 \oplus Z_2 \oplus Z_2 \oplus Z_2. \end{aligned}$$

The result on π_{25}^8 follows from this, (5.15) and Theorem 7.4.

By (12.5), we have an exact sequence

$$\pi_{27}^{17} \xrightarrow{\Delta} \pi_{25}^8 \xrightarrow{E} \pi_{26}^9 \longrightarrow 0.$$

$\pi_{27}^{17} = \{\eta_{17} \circ \mu_{18}\} \approx Z_2$. By Proposition 2.5 and by (5.16), we have that $\Delta(\eta_{17} \circ \mu_{18}) = E\sigma' \circ \eta_{15} \circ \mu_{16}$. It follows the result on π_{26}^9 and that the above Δ is an isomorphism into. By the exactness of (4.4), we have

(12.9). $E : \pi_{26}^8 \to \pi_{27}^9$ <u>is onto</u>.

<u>Proof of Theorem</u> 12.8.

The result on π_{20}^2 is a direct consequence of (5.2) and Theorem 12.7.

Consider the exact sequence

$$\pi_{22}^5 \xrightarrow{\Delta} \pi_{20}^2 \xrightarrow{E} \pi_{21}^3 \xrightarrow{H} \pi_{21}^5$$

of (4.4). By Lemma 5.2, $\Delta(\bar{\mu}_5) = \Delta(\eta_5 \circ \mu_6 \circ \sigma_{15}) = 0$. Since $E^2\bar{\varepsilon}' = E^2\nu' \circ \kappa_8 = 2\iota_5 \circ \nu_5 \circ \kappa_8$, by Lemma 12.3, then it follows from Lemma 5.7 that $E(\eta_2 \circ \bar{\varepsilon}') = 0$. Since $\pi^5_{22} = \{\nu_5 \circ \kappa_8\} \oplus \{\bar{\mu}_5\} \oplus \{\eta_5 \circ \mu_6 \circ \sigma_{15}\} \approx Z_4 \oplus Z_2 \oplus Z_2$, by Theorem 12.7, then it follows that

(12.10). $\Delta(\nu_5 \circ \kappa_8) = \pm(\eta_2 \circ \bar{\varepsilon}')$ <u>and the kernel of</u> $\Delta : \pi^5_{22} \to \pi^2_{20}$ <u>is</u> $\{\bar{\mu}_5\} \oplus \{\eta_5 \circ \mu_6 \circ \sigma_{15}\} \approx Z_2 \oplus Z_2$. Then

$E\pi^2_{20} = \{\eta_3 \circ \bar{\mu}_4\} \oplus \{\eta_3^2 \circ \mu_5 \circ \sigma_{14}\} \approx Z_2 \oplus Z_2$. $\pi^5_{21} = \{\mu_5 \circ \sigma_{14}\} \oplus \{\eta_5 \circ \bar{\varepsilon}_6\} \approx Z_2 \oplus Z_2$ by Theorem 12.6. We have

$$H(\mu' \circ \sigma_{14}) = \mu_5 \circ \sigma_{14} \quad \text{by Proposition 2.2 and (7.7)}$$

and $$H(\nu' \circ \bar{\varepsilon}_6) = \eta_5 \circ \bar{\varepsilon}_6 \quad \text{by Proposition 2.2 and (5.3).}$$

$2\mu' \circ \sigma_{14} = \eta_3^2 \circ \mu_5 \circ \sigma_{14}$ by (7.7) and $2(\nu' \circ \bar{\varepsilon}_6) = \nu' \circ 2\bar{\varepsilon}_6 = 0$. Then it follows from the exactness of the above sequence that

$$\pi^3_{21} = \{\mu' \circ \sigma_{14}\} \oplus \{\nu' \circ \bar{\varepsilon}_6\} \oplus \{\eta_3 \circ \bar{\mu}_4\} \approx Z_4 \oplus Z_2 \oplus Z_2.$$

The result on π^4_{22} is a consequence of this result, Theorem 10.5 and (5.6).

Next consider the exact sequence

$$\pi^9_{24} \xrightarrow{\Delta} \pi^4_{22} \xrightarrow{E} \pi^5_{23} \xrightarrow{H} \pi^9_{23} \xrightarrow{\Delta} \pi^4_{21}$$

of (4.4). By use of Proposition 2.5, we have

$$\Delta(\bar{\varepsilon}_9) = \Delta(\iota_9) \circ \bar{\varepsilon}_7 = E\nu' \circ \bar{\varepsilon}_7 \quad \text{by (5.8),}$$

$$\Delta(\sigma_9 \circ \bar{\nu}_{16}) = \Delta(\sigma_9) \circ \bar{\nu}_{14} = \nu_4 \circ \sigma' \circ \bar{\nu}_{12} + E(\varepsilon' \circ \bar{\nu}_{11}) \quad \text{by (7.16),}$$

$$\Delta(\sigma_9 \circ \varepsilon_{16}) = \Delta(\sigma_9) \circ \varepsilon_{14} = \nu_4 \circ \sigma' \circ \varepsilon_{12} + E(\varepsilon' \circ \varepsilon_{11}) \quad \text{by (7.16)}$$

and
$$\begin{aligned} \Delta(4\rho') &= \Delta(E^3\rho''') = \Delta(\iota_9) \circ E\rho''' \\ &= \pm(2\nu_4 \circ E\rho''' - E(\nu' \circ \rho''')) \quad \text{by (5.8)} \\ &= \pm(4\nu_4 \circ \rho'' - E(\nu' \circ \rho''')). \end{aligned}$$

The last relation shows that $\Delta(\rho')$ is an element of order 8. Thus $\Delta(\rho')$ is of the form $x\nu_4 \circ \rho'' +$ (linear combination of other generators) for an odd integer x. $\pi^9_{24} = \{\rho'\} \oplus \{\sigma_9 \circ \bar{\nu}_{16}\} \oplus \{\sigma_9 \circ \varepsilon_{16}\} \oplus \{\bar{\varepsilon}_9\} \approx Z_{16} \oplus Z_2 \oplus Z_2 \oplus Z_2$ by Theorem 10.5. Then it follows that

(12.11) <u>the kernel of</u> $\Delta : \pi^9_{24} \to \pi^4_{22}$ <u>is</u> $\{8\rho'\} \approx Z_2$ <u>and</u>

$$E\pi^4_{22} = \{\nu_5 \circ \bar{\varepsilon}_8\} \oplus \{E^2\mu' \circ \sigma_{16}\} \oplus \{\eta_5 \circ \bar{\mu}_6\} \approx Z_2 \oplus Z_4 \oplus Z_2 .$$

By (12.6), we have the following exact sequence :

$$0 \longrightarrow E\pi^4_{22} \longrightarrow \pi^5_{23} \xrightarrow{H} \{8\sigma_9^2\} .$$

By Proposition 2.2, Lemma 6.7 and (7.14),

$$H(\zeta_5 \circ \sigma_{16}) = 8\sigma_9^2 \quad \text{and} \quad 2(\zeta_5 \circ \sigma_{16}) = \pm E^2\mu' \circ \sigma_{16}.$$

Then it follows that

$$\pi_{23}^5 = \{\zeta_5 \circ \sigma_{16}\} \oplus \{\nu_5 \circ \bar{\varepsilon}_8\} \oplus \{\eta_5 \circ \bar{\mu}_6\} \approx Z_8 \oplus Z_2 \oplus Z_2$$

Consider the exact sequence

$$\pi_{24}^{11} \xrightarrow{\Delta} \pi_{23}^5 \xrightarrow{E} \pi_{24}^6 \xrightarrow{H} \pi_{24}^{11} \xrightarrow{\Delta} \pi_{22}^5$$

of (4.4). The kernel of $\Delta : \pi_{24}^{11} \to \pi_{22}^5$ is generated by $\theta' \circ \eta_{23} = H(\Delta(E\theta) \circ \eta_{23})$, since (12.7). Then it follows that

$$\pi_{24}^6 = E\pi_{23}^5 \oplus \{\Delta(E\theta) \circ \eta_{23}\} \approx E\pi_{23}^5 \oplus Z_2 .$$

We have, by use of Proposition 2.5,

$$\begin{aligned} \Delta(\kappa_{11}) &= \Delta(\iota_{11}) \circ \kappa_9 = \nu_5 \circ \eta_8 \circ \kappa_9 && \text{by (5.10)} \\ &= \nu_5 \circ \bar{\varepsilon}_8 && \text{by (10.23)} \end{aligned}$$

and

$$\begin{aligned} \Delta(\sigma_{11}^2) &= \Delta(\sigma_{11}) \circ \sigma_{16} \\ &= \nu_5 \circ (\bar{\nu}_8 + \varepsilon_8) \circ \sigma_{16} = 0 && \text{by (7.17) and Lemma 10.7.} \end{aligned}$$

Thus we have

$$\pi_{24}^6 = \{\Delta(E\theta) \circ \eta_{23}\} \oplus \{\zeta_6 \circ \sigma_{17}\} \oplus \{\eta_6 \circ \bar{\mu}_7\} \approx Z_2 \oplus Z_8 \oplus Z_2$$

and

(12.12). <u>the kernel of</u> $\Delta : \pi_{25}^{11} \to \pi_{23}^5$ <u>is</u> $\{\sigma_{11}^2\} \approx Z_{16}$.

In the exact sequence

$$\pi_{26}^{13} \xrightarrow{\Delta} \pi_{24}^6 \xrightarrow{E} \pi_{25}^7$$

of (4.4), E is onto by (12.8) and $\pi_{26}^{13} = \{E\theta \circ \eta_{25}\} \approx Z_2$ by Theorem 7.7. $\Delta(E\theta \circ \eta_{25}) = \Delta(E\theta) \circ \eta_{23}$ by Proposition 2.5. Thus we have

$$\pi_{25}^7 = \{\zeta_7 \circ \sigma_{18}\} \oplus \{\eta_7 \circ \bar{\mu}_8\} \approx Z_8 \oplus Z_2 ,$$

and that the homomorphism Δ is an isomorphism into. It follows from the exactness of (4.4) that

(12.13). $E : \pi_{25}^6 \to \pi_{26}^7$ <u>is onto</u>.

By (5.15) and Theorem 7.4,

$$\pi_{26}^8 = \{\sigma_8 \circ \zeta_{15}\} \oplus \{\zeta_8 \circ \sigma_{19}\} \oplus \{\eta_8 \circ \bar{\mu}_9\} \approx Z_8 \oplus Z_8 \oplus Z_2 .$$

In the exact sequence

$$\pi_{28}^{17} \xrightarrow{\Delta} \pi_{26}^8 \xrightarrow{E} \pi_{27}^9$$

of (4.4), E is onto by (12.9). $\pi_{28}^{17} = \{\zeta_{17}\} \approx Z_8$ by Theorem 7.4. By using the following Lemma 12.12, we have

$$\begin{aligned} \Delta(\zeta_{17}) &= \Delta(\iota_{17}) \circ \zeta_{15} = \pm (2\sigma_8 - E\sigma') \circ \zeta_{15} && \text{by (5.16)} \\ &= \pm (2\sigma_8 \circ \zeta_{15} - x\zeta_8 \circ \sigma_{19}) \end{aligned}$$

for an odd integer x. Then it follows that

$$\pi^9_{27} = \{\sigma_9 \circ \zeta_{16}\} \oplus \{\eta_9 \circ \bar{\mu}_{10}\} \approx Z_8 \oplus Z_2$$

and that the above Δ is an isomorphism onto. It follows from the exactness of (4.4) that

(12.14). $E : \pi^8_{27} \to \pi^9_{28}$ <u>is onto</u>.

<u>Lemma</u> 12.12. $\sigma' \circ \zeta_{14} = x\zeta_7 \circ \sigma_{18}$ <u>for an odd integer</u> x.

<u>Proof</u>. By Lemma 9.1, for an odd integer y, the element $y\sigma' \circ \zeta_{14}$ is contained in $\sigma' \circ \{8\iota_{14}, \nu_{14}, \sigma_{17}\}$. By Proposition 1.4,

$$\sigma' \circ \{8\iota_{14}, \nu_{14}, \sigma_{17}\} = \{\sigma', 8\iota_{14}, \nu_{14}\} \circ -\sigma_{18}.$$

Thus $y\sigma' \circ \zeta_{14} = -\alpha \circ \sigma_{18}$ for an element α of $\{\sigma', 8\iota_{14}, \nu_{14}\} \subset \pi^7_{18}$. $E^\infty\alpha \in \langle E^\infty\sigma', 8\iota, \nu \rangle = \langle 2\sigma, 8\iota, \nu \rangle$. By (9.2), $E^\infty\alpha = \zeta = E^\infty\zeta_7$. By Theorem 7.4, it follows that $\alpha = \zeta_7 + z\bar{\nu}_7 \circ \nu_{15}$ for $z = 0$ or 1. Since $\nu_{15} \circ \sigma_{18} = 0$, by (7.20), then we have the equality

$$y\sigma' \circ \zeta_{14} = -(\zeta_7 + z\bar{\nu}_7 \circ \nu_{15}) \circ \sigma_{18} = -\zeta_7 \circ \sigma_{18}.$$

By taking an odd integer x such that $xy \equiv -1 \pmod 8$, we obtain the equality of this lemma q.e.d.

<u>Proof of Theorem</u> 12.9.

The result on π^2_{21} is a direct consequence of Theorem 12.8 and (5.2).

Consider the exact sequence

$$\pi^5_{23} \xrightarrow{\Delta} \pi^2_{21} \xrightarrow{E} \pi^3_{22} \xrightarrow{H} \pi^5_{22} \xrightarrow{\Delta} \pi^2_{20}$$

of (4.4). By Lemma 5.2, $\Delta(\eta_5 \circ \bar{\mu}_6) = 0$ and $2\Delta(\zeta_5 \circ \sigma_{16}) = \Delta(\pm E^2\bar{\mu}' \circ \sigma_{16}) = 2(\eta_2 \circ \mu' \circ \sigma_{14})$. It follows that

$$\Delta(\zeta_5 \circ \sigma_{16}) \equiv \pm(\eta_2 \circ \mu' \circ \sigma_{15}) \bmod \{\eta_2 \circ \nu' \circ \bar{\varepsilon}_6\} + \{\eta^2_2 \circ \bar{\mu}_4\}.$$

By Proposition 2.5 and by Lemma 5.7, $\Delta(\nu_5 \circ \bar{\varepsilon}_8) = \eta_2 \circ \nu' \circ \bar{\varepsilon}_6$. Since π^5_{23} is generated by the elements $\zeta_5 \circ \sigma_{16}$, $\nu_5 \circ \bar{\varepsilon}_8$ and $\eta_5 \circ \bar{\mu}_6$, by Theorem 12.8, then it follows from the exactness of the above sequence that

$$E\pi^2_{21} = \{\eta^2_3 \circ \bar{\mu}_5\} \approx Z_2.$$

By Proposition 2.2 and (5.3),

$$H(\nu' \circ \mu_6 \circ \sigma_{15}) = H(\nu') \circ \mu_6 \circ \sigma_{15} = \eta_5 \circ \mu_6 \circ \sigma_{15}.$$

By Lemma 12.4,

$$H(\bar{\mu}') \equiv \mu_5 \bmod E^3\pi^2_{19}.$$

By Theorem 12.7 and by its proof, we have that

$$E^3\pi^2_{19} = \{\eta_5 \circ \mu_6 \circ \sigma_{15}\} + \{4\nu_5 \circ \kappa_8\} = \{\eta_5 \circ \mu_6 \circ \sigma_{15}\}.$$

Then it follows from the exactness of the above sequence and from (12.10) that the above two H-images span $H\pi_{22}^3 \approx Z_2 \oplus Z_2$. By Lemma 12.4, $2\bar{\mu}' = \eta_3^2 \circ \bar{\mu}_5$. Obviously, $2\nu' \circ \mu_6 \circ \sigma_{15} = \nu' \circ 2\mu_6 \circ \sigma_{15} = 0$. Therefore, we conclude from the above exact sequence that

$$\pi_{22}^3 = \{\bar{\mu}'\} \oplus \{\nu' \circ \mu_6 \circ \sigma_{15}\} \approx Z_4 \oplus Z_2 .$$

The result on π_{23}^4 is a consequence of this result, Theorem 12.6 and (5.6).

Next we have, by use of Proposition 2.5,

$$\begin{aligned}
\Delta(\sigma_9 \circ \nu_{16}^3) &= \Delta(\sigma_9) \circ \nu_{14}^3 \\
&= \nu_4 \circ \sigma' \circ \nu_{14}^3 + E\varepsilon' \circ \nu_{14}^3 && \text{by } (7.16) \\
&= \nu_4^2 \circ \sigma_{10} \circ \nu_{17}^2 + E(\varepsilon' \circ \nu_{13}^3) && \text{by } (7.19) \\
&= \nu_4 \circ \eta_7 \circ \bar{\varepsilon}_8 + E(\varepsilon' \circ \nu_{12}^3) && \text{by Lemma 12.10,} \\
\Delta(\sigma_9 \circ \mu_{16}) &= \Delta(\sigma_9) \circ \mu_{14} \\
&= \nu_4 \circ \sigma' \circ \mu_{14} + E(\varepsilon' \circ \mu_{13}) \\
\Delta(\sigma_9 \circ \eta_{16} \circ \varepsilon_{17}) &= \Delta(\sigma_9) \circ \eta_{14} \circ \varepsilon_{15} \\
&= \nu_4 \circ \sigma' \circ \eta_{14} \circ \varepsilon_{15} + E\varepsilon' \circ \eta_{14} \circ \varepsilon_{15} && \text{by } (7.16) \\
&= \nu_4 \circ E\zeta' + E(\varepsilon' \circ \eta_{13} \circ \varepsilon_{14}) && \text{by } (12.4)
\end{aligned}$$

and

$$\begin{aligned}
\Delta(\mu_9 \circ \sigma_{18}) &= \Delta(\iota_9) \circ \mu_7 \circ \sigma_{16} \\
&= E\nu' \circ \mu_7 \circ \sigma_{16} && \text{by } (5.8).
\end{aligned}$$

By Theorem 12.6, the image of $\Delta : \pi_{25}^9 \to \pi_{23}^4$ is generated by the above elements. It follows from the exactness of the sequence (4.4) that

$$E\pi_{23}^4 = \{\nu_5 \circ \mu_8 \circ \sigma_{17}\} + \{E^2\bar{\mu}'\} \approx Z_2 \oplus Z_4 .$$

By (12.11), we have the following exact sequence:

$$0 \longrightarrow E\pi_{23}^4 \longrightarrow \pi_{24}^5 \longrightarrow \{8\rho'\} \to 0$$

By Lemma 12.4, $H(\bar{\zeta}_5) = 8\rho'$ and $2\bar{\zeta}_5 = E^2\bar{\mu}'$. Thus we have

$$\pi_{24}^5 = \{\bar{\zeta}_5\} \oplus \{\nu_5 \circ \mu_8 \circ \sigma_{17}\} \approx Z_8 \oplus Z_2 .$$

Next we prove

(12.15). $\Delta(E^2\rho') = \Delta(\bar{\varepsilon}_{11}) = 0.$

$$\begin{aligned}
\Delta(\bar{\varepsilon}_{11}) &= \Delta(\iota_{11}) \circ \bar{\varepsilon}_9 = \nu_5 \circ \eta_8 \circ \bar{\varepsilon}_9 && \text{by } (5.10) \\
&= E\Delta(\sigma_9 \circ \nu_{16}^3) - E^2(\varepsilon' \circ \nu_{13}^3) \\
&= 2(\nu_5 \circ \sigma_8) \circ \nu_{13}^3 && \text{by } (7.10) \\
&= \nu_5 \circ \sigma_8 \circ 2\nu_{13}^3 = 0 .
\end{aligned}$$

By Proposition 2.5, Proposition 2.7, and Lemma 10.9,

$$H\Delta(2\rho_{13}) = H\Delta(E^4\rho') = H\Delta(\iota_{13}) \circ E^2\rho' = \pm 2E^2\rho' .$$

By Theorem 10.5, $\pi_{26}^{11} = \{E^2\rho'\} \oplus \{\bar{\varepsilon}_{11}\} \approx Z_{16} \oplus Z_2$. Since $H\Delta(2\rho_{13}) = \pm 2E^2\rho'$ is an element of order 8, then $H\Delta(\rho_{13})$ is an element of order 16. Thus $H\Delta(\rho_{13}) = xE^2\rho' + y\bar{\varepsilon}_{11}$ for some integers x and y such that $x \not\equiv 0 \pmod 2$. Let z be an integer such that $xz \equiv 1 \pmod{16}$, then

$$\Delta(E^2\rho') = \Delta(xzE^2\rho') = z\Delta H\Delta(\rho_{13}) - yz\Delta(\bar{\varepsilon}_{11}) = 0 .$$

Thus (12.15) is proved.

By this proof, we have that $\Delta\pi_{26}^{11} = 0$. It follows from the exactness of the sequence (4.4) and from (12.12) that the following sequence is exact.

$$0 \to \pi_{24}^5 \xrightarrow{E} \pi_{25}^6 \xrightarrow{H} \{\sigma_{11}^2\} \to 0.$$

By Lemma 12.5, $H(\bar{\sigma}_6)$ generates $\{\sigma_{11}^2\} \approx Z_{16}$. By Proposition 2.5, and Proposition 2.7, $H(\Delta(\sigma_{13}^2)) = \pm 2\sigma_{11}^2$ and it generates $\{2\sigma_{11}^2\}$. Thus $H(2\bar{\sigma}_6) = H(x\Delta(\sigma_{13}^2))$ for an odd integer x. By the exactness of the above sequence, we have that $2\bar{\sigma}_6 \equiv x\Delta(\sigma_{13}^2) \bmod E\pi_{24}^5$. Since $8\pi_{24}^5 = 0$, then it follows that $16\bar{\sigma}_6 = 8x\Delta(\sigma_{13}^2)$. By Proposition 2.5, and by (7.25), we have

$$16\bar{\sigma}_6 = 8x\Delta(\sigma_{13}) \circ \sigma_{18} = x\nu_6 \circ \mu_9 \circ \sigma_{18} = \nu_6 \circ \mu_9 \circ \sigma_{18}.$$

Therefore $\bar{\sigma}_6$ is of order 32 and we have that

$$\pi_{25}^6 = \{\bar{\sigma}_6\} \oplus \{\bar{\zeta}_6\} \approx Z_{32} \oplus Z_8 .$$

Next we prove

<u>Lemma</u> 12.13. $\nu_6 \circ \pi_{25}^9 = \{16\bar{\sigma}_6\}$.

<u>Proof</u>. By Theorem 12.6, it is sufficient to prove that the compositions of ν_6 with the elements $\sigma_9 \circ \nu_{16}^3$, $\sigma_9 \circ \mu_{16}$, $\sigma_9 \circ \eta_{16} \circ \varepsilon_{17}$ and $\mu_9 \circ \sigma_{18}$ are in $\{16\bar{\sigma}_6\}$. First,

$$\nu_6 \circ \mu_9 \circ \sigma_{18} = 16\bar{\sigma}_6$$

and

$$\nu_6 \circ \sigma_9 \circ \nu_{16}^3 = \bar{\varepsilon}_6 \circ \eta_{21} \circ \nu_{22} \quad \text{by Lemma 12.10}$$
$$= 0 \quad \text{by (5.9)}.$$

Next consider the secondary composition $\{\nu_5, 2\nu_8, \nu_{11}\}_1$. By Proposition 2.6 and by (5.13), we have $H\{\nu_5, 2\nu_8, \nu_{11}\}_1 = \nu_9^2$. Since $H(\nu_5 \circ \sigma_8) = \nu_9^2$, then it follows from Theorem 7.3 that any element of this secondary composition is of a form $x(\nu_5 \circ \sigma_8) + y\eta_5 \circ \mu_6$ for an odd integer x and an integer y. By (6.1) and Proposition 1.4,

$$\nu_5 \circ \varepsilon_8 \in \nu_5 \circ E\{2\nu_7, \nu_{10}, \eta_{13}\}$$
$$= \{\nu_5, 2\nu_8, \nu_{11}\}_1 \circ \eta_{15} .$$

Thus, $\nu_5 \circ \varepsilon_8 = \nu_5 \circ \sigma_8 \circ \eta_{15} + y\eta_5^2 \circ \mu_7$. $E^2(\nu_5 \circ \varepsilon_8) = E\Delta(\nu_{13}^2) = 0$, by (7.18) and $E^2(\nu_5 \circ \sigma_8 \circ \eta_{15}) = \nu_7 \circ \sigma_{10} \circ \eta_{17} = \nu_7 \circ \eta_{10} \circ \sigma_{11} = 0$ by (5.9). But $E^2(\eta_5^2 \circ \mu_7) = \eta_7^2 \circ \mu_9 \neq 0$. Thus $y \equiv 0 \pmod 2$ and we have that

$$\nu_5 \circ \varepsilon_8 = \nu_5 \circ \sigma_8 \circ \eta_{15} .$$

By Lemma 12.10 and (5.9),

$$\nu_6 \circ \sigma_9 \circ \eta_{16} \circ \varepsilon_{17} = \nu_6 \circ \varepsilon_9 \circ \varepsilon_{17} = \nu_6 \circ \eta_9 \circ \bar{\varepsilon}_{10} = 0 .$$

Finally,

$$\nu_6 \circ \sigma_9 \circ \mu_{16} \in \nu_6 \circ \sigma_9 \circ \{\eta_{16}, 2\iota_{17}, 8\sigma_{17}\}$$
$$\subset \{\nu_6 \circ \sigma_9 \circ \eta_{16}, 2\iota_{17}, 8\sigma_{17}\} \quad \text{by Proposition 1.2}$$

and $\nu_6 \circ E^6\{\varepsilon_3, 2\iota_{11}, 8\sigma_{11}\} \subset \{\nu_6 \circ \varepsilon_9, 2\iota_{17}, 8\sigma_{17}\}$ by Propositions 1.2 and 1.3.

Since $\nu_6 \circ \varepsilon_9 = \nu_8 \circ \sigma_9 \circ \eta_{16}$, then $\{\nu_6 \circ \varepsilon_9, 2\iota_{17}, 8\sigma_{17}\} = \{\nu_6 \circ \sigma_9 \circ \eta_{16}, 2\iota_{17}, 8\sigma_{17}\}$ and this secondary composition is a coset of

$$\begin{aligned} \nu_6 \circ \varepsilon_9 \circ \pi_{25}^{17} + \pi_{18}^6 \circ 8\sigma_{18} &= \{\nu_6 \circ \varepsilon_9 \circ \varepsilon_{17}\} + \{\nu_6 \circ \varepsilon_9 \circ \bar{\nu}_{17}\} + 8\pi_{18}^6 \circ \sigma_{18} \\ &= \{\nu_6 \circ \eta_9 \circ \bar{\varepsilon}_{10}\} + \{\nu_6 \circ \mu_9 \circ \sigma_{18}\} \quad \text{by Lemma 12.10} \\ &= \{16\bar{\sigma}_6\} . \quad \text{by (5.9)}. \end{aligned}$$

Thus $\nu_6 \circ \sigma_9 \circ \mu_{16}$ is an element of

$$\begin{aligned} \nu_6 \circ E^6\{\varepsilon_3, 2\iota_{11}, 8\sigma_{11}\} &\subset \nu_6 \circ E^6\pi_{19}^3 \\ &= \{\nu_6 \circ \eta_9 \circ \bar{\varepsilon}_{10}\} + \{\nu_6 \circ \mu_9 \circ \sigma_{18}\} = \{16\bar{\sigma}_6\} . \end{aligned}$$

Consequently the lemma is proved. q.e.d.

As a corollary we have

(12.16). $\Delta(\kappa_{13} + 4x\sigma_{13}^2) = 0$ <u>for some integer</u> x.

<u>Proof</u>. Since $2\kappa_{13} \equiv 0 \bmod \nu_{13} \circ \zeta_{17} = 8\sigma_{13}^2$ (see the proof of Theorem 10.3), then $2(\kappa_{13} + 4y\sigma_{13}^2) = E^3(2\iota_{10} \circ (\kappa_{10} + 4y\sigma_{10}^2)) = 0$ for some integer y. By Theorem 10.3, $E^3 : \pi_{24}^{10} \to \pi_{27}^{13}$ is an isomorphism onto, then it follows that $2\iota_{10} \circ (\kappa_{10} + 4y\sigma_{10}^2) = 0$.

$$\begin{aligned} \Delta(\kappa_{13} + 4y\sigma_{13}^2) &= \Delta(\iota_{13}) \circ (\kappa_{11} + 4y\sigma_{11}^2) \\ &\in \{\nu_6, \eta_9, 2\iota_{10}\} \circ (\kappa_{11} + 4y\sigma_{11}^2) \quad \text{by Lemma 5.10} \\ &= \nu_6 \circ \{\eta_9, 2\iota_{10}, \kappa_{10} + 4y\sigma_{11}^2\} \quad \text{by Proposition 1.4} \\ &\subset \nu_6 \circ \pi_{25}(S^9) . \end{aligned}$$

Since $\kappa_{11} + 4y\sigma_{11}^2$ is in the 2-primary component, then $\Delta(\kappa_{13} + 4y\sigma_{13}^2)$ is in $\nu_6 \circ \pi_{25}^9 = \{16\bar{\sigma}_6\} = \{8\Delta(\sigma_{13}^2)\}$, by Lemma 12.13. Thus $\Delta(\kappa_{13} + 4y\sigma_{13}^2) = 8z\Delta(\sigma_{13}^2)$ for an integer z. (12.16) holds for $x = y - 2z$. q.e.d.

Now, we consider the exact sequence

$$\pi_{27}^{13} \xrightarrow{\Delta} \pi_{25}^{6} \xrightarrow{E} \pi_{26}^{7}$$

of (4.4), where E is onto by (12.13). Since the result (12.16), as $\Delta(\sigma_{13}^2)$ is an element of order 16 and as σ_{13}^2 and κ_{13} generate π_{27}^{13} by Theorem, 10.3, then we have that the kernel of E is a subgroup of order 16. Then the result

$$\pi_{26}^{7} = \{\bar{\zeta}_7\} \oplus \{\bar{\sigma}_7\} \approx Z_8 \oplus Z_2$$

is proved by

(12.17). $2\bar{\sigma}_7 = 0$ <u>and thus</u> $2\bar{\sigma}_n = 2\bar{\sigma} = 0$ <u>for</u> $n \geq 7$.

For, $2\bar{\sigma}_7 \in 2E\{\nu_6, \bar{\nu}_9 + \varepsilon_9, \sigma_{17}\}_1$

$$\subset -2\{\nu_7, \bar{\nu}_{10} + \varepsilon_{10}, \sigma_{18}\}_1 \qquad \text{by Proposition 1.3}$$

$$= -\{\nu_7, 2(\bar{\nu}_{10} + \varepsilon_{10}), 5_{18}\}_1 \qquad \text{by Proposition 1.6}$$

$$= -\{\nu_7, 0, \sigma_{18}\}_1$$

$$= \nu_7 \circ E\pi_{25}^9 + \pi_{19}^7 \circ \sigma_{19} \qquad \text{by Proposition 1.2}$$

$$= \{E16\bar{\sigma}_6\} = 0 \quad \text{by Lemma 12.13 and Theorem 7.6.}$$

It follows from (5.15) and Theorem 7.6 that

$$\pi_{27}^{8} = \{\bar{\zeta}_8\} \oplus \{\bar{\sigma}_8\} \approx Z_8 \oplus Z_2$$

In the exact sequence

$$\pi_{29}^{17} \xrightarrow{\Delta} \pi_{27}^{8} \xrightarrow{E} \pi_{28}^{9}$$

of (4.4), $\pi_{29}^{17} = 0$ by Theorem 7.6 and E is onto by (12.14). Thus E is an isomorphism onto and

$$\pi_{28}^{9} = \{\bar{\zeta}_9\} \oplus \{\bar{\sigma}_9\} \approx Z_8 \oplus Z_2.$$

iii). <u>The groups</u> π_{n+16}^n <u>and</u> π_{n+17}^n <u>for</u> $n \geq 10$.

Choose elements $\eta^{*\prime} \in \pi_{31}^{15}$, $\eta_{16}^* \in \pi_{32}^{16}$, $\nu_{16}^* \in \pi_{34}^{16}$ and $\xi_{12} \in \pi_{30}^{12}$ from the following secondary compositions.

$$\eta^{*\prime} \in \{\sigma_{15}, 4\sigma_{22}, \eta_{29}\}_1,$$

$$\eta_{16}^* \in \{\sigma_{16}, 2\sigma_{23}, \eta_{30}\}_1,$$

$$\nu_{16}^* \in \{\sigma_{16}, 2\sigma_{23}, \nu_{30}\}_1,$$

and

$$\xi_{12} \in \{\sigma_{12}, \nu_{19}, \sigma_{22}\}_1.$$

Denote that

$$\eta_n^* = E^{n-16}\eta_{16}^* \quad \text{for } n \geq 16 \quad \text{and} \quad \eta^* = E^{\infty}\eta_{16}^*,$$

$$\nu_n^* = E^{n-16}\nu_{16}^* \quad \text{for } n \geq 16 \quad \text{and} \quad \nu^* = E^{\infty}\nu_{16}^*$$

and

$$\xi_n = E^{n-12}\xi_{12} \quad \text{for } n \geq 12 \quad \text{and} \quad \xi = E^{\infty}\xi_{12}.$$

Lemma 12.14. $H(\eta^{*\prime}) = \eta_{29}^2$, $H(\eta_{16}^*) = \eta_{31}$, $H(\nu_{16}^*) \equiv \nu_{31} \bmod 2\nu_{31}$ and $H(\xi_{12}) \equiv \sigma_{23} \bmod 2\sigma_{23}$. $2\eta^{*\prime} = 0$, $2\eta_n^* = 2\eta^* = 0$ for $n \geq 16$, $8\nu_{16}^* \equiv 0 \bmod \sigma_{16} \circ \zeta_{23}$ and $16\xi_{12} \equiv \sigma_{12} \circ \zeta_{19} \bmod 2\sigma_{12} \circ \zeta_{19}$.

Proof. By Proposition 2.6,

$$H\{\sigma_{16}, 2\sigma_{23}, \nu_{30}\}_1 = \Delta^{-1}(2\sigma_{15}^2) \circ \nu_{31}.$$

By the last relation of (10.10), we have that $\Delta^{-1}(2\sigma_{15}^2)$ is a coset of $2\iota_{31}^{31}$ containing ι_{31}. Then the relation $H(\nu_{16}^*) \equiv \nu_{31} \bmod 2\nu_{31}$ is proved. Similarly, the first two equalities of the lemma are proved. The relation $H(\xi_{12}) \equiv \sigma_{23} \bmod 2\sigma_{23}$ is also proved by use of Proposition 2.6 and the last formula of (7.21). We have

$$\begin{aligned}
2\eta_{16}^* &\in \{\sigma_{16}, 2\sigma_{23}, \eta_{30}\}_1 \circ 2\iota_{32} \\
&= \sigma_{16} \circ E\{2\sigma_{22}, \eta_{29}, 2\iota_{30}\} && \text{by Proposition 1.4} \\
&\supset \sigma_{16}^2 \circ E\{2\iota_{29}, \eta_{29}, 2\iota_{30}\} && \text{by Proposition 1.2} \\
&\subset \sigma_{16}^2 \circ \pi_{32}^{30} = \{\sigma_{16}^2 \circ \eta_{30}^2\} \\
&= \{\sigma_{16} \circ (\bar{\nu}_{23} + \varepsilon_{23}) \circ \eta_{31}\} && \text{by Lemma 6.4} \\
&= \{(\bar{\nu}_{16} + \varepsilon_{16}) \circ \sigma_{24} \circ \eta_{31}\} && \text{by Proposition 3.1} \\
&= 0 && \text{by Lemma 10.7.}
\end{aligned}$$

The secondary composition $\sigma_{16} \circ E\{2\sigma_{22}, \eta_{29}, 2\iota_{30}\}$ is a coset of

$$\sigma_{16} \circ E\pi_{31}^{22} \circ 2\iota_{32} = \sigma_{16} \circ E2\pi_{31}^{22} = 0.$$

Thus $2\eta_{16}^* = 0$ and $2\eta_n^* = 2\eta^* = 0$. Similarly, we have the relation $2\eta^{*\prime} = 0$. Next,

$$\begin{aligned}
8\nu_{16}^* &\in \{\sigma_{16}, 2\sigma_{23}, \nu_{30}\}_1 \circ 8\iota_{34} \\
&= \sigma_{16} \circ E\{2\sigma_{22}, \nu_{29}, 8\iota_{32}\} && \text{by Proposition 1.4} \\
&\subset \sigma_{16} \circ E\pi_{33}^{22} = \{\sigma_{16} \circ \zeta_{23}\}.
\end{aligned}$$

Thus $8\nu_{16}^* \equiv 0 \bmod \sigma_{16} \circ \zeta_{23}$. By Proposition 1.4,

$$\begin{aligned}
16\xi_{12} &\in \{\sigma_{12}, \nu_{19}, \sigma_{22}\} \circ 16\iota_{30} \\
&= \sigma_{12} \circ \{\nu_{19}, \sigma_{22}, 16\iota_{29}\}.
\end{aligned}$$

We may identify $\{\nu_{19}, \sigma_{22}, 16\iota_{29}\}$ to the stable $\langle \nu, \sigma, 16\iota \rangle$. Then, by (9.3), it contains $x\zeta_{19}$ for an odd integer x. $\sigma_{12} \circ \{\nu_{19}, \sigma_{22}, 16\iota_{29}\}$ is a coset of

$$\sigma_{12} \circ \pi_{30}^{19} \circ 16\iota_{30} = \sigma_{12} \circ 16\pi_{30}^{19} = 0.$$

Therefore $16\xi_{12} = x\sigma_{12} \circ \zeta_{19}$ and the last relation of the lemma is proved.

q.e.d.

Lemma 12.15. i). There exists an element ω_{14} of π_{30}^{14} such that

$$H(\omega_{14}) = \nu_{27}.$$

ii). There exists an element ε^*_{12} of π_{29}^{12} such that

$$H(\varepsilon^*_{12}) = \nu_{23}^2 \quad \text{and} \quad E^2\varepsilon^*_{12} = \omega_{14} \circ \eta_{30}.$$

Proof. By (10.21), $\Delta(\nu_{27}) = 0$. By the exactness of the sequence (4.4), it follows the existence of an element ω_{14} such that $H(\omega_{14}) = \nu_{27}$.

Next, apply Proposition 11.14 for the case $n = 11$, $\alpha = \omega_{14}$, $\beta = \nu_{22}$ and $\gamma = \eta_{25}$. Then there exists an element ε^*_{12} of π_{29}^{12} such that

$$E^2\varepsilon^*_{12} = \omega_{14} \circ \eta_{30} \quad \text{and} \quad H(\varepsilon^*_{12}) \in \{\eta_{23}, \nu_{24}, \eta_{27}\}.$$

By Lemma 5.12, it follows that $H(\varepsilon^*_{12}) = \nu_{23}^2$. Then the lemma has been proved. q.e.d.

Denote that

$$\omega_n = E^{n-14}\,\omega_{14} \quad \text{for} \quad n \geq 14 \quad \text{and} \quad \omega = E^{\infty}\,\omega_{14}$$

and

$$\varepsilon^*_n = E^{n-12}\,\varepsilon^*_{12} \quad \text{for} \quad n \geq 12 \quad \text{and} \quad \varepsilon^* = E^{\infty}\,\varepsilon^*_{12}.$$

By use of the elements $\eta^{*\prime}$, η^*_n, ω_n and ε^*_n, our results are stated as follows.

Theorem 12.16. $\pi_{26}^{10} = \{\Delta(\sigma_{21})\} \oplus \{\sigma_{10} \circ \mu_{17}\} \approx Z_{16} \oplus Z_2$,

$\pi_{n+16}^{n} = \{\sigma_n \circ \mu_{n+7}\} \approx Z_2$ for $n = 11, 12, 13$,

$\pi_{30}^{14} = \{\omega_{14}\} \oplus \{\sigma_{14} \circ \mu_{21}\} \approx Z_8 \oplus Z_2$,

$\pi_{31}^{15} = \{\eta^{*\prime}\} \oplus \{\omega_{15}\} \oplus \{\sigma_{15} \circ \mu_{22}\} \approx Z_2 \oplus Z_2 \oplus Z_2$,

$\pi_{32}^{16} = \{\eta^*_{16}\} \oplus \{E\eta^{*\prime}\} \oplus \{\omega_{16}\} \oplus \{\sigma_{16} \circ \mu_{23}\} \approx Z_2 \oplus Z_2 \oplus Z_2 \oplus Z_2$,

$\pi_{33}^{17} = \{\eta^*_{17}\} \oplus \{\omega_{17}\} \oplus \{\sigma_{17} \circ \mu_{24}\} \approx Z_2 \oplus Z_2 \oplus Z_2$,

$\pi_{n+16}^{n} = \{\omega_n\} \oplus \{\sigma_n \circ \mu_{n+7}\} \approx Z_2 \oplus Z_2$ for $n \geq 18$ and

$(G_{16};2) = \{\omega\} \oplus \{\sigma \circ \mu\} \approx Z_2 \oplus Z_2$.

Theorem 12.17.

$\pi_{n+17}^{n} = \{\sigma_n \circ \eta_{n+7} \circ \mu_{n+8}\} \oplus \{\nu_n \circ \kappa_{n+3}\} \oplus \{\bar{\mu}_n\} \approx Z_2 \oplus Z_2 \oplus Z_2$ for $n = 10, 11$,

$\pi_{n+17}^{n} = \{\varepsilon^*_n\} \oplus \{\sigma_n \circ \eta_{n+7} \circ \mu_{n+8}\} \oplus \{\nu_n \circ \kappa_{n+3}\} \oplus \{\bar{\mu}_n\} \approx Z_2 \oplus Z_2 \oplus Z_2 \oplus Z_2$ for $n = 12, 13, 14$,

$\pi_{32}^{15} = \{\eta^{*\prime} \circ \eta_{31}\} \oplus \{\varepsilon_{15}^*\} \oplus \{\sigma_{15} \circ \eta_{22} \circ \mu_{23}\} \oplus \{\nu_{15} \circ \kappa_{18}\} \oplus \{\bar{\mu}_{15}\}$
$\approx Z_2 \oplus Z_2 \oplus Z_2 \oplus Z_2 \oplus Z_2,$

$\pi_{33}^{16} = \{\eta_{16}^* \circ \eta_{32}\} \oplus \{E\eta^{*\prime} \circ \eta_{32}\} \oplus \{\varepsilon_{16}^*\} \oplus \{\sigma_{16} \circ \eta_{23} \circ \mu_{24}\}$
$\oplus \{\nu_{16} \circ \kappa_{19}\} \oplus \{\bar{\mu}_{16}\} \approx Z_2 \oplus Z_2 \oplus Z_2 \oplus Z_2 \oplus Z_2 \oplus Z_2,$

$\pi_{34}^{17} = \{\eta_{17}^* \circ \eta_{33}\} \oplus \{\varepsilon_{17}^*\} \oplus \{\sigma_{17} \circ \eta_{24} \circ \mu_{25}\} \oplus \{\nu_{17} \circ \kappa_{20}\} \oplus \{\bar{\mu}_{17}\}$
$\approx Z_2 \oplus Z_2 \oplus Z_2 \oplus Z_2 \oplus Z_2,$

$\pi_{35}^{18} = \{\Delta(\iota_{37})\} \oplus \{\varepsilon_{18}^*\} \oplus \{\sigma_{18} \circ \eta_{25} \circ \mu_{26}\} \oplus \{\nu_{18} \circ \kappa_{21}\} \oplus \{\bar{\mu}_{18}\}$
$\approx Z \oplus Z_2 \oplus Z_2 \oplus Z_2 \oplus Z_2,$

$\pi_{n+17}^{n} = \{\varepsilon_n^*\} \oplus \{\sigma_n \circ \eta_{n+7} \circ \mu_{n+8}\} \oplus \{\nu_n \circ \kappa_{n+3}\} \oplus \{\bar{\mu}_n\}$
$\approx Z_2 \oplus Z_2 \oplus Z_2 \oplus Z_2$ <u>for</u> $n \geq 19$ <u>and</u>

$(G_{17};2) = \{\varepsilon^*\} \oplus \{\sigma \circ \eta \circ \mu\} \oplus \{\nu \circ \kappa\} \oplus \{\bar{\mu}\} \approx Z_2 \oplus Z_2 \oplus Z_2 \oplus Z_2 .$

<u>Proof of Theorem</u> 12.16 <u>for</u> $n \leq 14$. By Proposition 2.5 and (7.1),

$$\Delta(\bar{\nu}_{19}) = \Delta(\iota_{19}) \circ \bar{\nu}_{17} = (\sigma_9 \circ \eta_{16} + \bar{\nu}_9 + \varepsilon_9) \circ \bar{\nu}_{17}$$
$$\equiv \sigma_9 \circ \nu_{16}^3 \mod E^3\pi_{22}^6$$

and $$\Delta(\varepsilon_{19}) \equiv \sigma_9 \circ \eta_{16} \circ \varepsilon_{17} \mod E^3\pi_{22}^6.$$

Then it follows from Theorem 7.1 and Theorem 12.6 that

(12.18). $\Delta : \pi_{27}^{19} \to \pi_{25}^{9}$ <u>is an isomorphism into and that</u>

$$\pi_{25}^9 / \Delta(\pi_{27}^{19}) = \{\sigma_9 \circ \mu_{16}\} \oplus \{\mu_9 \circ \sigma_{18}\} \approx Z_2 \oplus Z_2.$$

By (10.19), we have an exact sequence

$$0 \to \pi_{25}^9 / \Delta\pi_{27}^{19} \xrightarrow{E} \pi_{26}^{10} \longrightarrow \{2\sigma_{19}\} \; (\approx Z_8) .$$

We have $2\sigma_{19} = \pm H(\Delta(\sigma_{21}))$ by Propositions 2.5 and 2.7. By Proposition 3.1, $E(\sigma_{10} \circ \mu_{17}) = \sigma_8 \# \mu_3 = E(\mu_{10} \circ \sigma_{19})$. It follows from the exactness of the sequence (4.4) that $\sigma_{10} \circ \mu_{17} + \mu_{10} \circ \sigma_{19}$ is contained in $\Delta\pi_{28}^{21}$. Since σ_{21} generates $\pi_{28}^{21} \approx Z_{16}$, we have a relation $\sigma_{10} \circ \mu_{17} + \mu_{10} \circ \sigma_{19} = x\Delta(\sigma_{21})$ for some $x \not\equiv 0 \pmod{16}$. We have that $\pm 2x\sigma_{19} = H(\Delta(x\sigma_{21})) = HE(\sigma_9 \circ \mu_{16} + \mu_9 \circ \sigma_{18}) = 0$. Thus $2x \equiv 0 \pmod{16}$, and we have obtained a relation

$$\Delta(8\sigma_{21}) = \sigma_{10} \circ \mu_{17} + \mu_{10} \circ \sigma_{19}.$$

It follows from the exactness of the above sequence that

$$\pi_{26}^{10} = \{\Delta(\sigma_{21})\} \oplus \{\sigma_{10} \circ \mu_{17}\} \approx Z_{16} \oplus Z_2 .$$

By (10.20) we have an exact sequence

$$\pi_{28}^{21} \xrightarrow{\Delta} \pi_{26}^{10} \xrightarrow{E} \pi_{27}^{11} \longrightarrow 0 .$$

Then it follows that

(12.19) $\Delta : \pi^{21}_{28} \to \pi^{10}_{26}$ _is an isomorphism into_,

and $$\pi^{11}_{27} = \{\sigma_{11} \circ \mu_{18}\} \approx Z_2.$$

Next consider the exact sequence

$$\pi^{23}_{29} \xrightarrow{\Delta} \pi^{11}_{27} \xrightarrow{E} \pi^{12}_{28} \xrightarrow{H} \pi^{23}_{28}$$

of (4.4). $\pi^{23}_{28} = 0$ by Proposition 5.9. $\pi^{23}_{29} = \{\nu^2_{23}\}$ by Proposition 5.11. By ii) of Lemma 12.15, $\Delta(\nu^2_{23}) = \Delta H(\varepsilon^*_{12}) = 0$. Then it follows from the exactness of the sequence that E is an isomorphism onto and

$$\pi^{12}_{28} = \{\sigma_{12} \circ \mu_{19}\} \approx Z_2 .$$

$\pi^{25}_{29} = \pi^{25}_{30} = 0$ by Propositions 5.8 and 5.9. Then it follows from the exactness of (4.4) that $E : \pi^{12}_{28} \to \pi^{13}_{29}$ is an isomorphism onto and thus

$$\pi^{13}_{29} = \{\sigma_{13} \circ \mu_{20}\} \approx Z_2 .$$

$\pi^{27}_{31} = 0$ by Proposition 5.8. Then we have an exact sequence

$$0 \longrightarrow \pi^{13}_{29} \xrightarrow{E} \pi^{14}_{30} \xrightarrow{H} \pi^{27}_{30} .$$

By i) of Lemma 12.15, we have that H is onto and π^{14}_{30} is generated by ω_{14} and $\sigma_{14} \circ \mu_{21}$. Furthermore, $\pi^{14}_{30} \approx Z_8 \oplus Z_2$ if $8\omega_{14} = 0$. Consider the element $\Delta(\nu_{29})$. Since $H\Delta(\nu_{29}) = \pm 2\nu_{27} = H(\pm 2\omega_{14})$, then $\Delta(\nu_{29}) = \pm 2\omega_{14} + x\sigma_{14} \circ \mu_{21}$ for some integer x. We have then

$$8\omega_{14} = \pm \Delta(4\nu_{29}) + x(4\sigma_{14} \circ \mu_{21}) = \Delta(\eta^3_{29})$$

$$= \Delta(\eta^2_{29}) \circ \eta_{29} = \Delta H(\eta^{*\prime}) \circ \eta_{29} = 0 ,$$

by Lemma 12.14. Thus

$$\pi^{14}_{30} = \{\omega_{14}\} + \{\sigma_{14} \circ \mu_{21}\} \approx Z_8 \oplus Z_2.$$

We have obtained also

(12.20). _The kernel of_ $\Delta : \pi^{29}_{32} \to \pi^{14}_{30}$ _is_ $\{4\nu_{29}\} = \{\eta^3_{29}\}$.

Proof of Theorem 12.17 _for_ $n \leq 16$.

We prepare the following lemmas.

Lemma 12.18. _There exists an element_ λ _of_ π^{13}_{31} _such that_

$$H(\lambda) = \nu^2_{25} \ \text{and}\ E^3\lambda - 2\nu^*_{16} = \pm \Delta(\nu_{33}).$$

This is just the Lemma 11.17 of the case $n = 12$, and $\alpha = \nu^*_{16}$.

Lemma 12.19. _There are elements_ λ', $\xi' \in \pi^{11}_{29}$ _and_ λ'', $\xi'' \in \pi^{10}_{28}$ _such that_

$$E^2\lambda' = 2\lambda,\ H(\lambda') \equiv \varepsilon_{21} \bmod \bar\nu_{21} + \varepsilon_{21},$$

$$E\xi' - 2\xi_{12} = \pm \Delta(\sigma_{25}),\ H(\xi') = \bar\nu_{21} + \varepsilon_{21},$$

$$E\lambda'' = 2\lambda', \ H(\lambda'') \equiv \eta_{19} \circ \varepsilon_{20} \bmod \eta_{19} \circ \varepsilon_{20} + \nu_{19}^3$$

and
$$E\xi'' = 2\xi', \ H(\xi'') = \nu_{19}^3 + \eta_{19} \circ \varepsilon_{20} .$$

Proof. Consider Proposition 11.15 of the case that $n = 10$, $\alpha = \lambda$ and $\beta = \nu_{20}^2$. Then we have an element λ' of π_{29}^{11} such that

$$2\lambda = E^2\lambda' \ \text{and} \ H(\lambda') \in \{\eta_{21}, \nu_{22}^2, 2\iota_{28}\} .$$

By i) of Proposition 3.4 and by (7.6), the secondary composition $\{\eta_{21}, \nu_{22}^2, 2\iota_{28}\}$ is a coset of $\eta_{21} \circ \pi_{29}^{22} + 2\pi_{29}(S^{21})$ which contains ε_{21}. By Proposition 5.15, Theorem 7.1 and by Lemma 6.4, we have that $H(\lambda') \equiv \varepsilon_{21}$ mod $\bar{\nu}_{21} + \varepsilon_{21} = \eta_{21} \circ \sigma_{22}$.

Next consider Proposition 11.16 of the case that $n = 10$, $\alpha = \xi_{12}$ and $\beta \equiv \sigma_{20}$ mod $2\sigma_{20}$. Then we have that there is an element ξ' of π_{29}^{11} such that

$$E\xi' - 2\xi_{12} = \pm \Delta(\sigma_{25}) \ \text{and} \ H(\xi') \equiv \eta_{21} \circ \sigma_{22}$$

mod $2\pi_{29}^{21}$. By Theorem 7.1, $2\pi_{29}^{21} = 0$. Thus $H(\xi') = \eta_{21} \circ \sigma_{22} = \bar{\nu}_{21} + \varepsilon_{21}$ by Lemma 6.4.

Apply Proposition 11.13 to the case $n = 9$ and $\alpha = \lambda'$ or ξ'. Then we have the assertion for λ'' and ξ'', since $2E^2\pi_{26}^{17} = 0$ by Theorem 7.2. q.e.d.

Now, we prove

(12.21). $E : \pi_{n+17}^n \to \pi_{n+18}^{n+1}$ is an isomorphism into if $10 \leq n \leq 15$, and the kernel of E is generated by $\Delta(\mu_{19}) = \eta_9 \circ \mu_{10} \circ \sigma_{19} + \sigma_9 \circ \eta_{16} \circ \mu_{17}$ if $n = 9$.

Proof. We shall show that $\Delta\pi_{n+19}^{2n+1} = 0$ for $10 \leq n \leq 15$, then the first assertion of (12.21) is proved. By Propositions 5.8 and 5.9, $\Delta\pi_{n+19}^{2n+1} = 0$ for $n = 13$ and $n = 14$. By Lemma 12.14 $\Delta(\nu_{31}) = \Delta H(\nu_{16}^*) = 0$. Since ν_{31} generates π_{34}^{31}, we have that $\Delta\pi_{n+19}^{2n+1} = 0$ for $n = 15$. By Lemma 12.18, $\Delta(\nu_{25}^2) = \Delta H(\lambda) = 0$. Since ν_{25}^2 generates π_{31}^{25}, then $\Delta\pi_{n+19}^{2n+1} = 0$ for $n = 12$. By Lemma 12.14, $\Delta(\sigma_{23}) = \Delta H(\xi_{12}) = 0$. Since σ_{23} generates π_{30}^{23}, then $\Delta\pi_{n+19}^{2n+1} = 0$ for $n = 11$. By Lemma 12.19 and Theorem 7.1, the elements $H(\lambda')$ and $H(\xi')$ span π_{29}^{21}. It follows from the exactness of (4.4) that $\Delta\pi_{n+19}^{2n+1} = 0$ for $n = 10$. Similarly $H(\lambda'')$ and $H(\xi'')$ span the subgroup $\{\nu_{19}^3\} + \{\eta_{19} \circ \varepsilon_{20}\}$ of π_{28}^{19}. Since π_{28}^{19} is spanned by the subgroup and μ_{19} (Theorem 7.2), then $\Delta\pi_{28}^{19}$ is generated by $\Delta(\mu_{19})$ and it has at most two elements. By Theorem 3.1, we have the relation $E(\eta_9 \circ \mu_{10} \circ \sigma_{19}) = (\eta_2 \circ \mu_3) \# \sigma_8 = E(\sigma_9 \circ \eta_{16} \circ \mu_{17})$. By Theorem 12.7, $\eta_9 \circ \mu_{10} \circ \sigma_{19} + \sigma_9 \circ \eta_{16} \circ \mu_{17} \neq 0$ but it vanishes under E. Thus, by the exactness of (4.4), we have that

$$\Delta(\mu_{19}) = \eta_9 \circ \mu_{10} \circ \sigma_{19} + \sigma_9 \circ \eta_{16} \circ \mu_{17}. \qquad \text{q.e.d.}$$

By the second assertion of (12.21), (12.18) and by Theorem 12.7, we have that

$$\pi^{10}_{27} = E\pi^{9}_{26} = \{\sigma_{10} \circ \eta_{17} \circ \mu_{18}\} \oplus \{\nu_{10} \circ \kappa_{13}\} \oplus \{\bar{\mu}_{10}\} \approx Z_2 \oplus Z_2 \oplus Z_2 .$$

By (12.19) and (12.21) we have that $E : \pi^{10}_{27} \to \pi^{11}_{28}$ is an isomorphism onto. Thus we have the results for π^{11}_{28} in Theorem 12.17.

Consider the exact sequence

$$\pi^{11}_{28} \xrightarrow{E} \pi^{12}_{29} \xrightarrow{H} \pi^{23}_{29} = \{\nu^2_{23}\}$$

of (4.4), where E is an isomorphism into by (12.21). By Lemma 12.15, $H(\varepsilon^*_{12}) = \nu^2_{23}$ and $E^2\varepsilon^*_{12} = \omega_{14} \circ \eta_{30}$. By (12.21), the relation $2E^2\varepsilon^*_{12} = \omega_{14} \circ 2\eta_{30} = 0$ implies that $2\varepsilon^*_{12} = 0$. Thus,

$$\pi^{12}_{29} = E\pi^{11}_{28} \oplus \{\varepsilon^*_{12}\} \approx Z_2 \oplus Z_2 \oplus Z_2 \oplus Z_2.$$

$\pi^{25}_{30} = \pi^{27}_{31} = 0$ by Propositions 5.8 and 5.9. It follows from the exactness of (4.4) and from (12.21) that $E : \pi^{12}_{29} \to \pi^{13}_{30}$ and $E : \pi^{13}_{30} \to \pi^{14}_{31}$ are isomorphisms onto. Thus we have obtained the results for π^{13}_{30} and π^{14}_{31} in Theorem 12.17.

By (12.20) and (12.21), we have exact sequences

$$0 \to \pi^{14}_{31} \xrightarrow{E} \pi^{15}_{32} \xrightarrow{H} \{\eta^3_{29}\}$$

and

$$0 \to \pi^{15}_{32} \xrightarrow{E} \pi^{16}_{33} \xrightarrow{H} \{\eta^2_{31}\} .$$

By Lemma 12.14 and Proposition 2.2,

$$H(\eta^{*\prime} \circ \eta_{31}) = \eta^3_{29} \quad \text{and} \quad H(\eta^*_{16} \circ \eta_{32}) = \eta^2_{31} .$$

Obviously $2(\eta^{*\prime} \circ \eta_{31}) = 2(\eta^*_{16} \circ \eta_{32}) = 0$. Then we have

$$\pi^{15}_{32} = E\pi^{14}_{31} \oplus \{\eta^{*\prime} \circ \eta_{31}\} \approx Z_2 \oplus Z_2 \oplus Z_2 \oplus Z_2 \oplus Z_2$$

and

$$\pi^{16}_{33} = E\pi^{15}_{32} \oplus \{\eta^*_{16} \circ \eta_{32}\} \approx Z_2 \oplus Z_2 \oplus Z_2 \oplus Z_2 \oplus Z_2 \oplus Z_2 .$$

<u>Proof of Theorem</u> 12.16 <u>for</u> $n \geq 15$ <u>and Theorem</u> 12.17 <u>for</u> $n \geq 17$.

In the group $\pi^{14}_{30} = \{\omega_{14}\} \oplus \{\sigma_{14} \circ \mu_{21}\}$, the element $\Delta(\nu_{29})$ is of a form $\pm 2\omega_{14} + x\sigma_{14} \circ \mu_{21}$ for some integer x. If $x \not\equiv 0 \pmod 2$, then $\sigma_{15} \circ \eta_{22} \circ \mu_{23} = \sigma_{15} \circ \mu_{22} \circ \eta_{31} = \pm 2\omega_{15} \circ \eta_{31} - E\Delta(\nu_{29}) \circ \eta_{31} = 0$. But this contradicts the result on π^{15}_{32} in Theorem 12.17. Thus

$$\Delta(\nu_{29}) = \pm 2\omega_{14}.$$

Consider the exact sequence

$$\pi^{29}_{32} \xrightarrow{\Delta} \pi^{14}_{30} \xrightarrow{E} \pi^{15}_{31} \xrightarrow{H} \pi^{29}_{31}$$

of (4.4). π^{29}_{32} and π^{29}_{31} are generated by ν_{29} and η^2_{29}. By Lemma 12.14, $H(\eta^{*\prime}) = \eta^2_{29}$ and $2\eta^{*\prime} = 0$. Then it follows that

$$\pi^{15}_{31} = \{\eta^{*\prime}\} \oplus \{\omega_{15}\} \oplus \{\sigma_{15} \circ \mu_{22}\} \approx Z_2 \oplus Z_2 \oplus Z_2 .$$

In the exact sequence

$$\pi^{31}_{33} \xrightarrow{\Delta} \pi^{15}_{31} \xrightarrow{E} \pi^{16}_{32} \xrightarrow{H} \pi^{31}_{32} ,$$

π^{31}_{33} and π^{31}_{32} are generated by η^2_{31} and η_{31} respectively. By Lemma 12.14 and by Proposition 2.2, $H(\eta^*_{16}) = \eta_{31}$, $2\eta^*_{16} = 0$ and $\Delta(\eta^2_{31}) = \Delta H(\eta^*_{16} \circ \eta_{32}) = 0$. Thus we have

$$\pi^{16}_{32} = \{\eta^*_{16}\} \oplus \{E\eta^{*\prime}\} \oplus \{\omega_{16}\} \oplus \{\sigma_{16} \circ \mu_{23}\} \approx Z_2 \oplus Z_2 \oplus Z_2 \oplus Z_2 .$$

Consider the exact sequence

$$\pi^{33}_{35} \xrightarrow{\Delta} \pi^{16}_{33} \xrightarrow{E} \pi^{17}_{34} \xrightarrow{H} \pi^{33}_{34} \xrightarrow{\Delta} \pi^{16}_{32} \xrightarrow{E} \pi^{17}_{33} \xrightarrow{H} \pi^{33}_{33} \xrightarrow{\Delta} \pi^{16}_{31}$$

of (4.4). By Theorem 10.10, $\Delta : \pi^{33}_{33} \to \pi^{16}_{31}$ is an isomorphism into. Thus $E : \pi^{16}_{32} \to \pi^{17}_{33}$ is onto. By i) of Proposition 11.10, $\Delta(\eta_{33}) = E\beta$ for an element $\beta \in \pi^{15}_{31}$ such that $H(\beta) = \eta^2_{29} = H(\eta^{*\prime})$. Thus

$$\Delta(\eta_{33}) \equiv E\eta^{*\prime} \mod E^2\pi^{14}_{30}$$

and

$$\Delta(\eta^2_{33}) \equiv E\eta^{*\prime} \circ \eta_{32} \mod E^2\pi^{14}_{31} .$$

Then it follows from the exactness of the above sequence that

$$\pi^{17}_{33} = \{\eta^*_{17}\} \oplus \{\omega_{17}\} \oplus \{\sigma_{17} \circ \mu_{24}\} \approx Z_2 \oplus Z_2 \oplus Z_2$$

and $\pi^{17}_{34} = \{\eta^*_{17} \circ \eta_{33}\} \oplus \{\varepsilon^*_{17}\} \oplus \{\sigma_{17} \circ \eta_{24} \circ \mu_{25}\} \oplus \{\nu_{17} \circ \kappa_{20}\} \oplus \{\bar{\mu}_{17}\}$ $\approx Z_2 \oplus Z_2 \oplus Z_2 \oplus Z_2 \oplus Z_2$.

Next consider the exact sequence

$$\pi^{35}_{36} \xrightarrow{\Delta} \pi^{17}_{34} \xrightarrow{E} \pi^{18}_{35} \xrightarrow{H} \pi^{35}_{35} \xrightarrow{\Delta} \pi^{17}_{33} \xrightarrow{E} \pi^{18}_{34} \longrightarrow 0$$

of (4.4). By ii) of Proposition 11.10, $\Delta(\iota_{35}) = E\beta$ for an element $\beta \in \pi^{16}_{32}$ such that $H(\beta) = \eta_{31} = H(\eta^*_{16})$. Thus

$$\Delta(\iota_{35}) \equiv \eta^*_{17} \mod E^2\pi^{15}_{31}$$

and

$$\Delta(\eta_{35}) \equiv \eta^*_{17} \circ \eta_{33} \mod E\pi^{15}_{32} .$$

Then it follows from the exactness of the above sequence that

$$\pi^{18}_{34} = \{\omega_{18}\} \oplus \{\sigma_{18} \circ \mu_{25}\} \approx Z_2 \oplus Z_2$$

and $\pi^{18}_{35} = \{\Delta(\iota_{37})\} \oplus \{\varepsilon^*_{18}\} \oplus \{\sigma_{18} \circ \eta_{25} \circ \mu_{26}\} \oplus \{\nu_{18} \circ \kappa_{21}\} \oplus \{\bar{\mu}_{18}\}$ $\approx Z \oplus Z_2 \oplus Z_2 \oplus Z_2 \oplus Z_2$.

The results of Theorem 12.17 and Theorem 12.16 for the stable groups are proved easily.

Finally, we add the following relations.

Proposition 12.20. i). $\sigma_n \circ \mu_{n+7} = \mu_n \circ \sigma_{n+9} = \eta_n \circ \rho_{n+1} = \rho_n \circ \eta_{n+16}$ for $n \geq 13$ and $\eta_{12} \circ \rho_{13} = \sigma_{12} \circ \mu_{19}$.

ii). $\Delta(\iota_{35}) \equiv \eta^*_{17} + \omega_{17} \mod \sigma_{17} \circ \mu_{24}$, and $\eta^*_n \equiv \omega_n \mod \sigma_n \circ \mu_{n+7}$ for $n \geq 18$.

Proof. By Lemma 10.9 $\rho \in \langle \sigma, 2\sigma, 8\iota \rangle$. Thus, by (3.6),

$$\rho \circ \eta \in \langle \sigma, 2\sigma, 8\iota \rangle \circ \eta = \sigma \circ \langle 2\sigma, 8\iota, \eta \rangle.$$

By Lemma 6.5 and Lemma 5.14, $\mu \in \langle \eta, 2\iota, 8\sigma \rangle$. Then

$$\sigma \circ \mu \in \sigma \circ \langle \eta, 2\iota, 8\sigma \rangle$$
$$= \sigma \circ \langle 8\sigma, 2\iota, \eta \rangle \quad \text{by i) of (3.9)}$$
$$\subset \sigma \circ \langle 2\sigma, 8\iota, \eta \rangle .$$

The composition $\sigma \circ \langle 2\sigma, 8\iota, \eta \rangle$ is a coset of $\sigma \circ G_8 \circ \eta$ which is generated by $\sigma \circ \bar{\nu} \circ \eta$ and $\sigma \circ \varepsilon \circ \eta$. By Lemma 10.7, we have that $\sigma \circ G_8 \circ \eta = 0$ and hence $\sigma \circ \mu = \rho \circ \eta$. By the anti-commutativity of the composition operator, we have $\mu \circ \sigma = \sigma \circ \mu = \rho \circ \eta = \eta \circ \rho$. By Theorem 12.16, $E^\infty : \pi^{13}_{29} \to G_{16}$ and $E^\infty : \pi^{12}_{28} \to G_{16}$ are isomorphisms into. Then i) is proved immediately.

Next let $n \geq 18$. The secondary composition $\{\sigma_n, 2\sigma_{n+7}, \eta_{n+14}\}$ is a coset of the subgroup $\sigma_n \circ \pi^{n+7}_{n+16} + \pi^n_{n+15} \circ \eta_{n+15}$ which is generated by the compositions $\sigma_n \circ \nu^3_{n+7}$, $\sigma_n \circ \mu_{n+7}$, $\sigma_n \circ \eta_{n+7} \circ \varepsilon_{n+8}$, $\rho_n \circ \eta_{n+15}$ and $\bar{\varepsilon}_n \circ \eta_{n+15}$. By (7.20), $\sigma_n \circ \nu^3_{n+7} = 0$. $\sigma_n \circ \eta_{n+7} \circ \varepsilon_{n+8} = \sigma_n \circ \varepsilon_{n+7} \circ \eta_{n+15} = 0$ by Lemma 10.7. $\bar{\varepsilon}_n \circ \eta_{n+15} = \eta_n \circ \bar{\varepsilon}_{n+1} = \nu_n \circ \sigma_{n+3} \circ \nu^2_{n+10} = 0$ by Proposition 3.1, Lemma 12.10 and (7.20). Thus $\{\sigma_n, 2\sigma_{n+7}, \eta_{n+14}\} = \eta^*_n + \{\sigma_n \circ \mu_{n+7}\} = \eta^*_n + \{\rho_n \circ \eta_{n+15}\}$. By Theorem 12.6, $\eta^*_n \equiv x\omega_n \mod \sigma_n \circ \mu_{n+7}$ for some integer x. Assume that $x \equiv 0 \pmod 2$, then $\eta^*_n = y\sigma_n \circ \mu_{n+7}$ for some integer y. But this contradicts Theorem 8.4. Thus $x \not\equiv 0 \pmod 2$ and $\eta^*_n \equiv \omega_n \mod \sigma_n \circ \mu_{n+7}$ for $n \geq 18$. From Theorem 12.16, it follows that $\Delta(\iota_{35}) \equiv \omega_{17} + \eta^*_{17} \mod \sigma_{17} \circ \mu_{24}$. q.e.d

iv). The groups π^n_{n+18} and π^n_{n+19} for $n \geq 10$.

Lemma 12.21. There exists an element ω' of π^{12}_{31} such that $E^2\omega' = 2\omega_{14} \circ \nu_{30}$ and $H(\omega') \equiv \varepsilon_{23} \mod \bar{\nu}_{23} + \varepsilon_{23}$.

Proof. By Proposition 2.2 and Lemma 12.15, we have that $H(\omega_{14} \circ \nu_{30}) = \nu_{27}^2$. Apply Proposition 11.14 for the case that $n = 11$, $\alpha = \omega_{14} \circ \nu_{30}$ and $\gamma = 2\iota_{28}$. Then there exists an element ω' of π_{31}^{12} such that $E^2\omega' = 2\omega_{14} \circ \nu_{30}$ and $H(\omega') \equiv \varepsilon_{23} \bmod \bar{\nu}_{23} + \varepsilon_{23}$, by a similar way to the proof of Lemma 12.19. q.e.d.

By use of the elements ω', λ, λ', ξ', λ'', ξ'', our results are stated as follows.

Theorem 12.22. $\pi_{28}^{10} \approx Z_8 \oplus Z_2 \oplus Z_2$: generated by λ'', ξ'' and $\eta_{10} \circ \bar{\mu}_{11}$,

$\pi_{29}^{11} \approx Z_8 \oplus Z_4 \oplus Z_2$: generated by λ', ξ' and $\eta_{11} \circ \bar{\mu}_{12}$,

$\pi_{30}^{12} \approx Z_{32} \oplus Z_4 \oplus Z_4 + Z_2$: generated by ξ_{12}, $E\lambda'$, $E\xi'$ and $\eta_{12} \circ \bar{\mu}_{13}$,

$\pi_{n+18}^{n} = \{\xi_n\} \oplus \{E^{n-13}\lambda\} \oplus \{\eta_n \circ \bar{\mu}_{n+1}\} \approx Z_8 \oplus Z_8 \oplus Z_2$ for $n = 13, 14, 15$,

$\pi_{34}^{16} = \{\nu_{16}^*\} \oplus \{\xi_{16}\} \oplus \{E^3\lambda\} \oplus \{\eta_{16} \circ \bar{\mu}_{17}\} \approx Z_8 \oplus Z_8 \oplus Z_8 \oplus Z_2$,

$\pi_{35}^{17} = \{\nu_{17}^*\} \oplus \{\xi_{17}\} \oplus \{\eta_{17} \circ \bar{\mu}_{18}\} \approx Z_8 \oplus Z_8 \oplus Z_2$,

$\pi_{36}^{18} = \{\nu_{18}^*\} \oplus \{\nu_{18}^* + \xi_{18}\} \oplus \{\eta_{18} \circ \bar{\mu}_{19}\} \approx Z_8 \oplus Z_4 \oplus Z_2$,

$\pi_{37}^{19} = \{\nu_{19}^*\} \oplus \{\nu_{19}^* + \xi_{19}\} \oplus \{\eta_{19} \circ \bar{\mu}_{20}\} \approx Z_8 \oplus Z_2 \oplus Z_2$,

$\pi_{n+18}^{n} = \{\nu_n^*\} \oplus \{\eta_n \circ \bar{\mu}_{n+1}\} \approx Z_8 \oplus Z_2$ for $n \geq 20$, and

$(G_{18};2) = \{\nu^*\} \oplus \{\eta \circ \bar{\mu}\} \approx Z_8 \oplus Z_2$.

Theorem 12.23. $\pi_{29}^{10} = \{\bar{\sigma}_{10}\} \oplus \{\bar{\zeta}_{10}\} \approx Z_2 \oplus Z_8$,

$\pi_{30}^{11} = \{\lambda' \circ \eta_{29}\} \oplus \{\xi' \circ \eta_{29}\} \oplus \{\bar{\sigma}_{11}\} \oplus \{\bar{\zeta}_{11}\} \approx Z_2 \oplus Z_2 \oplus Z_2 \oplus Z_8$,

$\pi_{31}^{12} = \{\omega'\} \oplus \{\xi_{12} \circ \eta_{30}\} \oplus \{E\lambda' \circ \eta_{30}\} \oplus \{E\xi' \circ \eta_{30}\} \oplus \{\bar{\sigma}_{12}\} \oplus \{\bar{\zeta}_{12}\}$
$\approx Z_2 \oplus Z_2 \oplus Z_2 \oplus Z_2 \oplus Z_2 \oplus Z_8$,

$\pi_{32}^{13} = \{E\omega'\} \oplus \{\xi_{13} \circ \eta_{31}\} \oplus \{\bar{\sigma}_{13}\} \oplus \{\bar{\zeta}_{13}\} \approx Z_2 \oplus Z_2 \oplus Z_2 \oplus Z_8$,

$\pi_{33}^{14} = \{\omega_{14} \circ \nu_{30}\} \oplus \{\bar{\sigma}_{14}\} \oplus \{\bar{\zeta}_{14}\} \approx Z_4 \oplus Z_2 \oplus Z_8$,

$\pi_{n+19}^{n} = \{\omega_n \circ \nu_{n+16}\} \oplus \{\bar{\sigma}_n\} \oplus \{\bar{\zeta}_n\} \approx Z_2 \oplus Z_2 \oplus Z_8$ for $n = 15, 16, 17$,

$\pi_{39}^{20} = \{\Delta(\iota_{41})\} \oplus \{\bar{\sigma}_{20}\} \oplus \{\bar{\zeta}_{20}\} \approx Z \oplus Z_2 \oplus Z_8$,

$\pi_{n+19}^{n} = \{\bar{\sigma}_n\} \oplus \{\bar{\zeta}_n\} \approx Z_2 \oplus Z_8$ for $n \geq 18$ and $n \neq 20$, and

$(G_{19};2) = \{\bar{\sigma}\} \oplus \{\bar{\zeta}\} \approx Z_2 \oplus Z_8$.

First we prove

(12.22). $E : \pi_{n+19}^{n} \to \pi_{n+20}^{n+1}$ is an isomorphism into if $9 \leq n \leq 11$.

Proof. By the exactness of (4.4), it is sufficient to prove that $\Delta\pi^{2n+1}_{n+21} = 0$ for $9 \leq n \leq 11$. Apply i) of Proposition 11.11 for the case $n = 9$ and $\alpha = \eta_{18} \circ \varepsilon_{19}$. Then there exists an element β of π^{10}_{30} such that $H(\beta) \in \{2\iota_{19}, \eta_{19}, \eta_{20} \circ \varepsilon_{21}\}$. The group π^{19}_{30} is stable. Then it follows from Lemma 9.1 that $H(\beta)$ generates $\pi^{19}_{30} = \{\zeta_{19}\}$. Thus, $\Delta\pi^{19}_{30} = \Delta H\pi^{10}_{30} = 0$.

Next apply i) of Proposition 11.10 for the case $n = 10$ and $\alpha = \mu_{20}$. Then there exists an element β of π^{11}_{31} such that $H(\beta) = \eta_{21} \circ \mu_{22}$. Since $\eta_{21} \circ \mu_{22}$ generates π^{21}_{31}, by Theorem 7.3, then $\Delta\pi^{21}_{31} = \Delta H\pi^{11}_{31} = 0$.

Apply ii) of Proposition 11.11 for the case that $n = 11$ and $\alpha = 8\sigma_{22}$. Then there exists an element β of π^{12}_{32} such that $H(\beta) \in \{\eta_{23}, 2\iota_{24}, 8\sigma_{24}\}$. It is verified from Lemma 6.5 that

$$H(\beta) \equiv \mu_{23} \bmod \{\nu^3_{23}\}+\{\eta_{23} \circ \varepsilon_{24}\}.$$

By Proposition 2.2 and by Lemma 12.14, we have

$$H(\xi_{12} \circ \eta^2_{30}) = \sigma_{23} \circ \eta^2_{30} = \nu^3_{23} + \eta_{23} \circ \varepsilon_{24} .$$

By Proposition 2.2 and by Lemma 12.21, we have

$$H(\omega' \circ \eta_{31}) \equiv \eta_{23} \circ \varepsilon_{24} \bmod \nu^3_{23} + \eta_{23} \circ \varepsilon_{24}.$$

Since the elements ν^3_{23}, μ_{23} and $\eta_{23} \circ \varepsilon_{24}$ generate π^{23}_{32}, by Theorem 7.3, then we have that $\Delta\pi^{23}_{32} = \Delta H\pi^{12}_{32} = 0$. Consequently (12.22) is proved. q.e.d.

Next we prove

(12.23). $4\sigma_{10} \circ \zeta_{17} = 2\sigma_{11} \circ \zeta_{18} = \sigma_{13} \circ \zeta_{20} = 0$.

By Lemma 12.12 and by Lemma 5.14, we have an equality

$$2\sigma_9 \circ \zeta_{16} = x\zeta_9 \circ \sigma_{20} \quad \text{for } x \not\equiv 0 \pmod 2).$$

Since $4\zeta_5 = \eta^2_5 \circ \mu_7$ and $2\zeta_5 = \pm E^2\mu'$, by (7.14), then we have the following relations by use of Proposition 3.1.

$$4\sigma_{10} \circ \zeta_{17} = \sigma_8 \# (\eta^2_2 \circ \mu_{14}) = 4\zeta_{10} \circ \sigma_{19} = 4x\zeta_{10} \circ \sigma_{19}$$
$$= 8\sigma_{10} \circ \zeta_{17} = 0 ,$$
$$2\sigma_{11} \circ \zeta_{18} = \sigma_8 \# \mu' = 2\zeta_{11} \circ \sigma_{20} = 4\sigma_{11} \circ \zeta_{18} = 0$$

and $$\sigma_{13} \circ \zeta_{20} = \sigma_8 \# \zeta_5 = \zeta_{13} \circ \sigma_{22} = 2\sigma_{13} \circ \zeta_{20} = 0.$$

Thus (12.23) is proved.

Consider the exact sequence

$$\pi^9_{28} \xrightarrow{E} \pi^{10}_{29} \xrightarrow{H} \pi^{19}_{29} \xrightarrow{\Delta} \pi^9_{27} \xrightarrow{E} \pi^{10}_{28} \xrightarrow{H} \pi^{19}_{28} \xrightarrow{\Delta} \pi^9_{26}$$

of (4.4), where the first E is an isomorphism into by (12.22) and the image of last Δ is $\{\Delta(\mu_{19})\} \approx Z_2$ by (12.21). $\pi_{29}^{19} \approx Z_2$ by Theorem 7.3. By Theorem 12.8 and by (12.23), the homomorphism $E : \pi_{27}^9 \to \pi_{28}^{10}$ has a non-trivial kernel. Thus $\Delta : \pi_{29}^{19} \to \pi_{27}^9$ is an isomorphism into. It follows from the exactness of the above sequence that $E : \pi_{28}^9 \to \pi_{29}^{10}$ is an isomorphism onto,

$$\pi_{29}^{10} = \{\bar{\sigma}_{10}\} \oplus \{\bar{\zeta}_{10}\} \approx Z_2 \oplus Z_8 \quad \text{by Theorem 12.9}$$

and that $E\pi_{27}^9 = \{\sigma_{10} \circ \zeta_{17}\} \oplus \{\eta_{10} \circ \bar{\mu}_{11}\} \approx Z_4 \oplus Z_2$.

π_{28}^{19} is spanned by ν_{19}^3, μ_{19} and $\eta_{19} \circ \varepsilon_{20}$, by Theorem 7.2.

By Lemma 12.19, $H(\lambda'')$ and $H(\xi'')$ span the subgroup of π_{28}^{19} generated by ν_{23}^3 and $\eta_{23} \circ \varepsilon_{24}$. Then it follows from the exactness of the above sequence that the group π_{28}^{10} is spanned by $E\pi_{27}^9$, λ'' and ξ''. To prove the result

$$\pi_{28}^{10} = (\{\lambda''\} + \{\xi''\}) \oplus \{\eta_{10} \circ \bar{\mu}_{11}\} \approx (Z_8 \oplus Z_2) \oplus Z_2 ,$$

it is sufficient to prove that

(12.24). $\eta_{10} \circ \bar{\mu}_{11}$ <u>is not divisible by</u> 2 <u>and</u> $\sigma_{10} \circ \zeta_{17}$ <u>is divisible by</u> 2.

Assume that $\eta_{10} \circ \bar{\mu}_{11}$ is divisible by 2. Then $\eta_{10}^2 \circ \bar{\mu}_{12} = 0$. By Lemma 12.4, $4\bar{\zeta}_{10} = \eta_{10}^2 \circ \bar{\mu}_{12}$. But this contradicts the above result on π_{29}^{10}. Thus $\eta_{10} \circ \bar{\mu}_{11}$ is not divisible by 2.

The secondary composition $\{\sigma_{10}, 4\nu_{17}, \sigma_{20}\}$ is defined (cf. (7.20)). By Proposition 1.2,

$$2\iota_{10} \circ \{\sigma_{10}, 4\nu_{17}, \sigma_{20}\} \subset \{2\sigma_{10}, 4\nu_{17}, \sigma_{20}\}.$$

By Lemma 9.1 and Proposition 1.2,

$$\sigma_{10} \circ \zeta_{17} \in \sigma_{10} \circ \{2\iota_{17}, 4\nu_{17}, \sigma_{20}\}$$
$$\subset \{2\sigma_{10}, 4\nu_{17}, \sigma_{20}\}.$$

The secondary composition $\{2\sigma_{10}, 4\nu_{17}, \sigma_{20}\}$ is a coset of the subgroup $2\sigma_{10} \circ \pi_{28}^{17} + \pi_{21}^{10} \circ \sigma_{21}$. By Lemma 12.12 and Lemma 5.14, we have $2\sigma_{10} \circ \pi_{28}^{17} = \{2\sigma_{10} \circ \zeta_{17}\} = \{\zeta_{10} \circ \sigma_{21}\} = \pi_{21}^{10} \circ \sigma_{21}$. Therefore we conclude a relation

$$\sigma_{10} \circ \zeta_{17} \equiv 2\alpha \bmod 2\sigma_{10} \circ \zeta_{17}$$

for an element α of $\{\sigma_{10}, 4\nu_{17}, \sigma_{20}\}$. Thus $E(\sigma_{10} \circ \zeta_{17}) \equiv E(2\alpha) \bmod 2E(\sigma_{10} \circ \zeta_{17})$. Then the second assertion of (12.24) follows from the following.

(12.25). <u>The kernel of</u> $E : \pi_{28}^{10} \to \pi_{29}^{11}$ <u>is generated by</u> $\Delta(\mu_{21}) = 2\sigma_{10} \circ \zeta_{17}$.

By Proposition 2.2 and by Lemma 12.19, we have

$$H(\xi' \circ \eta_{29}) = \nu_{21}^3 + \eta_{21} \circ \varepsilon_{22}$$

and $$H(\lambda' \circ \eta_{29}) \equiv \eta_{21} \circ \varepsilon_{22} \bmod H(\xi' \circ \eta_{29}) .$$

Since $\pi_{30}^{21} = \{\nu_{21}^3\} \oplus \{\mu_{21}\} \oplus \{\eta_{21} \circ \varepsilon_{22}\} \approx Z_2 \oplus Z_2 \oplus Z_2$, by Theorem 7.2, then the cokernel of $H : \pi_{30}^{11} \to \pi_{30}^{21}$ has at most two elements. By (12.23), $2\sigma_{10} \circ \zeta_{17}$ is a non-trivial element which vanishes by E. Then it follows from the exactness of (4.4) the result (12.25), and the following two exact sequences :

$$0 \to \{2\sigma_{10} \circ \zeta_{17}\} \to \pi_{28}^{10} \xrightarrow{E} \pi_{29}^{11} \xrightarrow{H} \pi_{29}^{21} ,$$

$$\pi_{29}^{10} \to \pi_{30}^{11} \to \{H(\lambda' \circ \eta_{29})\} \oplus \{H(\xi' \circ \eta_{29})\} .$$

Obviously $2(\lambda' \circ \eta_{29}) = 2(\xi' \circ \eta_{29}) = 0$. Then the second sequence splits, where $E : \pi_{29}^{10} \to \pi_{30}^{11}$ is an isomorphism into by (12.22). Thus

$$\pi_{30}^{11} = \{\lambda' \circ \eta_{29}\} \oplus \{\xi' \circ \eta_{29}\} \oplus \{\bar{\sigma}_{11}\} \oplus \{\bar{\zeta}_{11}\} \approx Z_2 \oplus Z_2 \oplus Z_2 \oplus Z_8 .$$

By Lemma 12.19 and Theorem 7.1, the group π_{29}^{21} is spanned by $H(\lambda')$ and $H(\xi')$. By (12.24), the element $2\sigma_{10} \circ \zeta_{17}$ is divisible by 4. Thus it generates $4(\{\lambda''\} + \{\xi''\})$ and hence

$$E\pi_{28}^{10} = \{\eta_{11} \circ \bar{\mu}_{12}\} \oplus (\{E\lambda''\} + \{E\xi''\}) \approx Z_2 \oplus (Z_2 \oplus Z_4) .$$

Since $E\lambda' = 2\lambda'$ and $E\xi'' = 2\xi'$, by Lemma 12.19, then it follows from the above first exact sequence that

$$\pi_{29}^{11} = \{\eta_{11} \circ \bar{\mu}_{12}\} \oplus (\{\lambda'\} + \{\xi'\}) \approx Z_2 \oplus (Z_4 \oplus Z_8).$$

Remark that $4(\{\lambda'\} + \{\xi'\}) = \{\sigma_{11} \circ \zeta_{18}\}$.

We have

$$H(\xi_{12} \circ \eta_{30}) = \sigma_{23} \circ \eta_{30} \quad \text{by Proposition 2.2 and Lemma 12.4}$$

$$= \bar{\nu}_{23} + \varepsilon_{23} \quad \text{by Lemma 6.4}$$

and $H(\omega') \equiv \varepsilon_{23} \bmod \bar{\nu}_{23} + \varepsilon_{23}$.

Thus $H : \pi_{31}^{12} \to \pi_{31}^{23}$ is onto, by Theorem 7.1. Then it follows from the exactness of (4.4) and (12.22) that the following two sequences are exact.

$$0 \longrightarrow \pi_{29}^{11} \xrightarrow{E} \pi_{30}^{12} \xrightarrow{H} \pi_{30}^{23} ,$$

$$0 \longrightarrow \pi_{30}^{11} \xrightarrow{E} \pi_{31}^{12} \xrightarrow{H} \pi_{31}^{23} \longrightarrow 0 .$$

Consider the element $\sigma_{12} \circ \zeta_{19}$. By (12.23), $E(\sigma_{12} \circ \zeta_{19}) = 0$. Since $E : \pi_{29}^{11} \to \pi_{30}^{12}$ is an isomorphism into, then $\sigma_{12} \circ \zeta_{19} \neq 0$. By the exactness of (4.4) the element $\sigma_{12} \circ \zeta_{19}$ is in $\Delta\pi_{32}^{25} = \{\Delta(\sigma_{25})\}$. Thus,

$\sigma_{12} \circ \zeta_{19} = x\Delta(\sigma_{25})$ for an integer $x \not\equiv 0 \pmod{16}$. By Proposition 2.5 and Proposition 2.7, $\pm 2x\sigma_{23} = xH\Delta(\sigma_{25}) = HE(\sigma_{11} \circ \zeta_{18}) = 0$. Thus $2x \equiv 0 \pmod{16}$ and $\sigma_{12} \circ \zeta_{19} = 8\Delta(\sigma_{25})$. By Lemma 12.19, we have

$$\sigma_{12} \circ \zeta_{19} = 8\Delta(\sigma_{25}) = 8(E\xi' - 2\xi_{12}) = 16\xi_{12}.$$

This shows that ξ_{12} is an element of order 32. It follows from the exactness of the above first sequence that

$$\pi^{12}_{30} = \{\eta_{12} \circ \bar{\mu}_{13}\} \oplus (\{E\lambda'\} + \{E\xi'\} + \{\xi\})$$
$$\approx Z_2 \oplus (Z_4 \oplus Z_4 \oplus Z_{32}).$$

Consider the above second sequence. Obviously $2(\xi_{12} \circ \eta_{30}) = 0$. Then the result

$$\pi^{12}_{31} = E\pi^{11}_{30} \oplus \{\xi_{12} \circ \eta_{30}\} \oplus \{\omega'\} \approx Z_2 \oplus Z_2 \oplus Z_2 \oplus Z_8 \oplus Z_2 \oplus Z_2$$

of Theorem 12.23 is proved by the first assertion of the following

(12.26). $2\omega' = 0$. $\Delta\pi^{25}_{33} = \{E\lambda' \circ \eta_{30}\} + \{E\xi' \circ \eta_{30}\}$. $\Delta\pi^{27}_{34} = \{\xi_{13} \circ \eta_{31}\}$.

<u>Proof</u>. By Lemma 12.19, we have $E(E\lambda' \circ \eta_{30}) = 2\lambda \circ \eta_{31} = 0$ and $E(E\xi' \circ \eta_{30}) = \pm E\Delta(\sigma_{25}) \circ \eta_{31} + 2(\xi_{13} \circ \eta_{31}) = 0$. Thus the elements $E\lambda' \circ \eta_{30}$ and $E\xi' \circ \eta_{30}$ span a subgroup of 4 elements in the kernal of E. Since π^{25}_{33} has 4 elements, then it follows from the exactness of the sequence (4.4) that $\Delta\pi^{25}_{33}$ coincides with the subgroup. Thus the second assertion is proved. Next we have

$$\Delta(\sigma_{27}) = \Delta(\iota_{27}) \circ \sigma_{25} = E\theta \circ \sigma_{25} \quad \text{by (7.30)}$$
$$\in \{\sigma_{13}, \nu_{20}, \eta_{23}\} \circ \sigma_{25} \quad \text{by the definition of } \theta$$
$$\subset \{\sigma_{13}, \nu_{20}, \eta_{23} \circ \sigma_{24}\} \quad \text{by Proposition 1.2}$$

and $\xi_{13} \circ \eta_{31} \in \{\sigma_{13}, \nu_{20}, \sigma_{23}\} \circ \eta_{31}$

$$\subset \{\sigma_{13}, \nu_{20}, \eta_{23} \circ \sigma_{24}\} \quad \text{by Proposition 1.2.}$$

The secondary composition $\{\sigma_{13}, \nu_{20}, \eta_{23} \circ \sigma_{24}\}$ is a coset of

$$\sigma_{13} \circ \pi^{20}_{32} + \pi^{13}_{24} \circ \eta_{24} \circ \sigma_{25} = \sigma_{13} \circ \pi^{20}_{32} + \{\zeta_{13}\} \circ \eta_{24} \circ \sigma_{25}$$
$$\subset \sigma_{13} \circ \pi^{20}_{32} + E^8\pi^5_{17} \circ \sigma_{25} = 0 \quad \text{by Theorem 7.6.}$$

Thus $\Delta(\sigma_{27}) = \xi_{13} \circ \eta_{31}$. Since σ_{27} generates π^{27}_{34}, then the last assertion of (12.26) is proved. In the computation of π^{15}_{31}, we obtained a relation $\Delta(\nu_{29}) = \pm 2\omega_{14}$. Then we have

$$E^2(2\omega') = 2E^2\omega' = 4\omega_{14} \circ \nu_{30} \quad \text{by Lemma 12.21}$$
$$= \Delta(2\nu_{29}) \circ \nu_{30}$$
$$= \Delta(2\nu^2_{29}) = 0 \quad \text{by Proposition 2.5.}$$

By the exactness of (4.4), it follows that

$$2\omega' = x\xi_{12} \circ \eta_{30} + yE\lambda' \circ \eta_{30} + zE\xi' \circ \eta_{30} ,$$

for some integers x, y and z. Applying the homomorphism $H : \pi_{31}^{12} \to \pi_{31}^{23}$, we have the relation $0 = H(2\omega') = xH(\xi_{12} \circ \eta_{30}) = x(\bar{\nu}_{23} + \varepsilon_{23})$. It follows that $x \equiv 0 \pmod 2$. Let $H(\omega') = E^3\beta$ for $\beta \in \pi_{28}^{20}$. Then, by Proposition 11.16 and i) of Proposition 11.10, we have elements γ and γ' of π_{30}^{11} such that $\Delta(E^5\beta) \pm 2\omega' = E\gamma$, $H(\gamma) = \eta_{21} \circ E^2\beta$, $\Delta(E^5\beta) = E\gamma'$ and $H(\gamma') = \eta_{21} \circ E^2\beta$. By cancelling $\Delta(E^5\beta)$, we have that

$$2\omega' = \pm E(\gamma - \gamma') \quad \text{and} \quad H(\gamma - \gamma') = 0.$$

By (12.22), $E : \pi_{30}^{11} \to \pi_{31}^{12}$ is an isomorphism into. Then it follows that $\pm (\gamma - \gamma') = y\lambda' \circ \eta_{29} + z\xi' \circ \eta_{29}$.

Applying the homomorphism $H : \pi_{30}^{11} \to \pi_{30}^{21}$, we have

$$0 = yH(\lambda' \circ \eta_{29}) + zH(\xi' \circ \eta_{29}).$$

We know that $H(\lambda' \circ \eta_{29})$ and $H(\xi' \circ \eta_{29})$ are independent. Thus we have that $y \equiv z \equiv 0 \pmod 2$. Consequently we have obtained that

$$2\omega' = 0. \qquad \text{q.e.d.}$$

By a similar discussion to the proof of the fact $E^2(2\omega') = 0$, we have

(12.27). $$\Delta(\nu_{29}^2) = E^2\omega' = 2\omega_{14} \circ \nu_{30} .$$

Consider the homomorphism $\Delta : \pi_{32}^{25} \to \pi_{30}^{12}$. We have obtained that $8\Delta(\sigma_{25}) = 16\xi_{12} \neq 0$. Thus the homomorphism Δ is an isomorphism into. By Lemma 12.19, $\pm \Delta(\sigma_{25}) = E\xi' - 2\xi_{12}$. Then it follows from the exactness of the sequence (4.4) the result

$$E\pi_{30}^{12} = \{\eta_{13} \circ \bar{\mu}_{14}\} \oplus (\{\xi_{13}\} + \{E^2\lambda'\}) \approx Z_2 \oplus (Z_8 \oplus Z_4)$$

and the following exact sequence :

$$0 \longrightarrow E\pi_{30}^{12} \longrightarrow \pi_{31}^{13} \xrightarrow{H} \pi_{31}^{25} ,$$

$$0 \longrightarrow \Delta\pi_{33}^{25} \longrightarrow \pi_{31}^{12} \xrightarrow{E} \pi_{32}^{13} \longrightarrow 0 .$$

We note that $E^2\lambda'$ is the element of order 4, since $4E^2\lambda' \in \{E^2(\sigma_{11} \circ \zeta_{18})\} = 0$. Since π_{31}^{25} is generated by ν_{25}^2, then it follows from Lemma 12.18 and Lemma 12.19 that

$$\pi_{31}^{13} = \{\eta_{13} \circ \bar{\mu}_{14}\} \oplus \{\xi_{13}\} \oplus \{\lambda\} \approx Z_2 \oplus Z_8 \oplus Z_8 .$$

It follows from (12.26) that

$$\pi_{32}^{13} = \{E\omega'\} \oplus \{\xi_{13} \circ \eta_{31}\} \oplus \{\bar{\sigma}_{13}\} \oplus \{\bar{\zeta}_{13}\} \approx Z_2 \oplus Z_2 \oplus Z_2 \oplus Z_8.$$

By Proposition 2.2 and Lemma 12.15, we have

$$H(\omega_{14} \circ \nu_{30}) = \nu_{27}^2 .$$

Since ν_{27}^2 generates π_{33}^{27}, then it follows from the exactness of (4.4) that the following two sequences are axact.

$$0 \longrightarrow \pi_{31}^{13} \xrightarrow{E} \pi_{32}^{14} \xrightarrow{H} \pi_{32}^{27} ,$$

$$0 \longrightarrow \Delta\pi_{34}^{27} \longrightarrow \pi_{32}^{13} \xrightarrow{E} \pi_{33}^{14} \longrightarrow \{H(\omega_{14} \circ \nu_{30})\} .$$

By Proposition 5.9, $\pi_{32}^{27} = 0$. Thus

$$\pi_{31}^{13} \overset{E}{\approx} \pi_{32}^{14} = \{\eta_{14} \circ \bar{\mu}_{15}\} \oplus \{\varepsilon_{14}\} \oplus \{E\lambda\} \approx Z_2 \oplus Z_8 \oplus Z_8 .$$

By (12.27) and (12.26), we have

$$\pi_{33}^{14} = \{\omega_{14} \circ \nu_{30}\} \oplus \{\bar{\sigma}_{14}\} \oplus \{\bar{\zeta}_{14}\} \approx Z_4 \oplus Z_2 \oplus Z_8 .$$

By Propositions 5.8 and 5.9, $\pi_{33}^{29} = \pi_{34}^{29} = 0$. By Proposition 5.11, the group π_{35}^{29} is generated by ν_{29}^2. Then it follows from the exactness of (4.4) that $E : \pi_{33}^{14} \to \pi_{34}^{15}$ is an isomorphism onto and that the sequence

$$0 \longrightarrow \{\Delta(\nu_{29}^2)\} \longrightarrow \pi_{33}^{14} \xrightarrow{E} \pi_{34}^{15} \longrightarrow 0$$

is exact. By (12.27), we have that

$$\pi_{34}^{15} = \{\omega_{15} \circ \nu_{31}\} \oplus \{\bar{\sigma}_{15}\} \oplus \{\bar{\zeta}_{15}\} \approx Z_2 \oplus Z_2 \oplus Z_8 .$$

By Propositions 5.8 and 5.9, $\pi_{35}^{31} = \pi_{36}^{31} = 0$. Then we have the following two exact sequences

$$0 \longrightarrow \pi_{34}^{15} \xrightarrow{E} \pi_{35}^{16} \longrightarrow 0$$

$$0 \longrightarrow \pi_{33}^{15} \xrightarrow{E} \pi_{34}^{16} \xrightarrow{H} \pi_{34}^{31} .$$

Thus $E : \pi_{34}^{15} \to \pi_{35}^{16}$ is an isomorphism onto. Since ν_{31} generates π_{34}^{31}, then it follows from Lemma 12.14 that $H(\nu_{16}^*)$ generates π_{34}^{31}. We have also by Lemma 12.14 that $8\nu_{16}^* = 0$, since $\sigma_{16} \circ \zeta_{23} = 0$. Thus we have from the exactness of the last sequence that

$$\pi_{34}^{16} = \{\nu_{16}^*\} \oplus E\pi_{33}^{15} \approx Z_8 \oplus Z_8 \oplus Z_8 \oplus Z_2 .$$

Consider the homomorphism $\Delta : \pi_{36}^{33} \to \pi_{34}^{16}$. The image of the homomorphism is generated by $E^3\lambda - 2\nu_{16}^*$, since Lemma 12.18. Since $E^3\lambda$ is an element of order 8, then the homomorphism Δ is an isomorphism into. It follows from the exact sequence (4.4) that the sequences

$$0 \longrightarrow E\pi_{34}^{16} \longrightarrow \pi_{35}^{17} \longrightarrow \pi_{35}^{33} \xrightarrow{\Delta} \pi_{33}^{16}$$

and

$$\pi_{37}^{33} \xrightarrow{\Delta} \pi_{35}^{16} \xrightarrow{E} \pi_{36}^{17} \longrightarrow 0$$

are exact and

$$E\pi_{34}^{16} = \{\nu_{17}^*\} \oplus \{\varepsilon_{17}\} \oplus \{\eta_{17} \circ \bar{\mu}_{18}\} \approx Z_8 \oplus Z_8 \oplus Z_2 .$$

Since $\Delta(\eta_{33}^2) \neq 0$ and since η_{33}^2 generates $\pi_{35}^{33} \approx Z_2$, then $H : \pi_{35}^{17} \to \pi_{35}^{33}$ is trivial. It follows from the exactness of the above first sequence that

$$E\pi_{34}^{16} = \pi_{35}^{17}.$$

$\pi_{37}^{33} = 0$ by Proposition 5.8. Then it follows from the last sequence that $E : \pi_{35}^{16} \to \pi_{36}^{17}$ is an isomorphism onto.

Lemma 12.24. $\sigma \circ \zeta = \rho \circ \nu = 0$ and $\nu^* + \xi = 0$

Proof. We know already that $\sigma_{12} \circ \zeta_{19} = 8\Delta(\sigma_{25}) = 16\xi_{12}$ and hence $\sigma_{13} \circ \zeta_{20} = 0$. Thus $\sigma \circ \zeta = 0$. We have

$$\rho \circ \nu \in \langle \sigma, 2\sigma, 8\iota \rangle \circ \nu \quad \text{by Lemma 10.9}$$
$$= \sigma \circ \langle 2\sigma, 8\iota, \nu \rangle \quad \text{by (3.6)}$$

and $\sigma \circ \zeta \in \sigma \circ \langle 2\sigma, 8\iota, \nu \rangle$ by the definition of ζ and by (3.9). The secondary composition $\sigma \circ \langle 2\sigma, 8\iota, \nu \rangle$ is a coset of the subgroup $\sigma \circ G_8 \circ \nu = \{\sigma \circ \bar{\nu} \circ \nu\} + \{\sigma \circ \varepsilon \circ \nu\}$. By Lemma 10.7, $\sigma \circ G_8 \circ \nu = 0$. Thus we have that $\rho \circ \nu = \sigma \circ \zeta = 0$.

By the definition of ν^* and ξ, $\nu^* \in - \langle \sigma, 2\sigma, \nu \rangle$ and $\xi \in - \langle \sigma, \nu, \sigma \rangle$. Then we have

$$-(\nu^* + \xi) \in \langle \sigma, 2\sigma, \nu \rangle + \langle \sigma, \nu, \sigma \rangle$$
$$= \langle \nu, 2\sigma, \sigma \rangle + \langle \sigma, \nu, \sigma \rangle \quad \text{by (3.9), i)}$$
$$\equiv \langle \nu, \sigma, 2\sigma \rangle + \langle \sigma, \nu, \sigma \rangle \quad \text{by (3.5)}$$
$$\equiv 0 \bmod \sigma \circ G_{11} + \nu \circ G_{15} \quad \text{by (3.10)}.$$

Since G_{11} is generated by ζ and since $\sigma \circ \zeta = 0$, then $\sigma \circ G_{11} = 0$. G_{15} is generated by ρ and $\kappa \circ \eta$ since Theorem 10.10 and (10.23). Since $\rho \circ \nu = 0$ and $\eta \circ \nu = 0$, then $\nu \circ G_{15} = 0$. Therefore, we have that

$$\nu^* + \xi = 0. \qquad \text{q.e.d.}$$

Corollary 12.25. $4(\nu_{18}^* + \xi_{18}) = 2(\nu_{19}^* + \xi_{19}) = \nu_{20}^* + \xi_{20} = 0$, $\Delta(\eta_{35}^2) = 4(\nu_{17}^* + \xi_{17})$, $\Delta(\eta_{37}) = 2(\nu_{18}^* + \xi_{18})$ and $\Delta(\iota_{39}) = \nu_{19}^* + \xi_{19}$.

Proof. Since the group π_{38}^{20} is stable, then $\nu_{20}^* + \xi_{20} = 0$. Consider the exact sequences

$$\pi_{n+20}^{2n+1} \longrightarrow \pi_{n+18}^{n} \xrightarrow{E} \pi_{n+19}^{n+1}$$

for $n = 17, 18, 19$. Since $\pi_{37}^{35} \approx Z_2$ and $\pi_{38}^{37} \approx Z_2$, then the kernel of E has at most two elements for $n = 17$ and $n = 18$. Since $\Delta(\pm 2\iota_{39}) = \Delta H\Delta(\iota_{41}) = 0$, by Proposition 2.7, then the kernel of E has at most two elements for $n = 19$. Thus the kernel of $E^3 : \pi_{35}^{17} \to \pi_{38}^{20}$ has at most 8 elements. Since $\nu_{17}^* + \xi_{17}$ is of order 8 and $E^3(\nu_{17}^* + \xi_{17}) = \nu_{20}^* + \xi_{20}$

$= 0$, then it follows that the kernel of the E^3 is a subgroup generated by $\nu^*_{17} + \xi_{17}$. Then the corollary is verified easily by the above exact sequences. q.e.d.

It is checked in the results of Theorem 12.17 that the homomorphisms $\Delta : \pi^{35}_{36} \to \pi^{17}_{34}$ and $\Delta : \pi^{37}_{37} \to \pi^{18}_{35}$ are isomorphisms into. Then it follows from the exactness of (4.4) that the following sequences are exact:

$$\pi^{35}_{37} \xrightarrow{\Delta} \pi^{17}_{35} \xrightarrow{E} \pi^{18}_{36} \longrightarrow 0 ,$$

$$\pi^{37}_{38} \xrightarrow{\Delta} \pi^{18}_{36} \xrightarrow{E} \pi^{19}_{37} \longrightarrow 0 ,$$

$$\pi^{39}_{39} \xrightarrow{\Delta} \pi^{19}_{37} \xrightarrow{E} \pi^{20}_{38} \longrightarrow \pi^{39}_{38} = 0 .$$

Then, we have, by Corollary 12.25, that

$$\pi^{18}_{36} = \{\nu^*_{18}\} \oplus \{\nu^*_{18} + \xi_{18}\} \oplus \{\eta_{18} \circ \bar{\mu}_{19}\} \approx Z_8 \oplus Z_4 \oplus Z_2 ,$$

$$\pi^{19}_{37} = \{\nu^*_{19}\} \oplus \{\nu^*_{19} + \xi_{19}\} \oplus \{\eta_{19} \circ \bar{\mu}_{20}\} \approx Z_8 \oplus Z_2 \oplus Z_2 ,$$

$$\pi^{20}_{38} = \{\nu^*_{20}\} \oplus \{\eta_{20} \circ \bar{\mu}_{21}\} \approx Z_8 \oplus Z_2 .$$

The results for the stable groups on Theorem 12.22 follow directly.

We have also, from Corollary 12.25 and from the exactness of (4.4), the following exact sequences.

$$\pi^{35}_{38} \xrightarrow{\Delta} \pi^{17}_{36} \xrightarrow{E} \pi^{18}_{37} \longrightarrow 0,$$

$$\pi^{37}_{39} \xrightarrow{\Delta} \pi^{18}_{37} \xrightarrow{E} \pi^{19}_{38} \longrightarrow 0,$$

$$\pi^{39}_{40} \xrightarrow{\Delta} \pi^{19}_{38} \xrightarrow{E} \pi^{20}_{39} \xrightarrow{H} \{2\iota_{39}\},$$

$$\pi^{41}_{41} \xrightarrow{\Delta} \pi^{20}_{39} \xrightarrow{E} \pi^{21}_{40} \longrightarrow 0 .$$

By Proposition 2.5 and ii) of Proposition 12.20,

$$\Delta(\nu_{35}) \equiv \eta^*_{17} \circ \nu_{33} + \omega_{17} \circ \nu_{33} \bmod \sigma_{17} \circ \mu_{24} \circ \nu_{33}.$$

By Theorem 7.6, $\sigma_{17} \circ \mu_{24} \circ \nu_{33} \in \sigma_{17} \circ \pi^{24}_{36} = 0$. We have

$$\begin{aligned} \eta^*_{17} \circ \nu_{33} &\in \{\sigma_{17}, 2\sigma_{24}, \eta_{31}\} \circ \nu_{33} \\ &= \sigma_{17} \circ \{2\sigma_{24}, \eta_{31}, \nu_{33}\} && \text{by Proposition 1.4} \\ &\in \sigma_{17} \circ \pi^{24}_{36} = 0 && \text{by Theorem 7.6.} \end{aligned}$$

Thus we have obtained that

$$\Delta(\nu_{35}) = \omega_{17} \circ \nu_{33} .$$

Since ν_{35} generates π^{35}_{38}, we have from the above first sequence that

$$\pi^{18}_{37} = \{\bar{\sigma}_{18}\} \oplus \{\bar{\xi}_{18}\} \approx Z_2 \oplus Z_8 .$$

Apply i) of Proposition 11.10 for the case $n = 18$ and $\alpha = \eta_{36}$.

Then there exists an element β such that $H(\beta) = \eta_{37}^2$. Thus $\Delta\pi_{39}^{37} = \{\Delta(\eta_{37}^2)\} = \{\Delta H(\beta)\} = 0$. It follows from the exactness of the above second sequence that $E : \pi_{37}^{18} \to \pi_{38}^{19}$ is an isomorphism onto.

Apply ii) of Proposition 11.10 for the case $n = 19$ and $\alpha = \iota_{38}$. Then there exists an element β such that $H(\beta) = \eta_{39}$. Thus $\Delta\pi_{40}^{39} = \{\Delta(\eta_{39})\} = \{\Delta H(\beta)\} = 0$. By Proposit on 2.7, $H(\Delta(\iota_{41})) = \pm 2\iota_{39}$. Then it follows from the exactness of the above third sequence that the sequence splits and

$$\pi_{39}^{20} = \{\Delta(\iota_{41})\} \oplus \{\bar{\sigma}_{20}\} \oplus \{\bar{\zeta}_{20}\} \approx Z \oplus Z_2 \oplus Z_8$$

It follows from the exactness of the last sequence that

$$\pi_{40}^{21} = \{\bar{\sigma}_{21}\} \oplus \{\bar{\zeta}_{21}\} \approx Z_2 \oplus Z_8 .$$

The results for the stable groups in Theorem 12.23 follows immediately. Consequently, the proof of Theorem 12.22 and Theorem 12.23 is completed.

CHAPTER XIII

Odd Components

The purpose of this chapter is to compute p-primary components $\pi_{n+k}(S^n;p)$ for odd prime p, in order to complete our table of $\pi_{n+k}(S^n)$.

We shall introduce necessary results from [15] and [21].

First we mention Serre's isomorphism [15]

(13.1). $\pi_{i-1}(S^{2m-1};p) + \pi_i(S^{4m-1};p) \approx \pi_i(S^{2m};p)$

which is given by the correspondence $(\alpha, \beta) \to E\alpha + [\iota_{2m}, \iota_{2m}] \circ \beta$.

In the following, we shall devote to compute the groups $\pi_{n+k}(S^n;p)$ for the case that n is odd, then the case that n is even will be computed by (13.1) easily.

Let $\Omega^2(S^{n+2}) = \Omega(\Omega(S^{n+2}))$. The canonical injection of S^n into $\Omega(S^{n+1})$ and an injection $\Omega(S^{n+1}) \subset \Omega^2(S^{n+2})$ induced by the canonical injection of S^{n+1} define an injection of S^n into $\Omega^2(S^{n+2})$ such that the homomorphism $\pi_i(S^n) \to \pi_i(\Omega^2(S^{n+2}))$ induced by this injection is equivalent to $E^2 : \pi_i(S^n) \to \pi_{i+2}(S^{n+2})$. Then, by concerning homotopy exact sequence associated with the pair $(\Omega^2(S^{n+2}), S^n)$, we have the following exact sequence.

$$\cdots \longrightarrow \pi_i(S^n) \xrightarrow{E^2} \pi_{i+2}(S^{n+2}) \longrightarrow \pi_i(\Omega^2(S^{n+2}), S^n) \xrightarrow{\partial} \pi_{i-1}(S^n) \longrightarrow \cdots .$$

Since the groups $\pi_i(S^n)$ are finite except $\pi_n(S^n) \approx Z$ and since $E^2 : \pi_n(S^n) \longrightarrow \pi_{n+2}(S^{n+2})$ are isomorphisms onto, then the exact sequence induces the following exact sequence (n = odd).

(13.2). $$\cdots \longrightarrow \pi_i(S^n;p) \xrightarrow{E^2} \pi_{i+2}(S^{n+2};p) \longrightarrow \pi_i(\Omega^2(S^{n+2}), S^n;p) \xrightarrow{\partial} \pi_{i-1}(S^n;p) \longrightarrow \cdots .$$

The following proposition and corollary were proved in [21].

Proposition 13.1 [21 : Theorem (8.3)]. *Let n be odd and p be an odd prime. Then there exists an exact sequence*

$$\cdots \longrightarrow \pi_{i+2}(S^{p(n+1)+1};p) \xrightarrow{\Delta} \pi_i(S^{p(n+1)-1};p) \longrightarrow \pi_{i-1}(\Omega^2(S^{n+2}), S^n;p) \longrightarrow \pi_{i+1}(S^{p(n+1)+1};p) \longrightarrow \cdots$$

for $i > p(n+1)-1$ *such that* $\Delta \circ E^2 = f_{p*}$ for a mapping $f_p : S^{p(n+1)-1} \to S^{p(n+1)-1}$ of degree p.

Corollary 13.2. [21 : (8.7)']. *Let* n *be odd and* p *be an odd prime. Then we have an isomorphism*

$$\pi_{i-1}(\Omega^2(S^{n+2}), S^n;p) \approx \pi_i(S_f, S^{p(n+1)-1})$$

for $i < p^2(n+1)-2$, *where* S_f *is a mapping cylinder of* f_p.

The proposition is obtained from the exact sequence

$$\cdots \to \pi_i(\Omega^2(S^{n+2}), \Omega(S^{n+1}_{p-1})) \to \pi_{i-1}(\Omega(S^{n+1}_{p-1}), S^n) \to \pi_{i-1}(\Omega^2(S^{n+2}), S^n) \to \pi_{i-1}(\Omega^2(S^{n+2}), \Omega(S^{n+1}_{p-1})) \cdots$$

by two isomorphisms

$$\pi_i(\Omega^2(S^{n+2}), \Omega(S^{n+1}_{p-1});p) \approx \pi_{i+2}(S^{p(n+1)+1};p)$$

and
$$\pi_{i-1}(\Omega(S^{n+1}_{p-1}), S^n;p) \approx \pi_i(S^{p(n+1)-1};p) .$$

The first isomorphism is the compositions of isomorphisms Ω_0^{-1} : $\pi_{i-1}(\Omega^2(S^{n+2}), \Omega(S^{n+1}_{p-1})) \approx \pi_{i+1}(\Omega(S^{n+2}), S^{n+1}_{p-1})$, $i_* : \pi_{i+1}(S^{n+1}_\infty, S^{n+1}_{p-1}) \approx \pi_{i+1}(\Omega(S^{n+2}), S^{n+1}_{p-1})$, $h_{p*} : \pi_{i+1}(S^{n+1}, S^{n+1}_{p-1};p) \approx \pi_{i+1}(S^{p(n+1)};p)$ of Theorem 2.4 and $\Omega_1^{-1} : \pi_{i+1}(S^{p(n+1)}_\infty) \approx \pi_{i+2}(S^{p(n+1)+1})$, where $h_p : (S^{n+1}_\infty, S^{n+1}_{p-1}) \to (S^{p(n+1)}_\infty, e_0)$ is the combinatorial extension of a mapping $h'_p : (S^{n+1}_p, S^{n+1}_{p-1}) \to (S^{p(n+1)}, e_0)$ which shrinks S^{n+1}_{p-1} to e_0.

The proof of the second isomorphism is more complicated (see [21]). Here we explain a proof of the isomorphism for $i \leq p(p(n+1)-2)-3$. This is sufficient in the following applications.

First $H^*(S^{n+1}_{p-1}, Z_p)$ is a truncated polynomial ring generated by an element e of dimension $n+1$ with the relation $e^p = 0$. Then by spectral sequence arguments we have that $H^*(\Omega(S^{n+1}_{p-1}), Z_p)$ is isomorphic to the tensor product of $H^*(S^n, Z_p)$ and a subgroup isomorphic to $H^*(S^{p(n+1)-2}_\infty, Z_p)$. Furthermore, we may compute, by spectral sequence arguments, that $H^*(\Omega(\Omega(S^{n+1}_{p-1}), S^n), Z_p)$ is isomorphic to $H^*(S^{p(n+1)-3}, Z_p)$ up to dimension $p(p(n+1)-2)-3$. By generalized J. H. C. Whitehead's theorem [15], we have that $\pi_{i-2}(S^{p(n+1)-3};p)$ and $\pi_{i-2}(\Omega(\Omega(S^{n+1}_{p-1}), S^n);p) \approx \pi_{i-1}(\Omega(S^{n+1}_{p-1}), S^n;p)$ are isomorphic for $i \leq p(p(n+1)-2)-3$. Then the second isomorphism is proved by the fact that $E^2 : \pi_{i-2}(S^{p(n+1)-3};p) \to \pi_i(S^{p(n+1)-1};p)$ are isomorphic onto for $i \leq p(p(n+1)-2)-3$, which is proved by the exactness of (13.2) and by Corollary 13.2 for $p(n+1)-3$ in place of n.

The proof of the relation $\Delta \circ E^2 = f_{p*}$ and Corollary 13.2 is based on the fact that the p-primary component of the group $H^{p(n+1)-2}(\Omega^2(S^{n+1}))$ is Z_p.

For the case $n = 1$, we have the following

<u>Proposition</u> 13.3. <u>Let</u> p <u>be an odd prime</u>. <u>Then there exists an exact sequence</u>

$$\cdots \longrightarrow \pi_{i+2}(S^{2p+1};p) \xrightarrow{\Delta} \pi_i(S^{2p-1};p) \xrightarrow{G} \pi_{i+1}(S^3;p) \xrightarrow{H} \pi_{i+1}(S^{2p+1};p) \longrightarrow \cdots$$

<u>for</u> $i > 2p-1$, <u>such that</u> $\Delta \circ E^2 = f_{p*}$ <u>and</u> $G(\beta) = \alpha_0 \circ E\beta$ <u>for an arbitrary generator</u> α_0 <u>of</u> $\pi_{2p}(S^3;p) \approx Z_p$.

<u>Proof</u>. It is well known that

(13.3). $$\pi_{2p}(S^3;p) \approx Z_p .$$

This is proved by Corollary 13.2 or more precisely by use of 3-connective fibre space over S^3 (cf. [12]). By Theorem 2.4, we have an isomorphism

$$h_{p*} : \pi_{2p}(S^2_\infty, S^2_{p-1}) \approx \pi_{2p}(S^{2p}_\infty) \approx Z.$$

Let M be the complex projective space of complex (p-1) dimension. Then S^{2p-1} is a fibre bundle over M with the fibre S^1. Since $\pi_i(M) = 0$ for $2 < i < 2p-1$, the inclusion of S^2 into M is extendable over a mapping

$$f' : S^2_{p-1} \to M .$$

Let Y be the bundle induced by f and let $\bar{f} : Y \to S^{2p-1}$ the induced map. Remark that Y is a 2-connective fibre space over S^2_{p-1} and hence the projection p of Y onto S^2_{p-1} induces isomorphisms

$$\pi_i(Y) \approx \pi_i(S^2_{p-1}) \quad \text{for } i > 2.$$

It is seen that M and S^2_{p-1} has the same ring structure of the cohomology mod p. Then it follows that f' induces isomorphisms of the cohomology groups mod p and thus isomorphisms mod p of homotopy groups. Considering the bundle map $\bar{f}$, we have that $\bar{f}$ induces isomorphisms

$$\bar{f}_* : \pi_i(Y;p) \to \pi_i(S^{2p-1};p)$$

for all i and $\pi_{2p}(Y;p) \approx Z +$ (finite group of no p-torsion). Let $g : S^{2p-1} \to S^2_{p-1}$ be a mapping representing an element of $\pi_{2p-1}(S^2_{p-1})$ not divisible by p, then g induces isomorphisms of the p-primary component of the homotopy groups since the composition $f \circ g$ has an order not divisible by p and it induces automorphisms of $\pi_i(S^{2p-1};p)$.

Consider the exact sequence of the pair (S^2_∞, S^2_{p-1}) :

$$\cdots \longrightarrow \pi_i(S^2_{p-1}) \xrightarrow{i_*} \pi_i(S^2_\infty) \longrightarrow \pi_i(S^2_\infty, S^2_{p-1}) \longrightarrow \pi_{i-1}(S^2_{p-1}) \longrightarrow \cdots .$$

Since $\pi_{2p}(S^2_\infty) \approx \pi_{2p+1}(S^3)$ is finite, then we have an exact sequence

$$0 \longrightarrow Z \longrightarrow \pi_{2p-1}(S^2_{p-1}) \xrightarrow{\Omega_1^{-1} \circ i_*} \pi_{2p}(S^3) \longrightarrow 0 .$$

Then it follows from (13.3) that there exists an element α' of $\pi_{2p-1}(S^2_{p-1})$ such that $p\alpha'$ is the image of a generator of $\pi_{2p}(S^2_\infty, S^2_{p-1})$ and that $i_*(\alpha')$ is an element of order p.

Let $f : (CS^{2p-1}, S^{2p-1}) \to (S^2_\infty, S^2_{p-1})$ be a characteristic map of the cell $e^{2p} = S^2_p - S^2_{p-1}$ such that $f \mid S^{2p-1}$ represents $p\alpha'$.

The following diagram is commutative.

$$\begin{array}{ccccc}
\pi_i(CS^{2p-1}, S^{2p-1}) & & \xrightarrow[\approx]{\partial} & & \pi_{i-1}(S^{2p-1}) \\
\downarrow f_* & & & & \downarrow \pm E^2 \\
\pi_i(S^2_\infty, S^2_{p-1}) & \xrightarrow{h_{p*}} & \pi_i(S^{2p}_\infty) & \xleftarrow[\approx]{\Omega_1} & \pi_{i+1}(S^{2p+1})
\end{array} .$$

For an element β of $\pi_{i-1}(S^{2p-1};p)$, we have

$$h_{p*}^{-1}(\Omega_1 E^2\beta) = \pm f_* \partial^{-1}(\beta) = \pm (f|S^{2p-1})_* \partial \partial^{-1}(\beta)$$

$$= \pm p\alpha' \circ \beta .$$

Now we replace the p-primary parts of the above exact sequence of the pair (S^2_∞, S^2_{p-1}) by the isomorphisms

$$\Omega_1^{-1} \circ h_{p*} : \pi_i(S^2_\infty, S^2_{p-1};p) \approx \pi_{i+1}(S^{2p+1};p) ,$$

$$\Omega_1^{-1} : \pi_i(S^2_\infty ;p) \approx \pi_{i+1}(S^3;p)$$

$$g_*^{-1} : \pi_i(S^2_{p-1};p) \approx \pi_i(S^{2p-1};p), \quad i > 2 ,$$

where g is a representative of $\pm \alpha'$. Then we obtain the exact sequence of the proposition, where Δ and G are given by the commutativity of the following diagram :

$$\begin{array}{ccccc}
\pi_{i+2}(S^{2p+1};p) & \xrightarrow{\Delta} & \pi_i(S^{2p-1};p) & \xrightarrow{G} & \pi_{i+1}(S^3;p) \\
\downarrow \Omega_1 & & & & \\
\pi_{i+1}(S^{2p}_\infty ;p) & & \downarrow g_* & & \downarrow \Omega_1 \\
\uparrow h_{p*} & & & & \\
\pi_{i+1}(S^2_\infty, S^2_{p-1};p) & \xrightarrow{\partial} & \pi_i(S^2_{p-1};p) & \xrightarrow{i_*} & \pi_i(S^2_\infty ;p)
\end{array} .$$

It is verified that

$$\Delta(E^2\beta) = g_*^{-1}(\partial\, h_{p*}^{-1}\, \Omega_1 E^2\beta) = g_*^{-1}(\pm\, p\alpha' \circ \beta)$$

$$= p\iota_{2p-1} \circ \beta$$

and

$$G(\gamma) = \Omega_1^{-1}(i_*\, g_*(\gamma)) = \Omega_1^{-1}(\{i \circ g\} \circ \gamma)$$

$$= \alpha_0 \circ E\gamma\ ,$$

where α_0 is the class $\Omega_1^{-1}\{i \circ g\}$. Finally we remark that any element of order p in $\pi_{2p}(S^3)$ is of the form $r\alpha_0$ for $r \not\equiv 0 \pmod p$, and that the exactness of the sequence still holds when we replace α_0 by $r\alpha_0$.

q.e.d.

i). $\pi_{n+k}(S^n;p)$ <u>for</u> $k < 2p(p-1)-2$.

We recall the following results for stable groups from Theorem 4.15 of [25].

(13.4). $(G_{2i(p-1)-1};p) \approx Z_p$ *for* $i = 1, 2, \ldots, p-1$ *and* $(G_k;p) = 0$ *otherwise for* $k < 2p(p-1)-2$. *Let* α_1 *be a generator of* $(G_{2p-3};p)$, *then the secondary compositions*

$$\alpha_{i+1} = \langle\, \alpha_1,\ p\iota,\ \alpha_i \,\rangle$$

give generators of $(G_{2(i+1)(p-1)-1};p)$, $i + 1 < p$, *inductively*.

Theorem 13.4. *Let* p *be an odd prime*.

$\pi_{2m+1+2i(p-1)-2}(S^{2m+1};p) \approx Z_p$ *for* $1 \leq m < i$, *and* $i = 2, \ldots, p-1$

$\pi_{2m+1+2i(p-1)-1}(S^{2m+1};p) \approx Z_p$ *for* $i = 1, 2, \ldots, p-1$, $m \geq 1$,

$\pi_{2m+1+k}(S^{2m+1};p) = 0$ *otherwise for* $k < 2p(p-1)-2$.

Proof. By Corollary 13.2,

$$\pi_i(\Omega^2(S^{2m+1}),\ S^{2m-1};p) = 0 \quad \text{for} \quad i < 2mp - 2\ .$$

It follows from the exactness of the sequence (13.2) that

(13.5) $E^2 : \pi_i(S^{2m-1};p) \to \pi_{i+2}(S^{2m+1};p)$ is an isomorphism onto if $i < 2mp - 3$ and it is a homomorphism onto if $i = 2mp - 3$.

Then we see that the theorem is true for large values of m. In particular,

$$\pi_{2m+1+k}(S^{2m+1};p) = 0 \quad \text{for} \quad k < 2p - 3\ ,$$

since it is isomorphic to $\pi_{k+1}(S^1;p) = 0$. Also we have that

$$\pi_{2mp-1+k}(S^{2mp-1};p) \approx \begin{cases} Z_p & \text{for } k = 2i(p-1)-1, \quad 1 \leq i < p, \quad 1 \leq m, \\ 0 & \text{otherwise for } k < 2p(p-1)-2, \quad 1 \leq m. \end{cases}$$

Then by use of (13.5), Proposition 13.1 and by Corollary 13.2, we have that

(13.6).
$$\pi_{2m+k}(\Omega^2(S^{2m+1}), S^{2m-1};p) \approx \begin{cases} Z_p & \underline{\text{for}} \ k = 2m(p-1)-2 \\ Z_p & \underline{\text{for}} \ k = 2(m+i)(p-1)-3 \\ Z_p & \underline{\text{for}} \ k = 2(m+i)(p-1)-2 \\ 0 & \underline{\text{otherwise for}} \ k < 2(m+p)(p-1)-4, \end{cases}$$

<u>where</u> $2 \leq m$ <u>and</u> $1 \leq i < p$. We have also by use of Proposition 13.3,

(13.6)'.
$$\pi_{3+k}(S^3;p) \approx \begin{cases} Z_p & \underline{\text{for}} \ k = 2p-3, \ k = 2(i+1)(p-1)-2 \ \underline{\text{and}} \\ & \quad k = 2(i+1)(p-1)-1, \ i = 1, 2, \ldots, p-1 \\ 0 & \underline{\text{otherwise for}} \ k < 2(p+1)(p-1)-3. \end{cases}$$

Now let $k < 2p-3$ or $2i(p-1)-1 < k < 2(i+1)(p-1)-2$ $(1 \leq i \leq p)$. Then we have an exact sequence

$$\pi_{2m-1+k}(S^{2m-1};p) \xrightarrow{E^2} \pi_{2m+1+k}(S^{2m+1};p) \longrightarrow 0$$

from (13.2) and (13.6). If $\pi_{2m-1+k}(S^{2m-1};p) = 0$, then $\pi_{2m+1+k}(S^{2m+1};p) = 0$. Since $\pi_{2m+1+k}(S^{2m+1};p) = 0$ if $m = 1$, then it follows that

$$\pi_{2m+1+k}(S^{2m+1};p) = 0 \quad \text{for} \quad m = 1, 2, \ldots,$$

in this case.

Next consider the case that $k = 2i(p-1)-1$ for $1 \leq i < p$. We have exact sequences

$$0 \longrightarrow \pi_{2m-1+k}(S^{2m-1};p) \xrightarrow{E^2} \pi_{2m+1+k}(S^{2m+1};p)$$

from (13.2) and (13.6). Thus $E^\infty : \pi_{3+k}(S^3;p) \to (G_k;p)$ is an isomorphism into. Since the groups $\pi_{3+k}(S^3;p)$ and $(G_k;p)$ are isomorphic to Z_p, then the homomorphism E^∞ is an isomorphism onto. Then it follows that the above E^2 is an isomorphism onto for $m = 2, 3, \ldots,$ and that

$$\pi_{2m+1+k}(S^{2m+1};p) \approx Z_p$$

for $k = 2i(p-1)-1$, $1 \leq i < p$ and $m \geq 1$.

Finally consider the case that $k = 2i(p-1)-2$ and $2 \leq i < p$. Then it follows from the above results on E^2, (13.6) and from the exactness of (13.2) that the following sequences are exact.

$$0 \longrightarrow Z_p \xrightarrow{\Delta} \pi_{2i-1+k}(S^{2i-1};p) \xrightarrow{E^2} \pi_{2i+1+k}(S^{2i+1};p) \longrightarrow 0 ,$$

$$0 \longrightarrow Z_p \xrightarrow{\Delta} \pi_{2m-1+k}(S^{2m-1};p) \xrightarrow{E^2} \pi_{2m+1+k}(S^{2m+1};p) \longrightarrow Z_p \longrightarrow 0 ,$$

for $2 \le m < i$. The group $\pi_{2i+1+k}(S^{2i+1};p)$ is stable by (13.5), and hence it is trivial by (13.4). We see that

$$E^2\pi_{2m-1+k}(S^{2m-1};p) = 0 \quad \text{for} \quad 2 \le m \le i .$$

For, if one of the E^2-images is not trivial, then it follows from the result $\pi_{3+k}(S^3;p) \approx Z_p$ and from the exactness of the above sequences that $\pi_{2i+1+k}(S^{2i+1};p)$ is not trivial. But this is a contradiction. Thus $E^2 = 0$ and we have isomorphisms

$$\Delta : Z_p \approx \pi_{2m-1+2i(p-1)-2}(S^{2m-1};p) \quad \text{for} \quad 2 \le m < i .$$

Consequently the theorem is proved. q.e.d.

<u>Lemma</u> 13.5. There exists a sequence $\{\alpha_i(3)\}$ of elements $\alpha_i(3) \in \pi_{2i(p-1)+2}(S^3;p)$ <u>such that</u> $\alpha_1(3)$ <u>is a generator of</u> $\pi_{2p}(S^3;p)$, $p\alpha_i(3) = 0$ <u>and</u>

$$\alpha_{i+1}(3) \in \{\alpha_i(3),\ p\iota_{2i(p-1)+2},\ E^{2i(p-1)-1}\alpha_1(3)\}_1 .$$

<u>Proof</u>. Obviously $\alpha_1(3)$ exists. Assume that there exists an element $\alpha_i(3)$ of $\pi_{2i(p-1)+2}(S^3;p)$ such that $p\alpha_i(3) = 0$. Then the secondary composition $\{\alpha_i(3),\ p\iota_{2i(p-1)+2},\ E^{2i(p-1)-1}\alpha_1(3)\}_1$ is defined.

Let $\alpha_{i+1}(3)$ be an element of the secondary composition. Then

$$\begin{aligned} p\alpha_{i+1}(3) &\in p\{\alpha_i(3),\ p\iota_{2i(p-1)+2},\ E^{2i(p-1)-1}\alpha_1(3)\}_1 \\ &= \alpha_i(3)\circ E\{p\iota_{2i(p-1)+1},\ E^{2i(p-1)-2}\alpha_1(3),\ p\iota_{2(i+1)(p-1)}\} \\ &\subset \alpha_i(3)\circ E\pi_{2(i+1)(p-1)+1}(S^{2i(p-1)+1};p) && \text{by (4.6)} \\ &= 0 && \text{by Theorem 13.4.} \end{aligned}$$

By induction on i, we have the existence of the sequence $\{\alpha_i(3)\}$.

q.e.d.

Denote that

$$\alpha_i(n) = E^{n-3}\alpha_i(3) \quad \text{for} \quad n \ge 3 \quad \text{and} \quad \alpha_i = E^\infty\alpha_i(3) .$$

Then

$$\alpha_{i+1} \in \langle \alpha_i,\ p\iota,\ \alpha_1 \rangle = \langle \alpha_1,\ p\iota,\ \alpha_i \rangle$$

by i) of (3.9). We may consider that this α_i coincides with the α_i in Theorem 4.15 of [25]. The following proposition is easily verified from the above discussions.

Proposition 13.6. $\pi_{2m+1+2i(p-1)-1}(S^{2m+1};p) = \{\alpha_i(2m+1)\} \approx Z_p$ for $i \leq i < p$ and $m \geq 1$. $\pi_{2i(p-1)+1}(S^3;p) = \{\alpha_1(3) \circ \alpha_{i-1}(2p)\} \approx Z_p$ for $2 \leq i < p$.

Since $E^2\pi_{2i(p-1)+1}(S^3;p) = 0$ for $2 \leq i < p$, we have

(13.7). $\alpha_i(n) \circ \alpha_j(n+2i(p-1)-1) = 0$ for $m \geq 5$ and $i+j < p$.

Since $E^\infty : \pi_{2m+1+2i(p-1)-1}(S^{2m+1};p) \to (G_{2i(p-1)-1};p)$ is an isomorphism onto if $1 \leq i < p$ and $m \geq 1$, then it follows from the result

$$(i/(i+j))\alpha_{i+j} \in < \alpha_i,\ \alpha_j,\ p\iota >$$

of Proposition 4.17 of [25] that

(13.8). $(i/(i+j))\alpha_{i+j}(n) \in \{\alpha_i(n),\ \alpha_j(n+2i(p-1)-1), p\iota_{2(i+j)(p-1)-2}\}_t$ for $i+j < p$, $n \geq 5$ and $t \leq n+2i(p-1)-4$.

ii). $\pi_{n+k}(S^n;3)$ for $k \leq 19$.

First we show some additional properties of the exact sequence of Proposition 13.3. It is directly verified from the definition of the exact sequence in Proposition 13.3 that the homomorphism $H : \pi_{i+1}(S^3;p) \to \pi_{i+1}(S^{2p+1};p)$ is defined by the formula

$$H = \Omega_1^{-1} \circ h_{p*} \circ \Omega_1 : \pi_{i+1}(S^3;p) \longrightarrow \pi_i(S^2_\infty;p) \longrightarrow \pi_i(S^{2p}_\infty;p) \longrightarrow \pi_{i+1}(S^{2p+1};p) .$$

Then the proof of the following formula (13.9) is similar to that of Proposition 2.2.

(13.9). $$H(\alpha \circ E\beta) = H(\alpha) \circ E\beta .$$

Remark that this homomorphism H is generalized to

$$H : \pi_{i+1}(S^{2m+1}) \longrightarrow \pi_{i+1}(S^{2pm+1})$$

and that it coincides with the composition of the homomorphism $\pi_{i+1}(S^{2m+1};p) \to \pi_{i-1}(\Omega^2(S^{2m+1}),\ S^{2m-1};p)$ in (13.2) and the homomorphism $\pi_{i-1}(\Omega^2(S^{2m+1}),\ S^{2m-1};p) \to \pi_{i+1}(S^{2pm+1};p)$ in Proposition 13.1 on the p-primary components. (13.9) holds for the general case.

We see in the proof of Theorem 13.4 that

(13.10). $H(\alpha_i(3)) = x\alpha_{i-1}(2p+1)$ for $1 < i < p$ and for some integer $x \not\equiv 0 \pmod p$.

Remark that (13.10) is true for all $i > 1$ and the coefficient x is constant.

Lemma 13.7. Assume that $E^2\beta = 0$ for $\beta \in \pi_i(S^{2p-1};p)$, then there exists an element $\gamma \in \pi_{i+2}(S^{2p+1};p)$ such that $\Delta\gamma = \beta$.

Proof. By Proposition 13.3, $EG(\beta) = E\alpha_0 \circ E^2\beta = 0$. Since $E : \pi_{i+1}(S^3) \to \pi_{i+2}(S^4)$ is an isomorphism into, it follows that $G(\beta) = 0$. By the exactness of the sequence of Proposition 13.3, there exists an element γ of $\pi_{i+2}(S^{2p+1};p)$ such that $\Delta\gamma = \beta$. q.e.d.

In the following, we consider the case $p = 3$.

Lemma 13.8. *Let $p = 3$. Let $\beta_1(5)$ be an element of $\{\alpha_1(5), \alpha_1(8), \alpha_1(11)\}_1$, then*

$$3\beta_1(5) = -\alpha_1(5) \circ \alpha_2(8) .$$

Proof. We have

$$3\beta_1(5) \in \{\alpha_1(5), \alpha_1(8), \alpha_1(11)\}_1 \circ 3\iota_{15}$$
$$= \alpha_1(5) \circ E\{\alpha_1(7), \alpha_1(10), 3\iota_{13}\} \text{ by Proposition 1.4}$$

and $-\alpha_1(5) \circ \alpha_2(8) \in \alpha_1(5) \circ E\{\alpha_1(7), \alpha_1(10), 3\iota_{13}\}$ by (13.8). The composition $\alpha_1(5) \circ E\{\alpha_1(7), \alpha_1(10), 3\iota_{13}\}$ is a coset of

$$\alpha_1(5) \circ E\pi_{14}(S^7) \circ 3\iota_{15} = 3\alpha_1(5) \circ E\pi_{14}(S^7) = 0 .$$

Thus we have that $3\beta_1(5) = -\alpha_1(5) \circ \alpha_2(8)$. q.e.d.

Denote that

$$\beta_1(n) = E^{n-5}\beta_1(5) \quad \text{for} \quad n \geq 5 .$$

Now we recall the results on the stable groups $(G_k;3)$ from Theorem 4.15 of [25]:

(13.11).
$$(G_{10};3) = \{\beta_1\} \approx Z_3,$$
$$(G_{11};3) = \{\alpha_3'\} \approx Z_9, \quad 3\alpha_3' = \alpha_3,$$
$$(G_{13};3) = \{\alpha_1 \circ \beta_1\} \approx Z_3,$$
$$(G_{15};3) = \{\alpha_4\} \approx Z_3,$$
$$(G_{19};3) = \{\alpha_5\} \approx Z_3,$$
and
$$(G_k;3) = 0 \quad \text{for} \quad k = 12, 14, 16, 17, 18.$$

Theorem 13.9. *There exists an element $\alpha_3'(5)$ such that $3\alpha_3'(5) = \alpha_3(5)$. Denote that $\alpha_3'(n) = E^{n-5}\alpha_3'(5)$ for $n \geq 5$. Then*

$$\pi_{13}(S^3;3) = \{\alpha_1(3) \circ \alpha_2(6)\} \approx Z_3,$$
$$\pi_{15}(S^5;3) = \{\beta_1(5)\} \approx Z_9,$$
$$\pi_{2m+1+10}(S^{2m+1};3) = \{\beta_1(2m+1)\} \approx Z_3 \quad \text{for} \quad m \geq 3,$$
$$\pi_{14}(S^3;3) = \{\alpha_3(3)\} \approx Z_3,$$
$$\pi_{2m+1+11}(S^{2m+1};3) = \{\alpha_3'(2m+1)\} \approx Z_9 \quad \text{for} \quad m \geq 2.$$
and
$$\pi_{2m+1+12}(S^{2m+1};3) = 0 \quad \text{for} \quad m \geq 1.$$

Proof. By (13.6)', $\pi_{13}(S^3;3) \approx \pi_{14}(S^3;3) \approx Z_3$ and $\pi_{15}(S^3;3) = 0$. Furthermore, the group $\pi_{13}(S^3;3)$ is the image of $\pi_{12}(S^5;3) = \{\alpha_2(5)\}$ under the homomorphism G. Thus $\pi_{13}(S^3;3)$ is generated by $G(\alpha_2(5)) = \alpha_0 \circ E\alpha_2(5) = \alpha_0 \circ \alpha_2(6)$. Since α_0 is an element of order 3 in $\pi_6(S^3;3) = \{\alpha_1(3)\} \approx Z_3$, then

$$\pi_{13}(S^3;3) = \{\alpha_1(3) \circ \alpha_2(6)\} \approx Z_3.$$

Since $E^\infty\alpha_3(3) = \alpha_3 \neq 0$ in $(G_{11};3)$, by (13.11), then $\alpha_3(3) \neq 0$, and

$$\pi_{14}(S^3;3) = \{\alpha_3(3)\} \approx Z_3$$

By (13.2) and (13.6), we have the following two exact sequences.

$$\pi_{15}(S^3;3) \xrightarrow{E^2} \pi_{17}(S^5;3) \to 0 \to \pi_{14}(S^3;3) \xrightarrow{E^2} \pi_{16}(S^5;3) \to Z_3 \to \pi_{13}(S^3;3) \xrightarrow{E^2} \pi_{15}(S^5;3) \to Z_3,$$

$$0 \longrightarrow \pi_{16}(S^5;3) \xrightarrow{E^2} \pi_{18}(S^7;3) \to Z_3 \to \pi_{15}(S^5;3) \xrightarrow{E^2} \pi_{17}(S^7;3) \longrightarrow 0 .$$

By (13.5) and by (13.11), we have that $\pi_{17}(S^7;3) \approx Z_3$, $\pi_{18}(S^7;3) \approx Z_9$ and $\pi_{19}(S^7;3) = 0$. It follows from the exactness of the above first sequence that $\pi_{15}(S^3;3) = 0$ implies $\pi_{17}(S^5;3) = 0$.

By Theorem 2.11 of [24], the mod 3 Hopf invariant

$$H_3 : \pi_{18}(S^7) \to Z_3$$

is trivial. By Proposition 2.2 and the diagram (2.4) of [24], we see that the sequence $\pi_{16}(S^5) \xrightarrow{E^2} \pi_{18}(S^7) \xrightarrow{H_3} Z_3$ is an exact sequence of the 3-primary components. Thus $E^2 : \pi_{16}(S^5;3) \to \pi_{18}(S^7;3)$ is onto. It follows from the exactness of the above second sequence that

$$\pi_{16}(S^5;3) \approx Z_9 \quad \text{and} \quad \pi_{15}(S^5;3) \approx Z_9 \text{ or } Z_3 + Z_3.$$

Furthermore, by the exactness of the above first sequence, we have that $E^2 : \pi_{13}(S^3;3) \to \pi_{15}(S^5;3)$ is an isomorphism into. Thus $\alpha_1(5) \circ \alpha_2(8) \neq 0$. It follows from Lemma 13.8 that $\beta_1(5)$ is an element of order 9 and hance it generates $\pi_{15}(S^5;3) \approx Z_9$. We see also that $E^2 : \pi_{14}(S^3;3) \to \pi_{16}(S^5;3)$ is an isomorphism into. Then $E^2\alpha_3(3) = \alpha_3(5)$ is divisible by 3. Let $\alpha_3'(5)$ be an element which divides $\alpha_3(5)$, then $\pi_{16}(S^5;3)$ is generated by $\alpha_3'(5)$. Since $E^2 : \pi_{16}(S^5;3) \to \pi_{18}(S^7;3)$ and $E^2 : \pi_{15}(S^5;3) \to \pi_{17}(S^7;3)$ are onto, then the groups $\pi_{18}(S^7;3)$ and $\pi_{17}(S^7;3)$ are generated by $\alpha_3'(7)$ and $\beta_1(7)$ respectively.

For the case that $m \geq 4$, the theorem is proved by the above results and by (13.5). q.e.d.

Theorem 13.10.

i). $\pi_{2m+1+13}(S^{2m+1};3) = \{\alpha_1(2m+1) \circ \beta_1(2m+4)\} \approx Z_3$ for $m \geq 1$.

ii). $\pi_{2m+1+14}(S^{2m+1};3) \approx \begin{cases} Z_3 & \text{for } 3 \geq m \geq 1, \\ 0 & \text{for } m \geq 4. \end{cases}$

iii). $\pi_{2m+1+15}(S^{2m+1};3) = \{\alpha_4(2m+1)\} \approx Z_3$ for $m \geq 1$.

iv). $\pi_{2m+1+16}(S^{2m+1};3) \approx \begin{cases} Z_3 & \text{for } m = 1, \\ 0 & \text{for } m \geq 2. \end{cases}$

v). $\pi_{2m+1+17}(S^{2m+1};3) \approx \begin{cases} Z_3 & \text{for } m = 1, \\ 0 & \text{for } m \geq 2. \end{cases}$

vi). $\pi_{2m+1+18}(S^{2m+1};3) \approx \begin{cases} Z_3 & \text{for } 4 \geq m \geq 1, \\ 0 & \text{for } m \geq 5. \end{cases}$

vii). $\pi_{2m+1+19}(S^{2m+1};3) = \{\alpha_5(2m+1)\} \approx Z_3$ for $m \geq 1$.

Proof. For the simplicity we shall use the following notations:

$$\pi^n_{n+k} = \pi_{n+k}(S^n;3).$$

First we mention that the theorem is true for the stable groups π^{2m+1}_{2m+1+k} of $m \geq (k-1)/4$, by (13.5) and (13.11).

The case that $m \geq 5$, the groups are stable and the theorem is true.

Consider the case that $m = 4$. Then the unstable groups are π^9_{27} and π^9_{28}. By (13.2) and (13.6), we have an exact sequence

$$0 \longrightarrow \pi^9_{28} \xrightarrow{E^2} \pi^{11}_{30} \longrightarrow Z_3 \longrightarrow \pi^9_{27} \xrightarrow{E^2} \pi^{11}_{29} .$$

Since $E^2\alpha_5(9) = \alpha_5(11)$ generates π^{11}_{30}, then $E^2 : \pi^9_{28} \to \pi^{11}_{30}$ is onto, and hence it is an isomorphism onto. Since $\pi^{11}_{29} = 0$, it follows that $\pi^9_{27} \approx Z_3$. Therefore, we see that the theorem is true for the case $m = 4$.

Next consider the case that $m = 3$ and $k \leq 16$. By (13.2) and (13.6), we have isomorphisms

$$E^2 : \pi^7_{23} \approx \pi^9_{25}, \quad E^2 : \pi^7_{20} \approx \pi^9_{22}$$

and an exact sequence

$$0 \longrightarrow \pi^7_{22} \xrightarrow{E^2} \pi^9_{24} \longrightarrow Z_3 \longrightarrow \pi^7_{21} \xrightarrow{E^2} \pi^9_{23} \longrightarrow 0.$$

Since $E^2\alpha_4(7) = \alpha_4(9)$ generates π^9_{24}, then $E^2 : \pi^7_{22} \to \pi^9_{24}$ is an isomorphism onto. Since $\pi^9_{23} = 0$, it follows that $\pi^7_{21} \approx Z_3$. Therefore, we see that the theorem is true for the case $m = 3$ and $k \leq 16$.

Consider the case that $m = 2$ and $k \leq 16$. By (13.2) and (13.6), we have an isomorphism

$$E^2 : \pi^5_{21} \approx \pi^7_{23}$$

and an exact sequence

$$0 \longrightarrow \pi^5_{20} \xrightarrow{E^2} \pi^7_{22} \longrightarrow Z_3 \longrightarrow \pi^5_{19} \xrightarrow{E^2} \pi^7_{21} \longrightarrow Z_3 \longrightarrow \pi^5_{18} \xrightarrow{E^2} \pi^7_{20} \longrightarrow 0 .$$

Since $E^2\alpha_4(5) = \alpha_4(7)$ generates π^7_{22}, then $E^2 : \pi^5_{20} \to \pi^7_{22}$ is an isomorphism onto. Then we see that the theorem is true for $m = 2$, $k = 15$ and $k = 16$. It follows from the exactness of the above sequence that

(13.12). _If_ $E^2\pi^5_{19} = 0$, _then the groups_ π^5_{19} _and_ π^5_{18} _are isomorphic to_ Z_3. _If_ $E^2\pi^5_{19} \neq 0$, _namely if_ $E^2\pi^5_{19} = \pi^7_{21}$, _then the groups_ π^5_{19} _and_ π^5_{18} _are isomorphic to_ Z_9 _or_ $Z_3 + Z_3$.

Consider the homomorphism $\Delta : \pi^7_{17} \to \pi^5_{15}$ and the relation $\Delta \circ E^2 = f_{3*}$ of Proposition 13.3. Remark that

(13.13). $f_{p*}(2) = p\alpha$ _for_ $\alpha \in \pi_i(S^{2m-1};p)$ _and for a mapping_ $f_p : S^{2m-1}$ S^{2m-1} _of degree_ p (p : _odd prime_).

For, $E(f_{p*}\alpha) = E(p\iota_{2m-1} \circ \alpha) = E(p\alpha)$ and this implies $f_{p*}(\alpha) = p\alpha$, by virtue of (13.1).

Since $\pi^7_{17} \approx Z_3$, $\pi^5_{15} \approx Z_9$ and since $E^2 : \pi^5_{15} \twoheadrightarrow \pi^7_{17}$ is onto, then it is an algebraic consequence that $\Delta : \pi^7_{17} \to \pi^5_{15}$ is an isomorphism into. Similarly, it follows from the isomorphism $E^2 : \pi^5_{16} \approx \pi^7_{18} \approx Z_9$ and from the formula $\Delta \circ E^2 = f_{3*}$ that $\Delta : \pi^7_{18} \to \pi^5_{16}$ is a homomorphism of degree 3. Then it follows from the exactness of the sequence in Proposition 13.3 and from $\pi^7_{16} = \pi^5_{17} = 0$ that

$$\pi^3_{16} \approx \pi^3_{17} \approx \pi^3_{18} \approx Z_3 .$$

Since $\alpha_1(3) \circ \beta_1(6)$ and $\alpha_4(3)$ do not vanish under E^∞, then they generate π^3_{16} and π^3_{18} respectively. Therefore, we see that the theorem is true for $m = 1$ and $k \leq 15$.

By (13.2) and (13.6), we have an exact sequence

$$\pi^3_{16} \xrightarrow{E^2} \pi^5_{18} \longrightarrow 0 .$$

Since $\pi^3_{16} \approx Z_3$, then π^5_{18} has at most three elements. By combining this fact with (13.12), we have that

$$E^2\pi^5_{19} = 0 \quad \text{and} \quad \pi^5_{19} \approx \pi^5_{18} \approx Z_3 .$$

Then the theorem is true for the case $m = 2$ and $k \leq 16$.

Apply Lemma 13.7 to the above result $E^2\pi^5_{19} = 0$, then we have that $\Delta : \pi^7_{21} \to \pi^5_{19}$ is an isomorphism onto. Since $E^2 : \pi^5_{18} \to \pi^7_{20}$ and $E^2 : \pi^5_{20} \to \pi^7_{22}$ are isomorphisms onto, then it follows from the relation $\Delta \circ E^2 = f_{3*}$ that $\Delta : \pi^7_{20} \to \pi^5_{18}$ and $\Delta : \pi^7_{22} \to \pi^5_{20}$ are trivial. Then it follows from the exactness of the sequence in Proposition 13.3 and from $\pi^5_{21} = \pi^7_{19} = 0$ that

$$\pi^3_{22} \approx \pi^3_{21} \approx \pi^3_{20} \approx \pi^3_{19} \approx Z_3 .$$

Since $E^\infty\alpha_5(3) \neq 0$, then $\alpha_5(3)$ generates π^3_{19}. Therefore, we see that the theorem is true for the case $m = 1$.

Next we compute the groups π^5_{24} and π^7_{26}. By (3.2) and (3.6), we have exact sequences

$$0 \longrightarrow \pi^7_{26} \xrightarrow{E^2} \pi^9_{28}$$

and

$$0 \longrightarrow \pi^5_{24} \xrightarrow{E^2} \pi^7_{26} .$$

Since $E^4\alpha_5(5) = \alpha_5(9)$ generates $\pi^9_{28} \approx Z_3$, then it follows that these E^2 are isomorphisms onto and that

$$\pi^5_{24} = \{\alpha_5(5)\} \approx Z_3 \quad \text{and} \quad \pi^7_{26} = \{\alpha_5(7)\} \approx Z_3 .$$

Now we shall prove that

$$(13.14). \qquad E^2\pi^3_{20} = 0 .$$

Consider the composition $\alpha_2(3) \circ \beta_1(10) \in \pi^3_{20}$. Then

$$\begin{aligned} H(\alpha_2(3) \circ \beta_1(10)) &= H(\alpha_2(3)) \circ \beta_1(10) && \text{by (13.9)} \\ &= \pm\, \alpha_1(7) \circ \beta_1(10) && \text{by (13.10)} \end{aligned}$$

Thus $\alpha_2(3) \circ \beta_1(10) \neq 0$ and it generates π^3_{20}, since $\pi^3_{20} \approx Z_3$.
Next we have

$$\begin{aligned} \alpha_2(5) \circ \beta_1(12) &\in -\{\alpha_0(5), \alpha_1(8), 3\iota_{11}\}_3 \circ \beta_1(12) && \text{by (13.8)} \\ &= -\alpha_1(5) \circ E^3\{\alpha_1(5), 3\iota_8, \beta_1(8)\} && \text{by Proposition 1.4} \\ &\subset \alpha_1(5) \circ E^3\pi^5_{14} && \text{by (4.6).} \end{aligned}$$

Since $E^2\pi^5_{14} = 0$, then it follows that $E^2(\alpha_2(3) \circ \beta_1(10)) = \alpha_2(5) \circ \beta_1(12) = 0$ and hence $E^2\pi^3_{20} = 0$.

It is verified from Proposition 13.1 that

$$\pi_i(\Omega^2(S^5), S^3) \approx \begin{cases} Z_3 & \text{for } i = 21 \text{ and } 23, \\ Z_9 \text{ or } Z_3 + Z_3 & \text{for } i = 22. \end{cases}$$

Then we have from (13.2) and (13.14) the following exact sequences.

$$0 \xrightarrow{E^2} \pi^5_{22} \longrightarrow Z_3 \longrightarrow \pi^3_{19} \xrightarrow{E^2} \pi^5_{21} ,$$

$$\pi^3_{22} \xrightarrow{E^2} \pi^5_{24} \to Z_3 \to \pi^3_{21} \xrightarrow{E^2} \pi^5_{23} \to \pi_{22}(\Omega^2(S^5), S^3) \to \pi^3_{20} \to 0 .$$

Since $\pi^5_{21} = 0$ and $\pi^3_{19} \approx Z_3$, then we have that $\pi^5_{22} = 0$. Since $E^2 : \pi^3_{22} = \{\alpha_5(3)\} \to \pi^5_{24} = \{\alpha_5(5)\}$ is an isomorphism onto and since $\pi^3_{20} \approx Z_3$, we have an exact sequence

$$0 \to Z_3 \to \pi^3_{21} \xrightarrow{E^2} \pi^5_{23} \to Z_3 \to 0 .$$

Since $\pi^3_{21} \approx Z_3$, it follows that $\pi^5_{23} \approx Z_3$.

By (13.2) and (13.6), we have an exact sequence

$$\pi^5_{24} \xrightarrow{E^2} \pi^7_{26} \to Z_3 \to \pi^5_{23} \xrightarrow{E^2} \pi^7_{25} \to Z_3 \to \pi^5_{22} \xrightarrow{E^2} \pi^7_{24} \to 0 .$$

Since $E^2 : \pi^5_{24} \to \pi^7_{26}$ is an isomorphism onto and since $\pi^5_{22} = 0$, we have that $\pi^7_{24} = 0$ and that the following sequence is exact:

$$0 \to Z_3 \to \pi^5_{23} \to \pi^7_{25} \to Z_3 \to 0 .$$

Since $\pi^5_{23} \approx Z_3$, it follows that $\pi^7_{25} \approx Z_3$.

Consequently, we see that the proof of the theorem is established.

CHAPTER XIV

Tables.

The following tables of the homotopy groups of spheres are given by compiling the results in Propositions 5.1, 5.3, 5.6, 5.8, 5.9, 5.11, 5.15, Theorems 7.1, 7.2, 7.3, 7.4, 7.6, 7.7, 10.3, 10.5, 10.10, 12.6, 12.7, 12.8, 12.9, 12.16, 12.17, 12.22, 12.23, 13.4, 13.9, 13.10 and (13.1).

In the table, an integer n indicates a cyclic group Z_n of order n, the symbol "∞" an infinite cyclic group Z, the symbol "+" the direct sum of the groups and $(2)^k$ indicates the direct sum of k-copies of Z_2.

Table of $\pi_{n+k}(S^n)$, I.

	n = 2	n = 3	n = 4	n = 5	n = 6	n = 7	n = 8	n = 9	n = 10	n = 11	n > k+1
k = 1	∞	2	...	...	...	...	...	...	...	...	2
k = 2	2	2	2	..	...	...	...	...	...	...	2
k = 3	2	12	∞ + 12	24	...	...	...	...	...	...	24
k = 4	12	2	$(2)^2$	2	0	...	...	...	...	...	0
k = 5	2	2	$(2)^2$	2	∞	0	...	...	...	...	0
k = 6	2	3	24 + 3	2	2	2	2	...	...	...	2
k = 7	3	15	15	30	60	120	∞ + 120	240	...	...	240
k = 8	15	2	2	2	24 + 2	$(2)^3$	$(2)^4$	$(2)^3$	$(2)^2$	...	$(2)^2$
k = 9	2	$(2)^2$	$(2)^3$	$(2)^3$	$(2)^3$	$(2)^4$	$(2)^5$	$(2)^4$	∞ + $(2)^3$	$(2)^3$	$(2)^3$
k = 10	$(2)^2$	12 + 2	120 + 12 + 2	72 + 2	72 + 2	24 + 2	24 + 24 + 2	24 + 2	12 + 2	6 + 2	6

Table of $\pi_{n+k}(S^n)$, II.

	n = 2	n = 3	n = 4	n = 5	n = 6	n = 7	n = 8	n = 9	n = 10	n = 11
k = 11	$12 + 2$	$84 + (2)^2$	$84 + (2)^5$	$504 + (2)^2$	$504 + 4$	$504 + 2$	$504 + 2$	$504 + 2$	504	504
k = 12	$84 + (2)^2$	$(2)^2$	$(2)^6$	$(2)^3$	240	0	0	0	12	2
k = 13	$(2)^2$	6	$24 + 6 + 2$	$6 + 2$	6	6	$6 + 2$	6	6	$6 + 2$
k = 14	6	30	$2520 + 6 + 2$	$6 + 2$	$12 + 2$	$24 + 4$	$240 + 24 + 4$	$16 + 4$	$16 + 2$	$16 + 2$
k = 15	30	30	30	$30 + 2$	$60 + 6$	$120 + (2)^3$	$120 + (2)^5$	$240 + (2)^3$	$240 + (2)^2$	$240 + 2$
k = 16	30	$6 + 2$	$6 + 6 + 2$	$(2)^2$	$504 + (2)^2$	$(2)^4$	$(2)^7$	$(2)^4$	$240 + 2$	2
k = 17	$6 + 2$	$12 + (2)^2$	$24 + 12 + 4 + (2)^2$	$4 + (2)^2$	$(2)^4$	$(2)^4$	$6 + (2)^4$	$(2)^4$	$(2)^3$	$(2)^3$
k = 18	$12 + (2)^2$	$12 + (2)^2$	$120 + 12 + (2)^5$	$24 + (2)^2$	$24 + 6 + 2$	$24 + 2$	$504 + 24 + 2$	$24 + 2$	$24 + (2)^2$	$8 + 4 + 2$
k = 19	$12 + (2)^2$	$132 + 2$	$132 + (2)^5$	$264 + 2$	$1056 + 8$	$264 + 2$	$264 + 2$	$264 + 2$	$264 + 6$	$264 + (2)^3$

Table of $\pi_{n+k}(S^n)$, III.

	n = 12	n = 13	n = 14	n = 15	n = 16	n = 17	n = 18	n = 19	n = 20	n > k+1
k = 11	$\infty + 504$	504	...	...	...	...	...	...	...	504
k = 12	$(2)^2$	2	0	...	...	...	...	...	...	0
k = 13	$6 + 2$	6	$\infty + 3$	3	...	...	...	...	...	3
k = 14	$48 + 4 + 2$	$16 + 2$	$8 + 2$	$4 + 2$	$(2)^2$	...	...	...	...	$(2)^2$
k = 15	$240 + 2$	$480 + 2$	$480 + 2$	$480 + 2$	$\infty + 480 + 2$	$480 + 2$	...	...	...	$480 + 2$
k = 16	2	2	$24 + 2$	$(2)^3$	$(2)^4$	$(2)^3$	$(2)^2$	...	...	$(2)^2$
k = 17	$(2)^4$	$(2)^4$	$(2)^4$	$(2)^5$	$(2)^6$	$(2)^5$	$\infty + (2)^4$	$(2)^4$	...	$(2)^4$
k = 18	$480 + 4 + 4 + 2$	$8 + 8 + 2$	$8 + 8 + 2$	$8 + 8 + 2$	$24 + 8 + 8 + 2$	$8 + 8 + 2$	$8 + 4 + 2$	$8 + (2)^2$	$8 + 2$	$8 + 2$
k = 19	$264 + (2)^5$	$264 + (2)^3$	$264 + 4 + 2$	$264 + (2)^2$	$264 + (2)^2$	$264 + (2)^2$	$264 + 2$	$264 + 2$	$\infty + 264 + 2$	$264 + 2$

For the stable group $G_k \approx \pi_{n+k}(S^n)$, $n > k+1$, we have the following table of generators.

	$k = 0$	$k = 1$	$k = 2$	$k = 3$	$k = 6$	$k = 7$
$G_k \approx$	Z	Z_2	Z_2	Z_8+Z_3	Z_2	$Z_{16}+Z_3+Z_5$
Generators	ι	η	η^2	ν, α_1	ν^2	$\sigma, \alpha_2, \alpha_{1,5}$

$k = 8$	$k = 9$	$k = 10$	$k = 11$	$k = 13$
Z_2+Z_2	$Z_2+Z_2+Z_2$	Z_2+Z_3	$Z_8+Z_9+Z_7$	Z_3
$\bar{\nu}, \varepsilon$	$\nu^3, \mu, \eta\circ\varepsilon$	$\eta\circ\mu, \beta_1$	$\zeta, \alpha_3', \alpha_{1,7}$	$\alpha_1\circ\beta_1$

$k = 14$	$k = 15$	$k = 16$	$k = 17$
Z_2+Z_2	$Z_{32}+Z_2+Z_3+Z_5$	Z_2+Z_2	$Z_2+Z_2+Z_2+Z_2$
σ^2, κ	$\rho, \eta\circ\kappa, \alpha_4, \alpha_{2,5}$	$\eta^*, \eta\circ\rho$	$\eta\circ\eta^*, \nu\circ\kappa, \eta^2\circ\rho, \bar{\mu}$

$k = 18$	$k = 19$
Z_8+Z_2	$Z_8+Z_2+Z_3+Z_{11}$
$\nu^*, \eta\circ\bar{\mu}$	$\bar{\zeta}, \bar{\sigma}, \alpha_5, \alpha_{1,11}$

We mention the following relations in secondary compositions, which are taken as the definitions of the corresponding elements.

$$\bar{\nu} \in \langle \nu, \eta, \nu \rangle ,$$
$$\varepsilon \in \langle \nu^2, 2\iota, \eta \rangle ,$$
$$\mu \in \langle 8\sigma, 2\iota, \eta \rangle ,$$
$$\zeta \in \langle 2\sigma, 8\iota, \nu \rangle ,$$
$$\eta\circ\kappa \in \langle \varepsilon, 2\iota, \nu^2 \rangle ,$$
$$\eta^* \in \langle \sigma, 2\sigma, \eta \rangle ,$$
$$\bar{\mu} \in \langle \mu, 2\iota, 8\sigma \rangle ,$$
$$\nu^* \in \langle \sigma, 2\sigma, \nu \rangle = - \langle \sigma, \nu, \sigma \rangle ,$$
$$\bar{\zeta} \in \langle \zeta, 8\iota, 2\sigma \rangle ,$$
$$\bar{\sigma} \in \langle \nu, \bar{\nu} + \varepsilon, \sigma \rangle ,$$

Let G denote the direct sum of all G_k. Then G is a graded ring with respect to the composition as the multiplication in G. The ring is anti-commutative by (3.4). The class ι of the identity acts as the identity of the ring. The following theorem clarifies the multiplicative structure in G_k for $k \leq 19$.

Theorem 14.1.

i). $\eta^3 = 4\nu$, $\eta \circ \sigma = \bar{\nu} + \varepsilon$, $\eta \circ \bar{\nu} = \nu^3$, $\eta^2 \circ \mu = 4\zeta$, $\eta^2 \circ \eta^* = 4\nu^*$, $\eta^2 \circ \bar{\mu} = 4\bar{\zeta}$ and

$$\eta \circ \nu = \eta^2 \circ \varepsilon = \eta \circ \zeta = \eta \circ \sigma^2 = \eta^2 \circ \kappa = \eta \circ \nu^* = 0.$$

ii). $\nu \circ \sigma = \nu \circ \varepsilon = \nu \circ \bar{\nu} = \nu \circ \mu = \nu \circ \zeta = \nu \circ \rho = \nu \circ \eta^* = 0.$

iii). $\sigma \circ \mu = \eta \circ \rho$ and $\sigma \circ \varepsilon = \sigma \circ \bar{\nu} = \sigma \circ \zeta = 0$.

iv). $\varepsilon \circ \mu = \eta^2 \circ \rho$ and $\varepsilon \circ \varepsilon = \bar{\nu} \circ \varepsilon = \bar{\nu} \circ \bar{\nu} = \bar{\nu} \circ \mu = \varepsilon \circ \zeta = \bar{\nu} \circ \zeta = 0$.

v). $\mu \circ \mu = \eta \circ \bar{\mu}$.

Proof. i). By (5.5), $4\nu = \eta^3$. By Lemma 6.4, $\eta \circ \sigma = \bar{\nu} + \varepsilon$. By Lemma 6.3, $\eta \circ \bar{\nu} = \nu^3$. By (7.14), $\eta^2 \circ \mu = 4\zeta$. We have

$$\eta^2 \circ \eta^* = \eta^* \circ \eta^2 \in \langle \sigma, 2\sigma, \eta \rangle \circ \eta^2$$
$$\subset \langle \sigma, 2\sigma, \eta^3 \rangle = \langle \sigma, 2\sigma, 4\nu \rangle$$

and $$4\nu^* = \langle \sigma, 2\sigma, \nu \rangle \circ 4\iota \subset \langle \sigma, 2\sigma, 4\nu \rangle .$$

The secondary composition $\langle \sigma, 2\sigma, 4\nu \rangle$ is a coset of

$$\sigma \circ G_{11} + 4\nu \circ G_{15} = \{\sigma \circ \zeta\} + \{4\nu \circ \eta \circ \kappa\} + \{4\nu \circ \rho\}$$
$$= \{\sigma \circ \zeta\} + \{4\nu \circ \rho\} = 0 \quad \text{by Lemma 12.24.}$$

Thus, $\eta^2 \circ \eta^* = 4\nu^*$. By Lemma 12.4, $\eta^2 \circ \bar{\mu} = 4\bar{\zeta}$. Since $G_4 = 0$, then $\eta \circ \nu = 0$. By (7.10) and (7.20), $\eta^2 \circ \varepsilon = 4\nu \circ \sigma = 0$. Since $G_{12} = 0$, then $\eta \circ \zeta = 0$. We have

$$\eta \circ \sigma^2 = (\bar{\nu} + \varepsilon) \circ \sigma = \bar{\nu} \circ \sigma + \varepsilon \circ \sigma = 0 \quad \text{by Lemma 10.7,}$$
$$\eta^2 \circ \kappa = \eta \circ \bar{\varepsilon} = \nu \circ \sigma \circ \nu^2 \quad \text{by Lemma 12.10 and (10.23)}$$
$$= 0 \quad \text{by (7.20)}$$

and $$\eta \circ \nu^* = \nu^* \circ \eta \in \langle \sigma, 2\sigma, \nu \rangle \circ \eta$$
$$= \sigma \circ \langle 2\sigma, \nu, \eta \rangle \subset \sigma \circ G_{12} = 0.$$

ii). By (7.20), $\nu \circ \sigma = 0$. By (7.17) and (7.18), $\nu \circ \bar{\nu} = \nu \circ \varepsilon = E^\infty \Delta(\nu_{13}^2) = 0$. Since $G_{12} = 0$, then $\nu \circ \mu = 0$. By (10.7), $\nu \circ \zeta = 8\sigma^2 = 0$. By Lemma 12.24, $\nu \circ \rho = 0$. We have

$$\nu \circ \eta^* = \eta^* \circ \nu \in \langle \sigma, 2\sigma, \eta \rangle \circ \nu$$
$$= \sigma \circ \langle 2\sigma, \eta, \nu \rangle \subset \sigma \circ G_{12} = 0.$$

iii). By i) of Proposition 12.20, $\sigma \circ \mu = \eta \circ \rho$. By Lemma 10.7, $\sigma \circ \bar{\nu} = \sigma \circ \varepsilon = 0$. By Lemma 12.24, $\sigma \circ \zeta = 0$.

iv). By Lemma 12.10, $\varepsilon \circ \varepsilon = \bar{\nu} \circ \varepsilon = \nu \circ \sigma \circ \nu^2 = 0$. Then $\bar{\nu} \circ \bar{\nu} = \eta \circ \sigma \circ \bar{\nu} + \varepsilon \circ \bar{\nu} = 0$. We see that the secondary composition $\{\eta_6, \nu_7, \mu_{10}\}$

can be defined. Let α be an element of $\{\eta_6, \nu_7, \mu_{10}\} \subset \pi_{20}(S^6)$. By Theorem 10.3, we have that $E\pi_{20}(S^6;2) \subset 2\pi_{21}(S^7)$. Since $G_{14} \approx Z_2 + Z_2$ it follows that $E^\infty\alpha \in E^\infty\pi_{20}(S^6) = 0$. Then we have

$$0 = \nu \circ (-E^\infty\alpha) \in \nu \circ < \eta, \nu, \mu > = < \nu, \eta, \nu > \circ \mu \ni \bar{\nu} \circ \mu.$$

Since $< \nu, \eta, \nu > \circ \mu$ is a coset of $\nu \circ G_5 \circ \mu = 0$, then we have that $\bar{\nu} \circ \mu = 0$ and $\varepsilon \circ \mu = \eta \circ \sigma \circ \mu + \bar{\nu} \circ \mu = \eta^2 \circ \rho$. We have that

$$\varepsilon \circ \zeta \in < \nu^2, 2\iota, \eta > \circ \zeta = \nu^2 \circ < 2\iota, \eta, \zeta >$$
$$\subset \nu^2 \circ G_{13} = 0.$$

Then $\bar{\nu} \circ \zeta = \eta \circ \sigma \circ \zeta + \varepsilon \circ \zeta = 0$.

v). By ii) or (3.9), $\mu \in < 8\sigma, 2\iota, \eta > \equiv < 2\iota, \eta, 8\sigma > + < \eta, 8\sigma, 2\iota > \bmod 2G_9 + \eta \circ G_8 = \eta \circ G_8$. Since $2 < 2\iota, \eta, 2\iota > \subset 2G_2 = 0$, then $0 \in < 2\iota, \eta, 2\iota > \circ 4\sigma \subset < 2\iota, \eta, 8\sigma >$, by (3.5). It follows then that μ is an element of $< \eta, 8\sigma, 2\iota > = < 2\iota, 8\sigma, \eta >$. Now we have

$$\mu \circ \mu \in \mu \circ < 2\iota, 8\sigma, \eta > = < \mu, 2\iota, 8\sigma > \circ \eta$$

and

$$\eta \circ \bar{\mu} = \bar{\mu} \circ \eta \in < \mu, 2\iota, 8\sigma > \circ \eta .$$

The composition $< \mu, 2\iota, 8\sigma > \circ \eta$ is a coset of

$$\mu \circ G_8 \circ \eta = \{\mu \circ \varepsilon \circ \eta\} + \{\mu \circ \bar{\nu} \circ \eta\} = \{\eta^3 \circ \rho\} = \{4\nu \circ \rho\} = 0.$$

Thus we have $\mu \circ \mu = \eta \circ \bar{\mu}$. q.e.d.

BIBLIOGRAPHY

[1]. J. F. Adams, On the non-existence of elements of Hopf invariant one, Ann. of Math., 72 (1960), 20-104.

[2]. J. Adem, The relations on Steenrod powers of cohomology classes, Algebraic geometry and topology, Princeton.

[3]. M. G. Barratt, Track groups I, II, Proc. London Math. Soc. 5 (1955), 71-106, 285-329.

[4]. M. G. Barratt, Note on a formual due to Toda, Jour. London Math. Soc., 36 (1961), 95-96.

[5]. M. G. Barratt and P. J. Hilton, On join operations of homotopy groups, Proc. London Math. Soc., 3 (1953), 430-445.

[6]. A. L. Blakers and W. S. Massey, The homotopy groups of a triad II, Ann. of Math., 55 (1952), 192-201.

[7]. H. Freudenthal, Uber die Klassen der Sphärenabbildungen, Comp. Math., 5 (1937).

[8]. H. Hopf, Uber die Abbildungen von Sphären auf Sphären niedrigerer Dimension, Fund. Math., 25 (1935).

[9]. I. M. James, Reduced product spaces, Ann. of Math., 62 (1955), 170-197.

[10]. I. M. James, Suspension triad of a sphere, Ann. of Math., 63 (1956), 407-429.

[11]. I. M. James, The intrinsic join : A study of the homotopy groups of Stiefel manifold, Proc. London Math. Soc., 8 (1958), 507-535.

[12]. J. C. Moore, Some applications of homology theory to homotopy problems, Ann. of Math., 58(1953), 325-350.

[13]. J.-P. Serre, Homologie singulière des espaces fibrés, Ann. of Math., 54 (1951), 425-505.

[14]. J.-P. Serre, Cohomologie modulo 2 des complexes d'Eilenberg-MacLane, Comm. Math. Helv., 27 (1953), 198-231.

[15]. J.-P. Serre, Groups d'homotopie et classes groupes abéliens, Ann. of Math., 58 (1953), 258-294.

[16]. J.-P. Serre, Quelques calculs de groupes d'homotopie, Comptes Rendus Paris 236 (1953), 2475-2477.

[17]. N. E. Steenrod, Topology of fibre bundles, Princeton.

[18]. N. E. Steenrod, Cohomology invariants of mappings, Ann. of Math., 50 (1949), 954-988.

[19]. N. E. Steenrod, Cyclic reduced powers of cohomology classes, Proc. Nat. Acad. Sci. U. S. A., 39 (1953), 217-223.

[20]. H. Toda, Generalized Whitehead products and homotopy groups of spheres, Jour. of Inst. Poly. Osaka City Univ., 3 (1952), 43-82.

[21]. H. Toda, On double suspension E^2, Jour. of Inst. Poly. Osaka City Univ., 7 (1956), 103-145.

[22]. H. Toda, Reduced join and Whitehead product, Jour. of Inst. Poly. Osaka City Univ., 8 (1957) 15-30.

[23]. H. Toda, On exact sequences in Steenrod algebra mod. 2, Memoirs Univ. of Kyoto, 31 (1958), 33-64.

[24]. H. Toda, p-primary components of homotopy groups II. mod p Hopf invariant, Memoirs Univ. of Kyoto, 31 (1958), 143-160.

[25]. H. Toda, p-primary components of homotopy groups IV, Compositions and toric constructions, Memoirs Univ. of Kyoto, 32 (1959), 297-332.

[26]. G. W. Whitehead, A generalization of the Hopf invariant, Ann. of Math., 51 (1950), 192-237.

[27]. G. W. Whitehead, The $(n+2)^{nd}$ homotopy group of the n-sphere, Ann. of Math., 52 (1950), 245-247.

[28]. J. H. C. Whitehead, On adding relations to homotopy groups, Ann. of Math., 42 (1941), 409-428.

[29]. J. H. C. Whitehead, On the groups $\pi_r(V_{m,n})$ and sphere bundles, Proc. London Math. Soc., 48 (1944) 243-291.

The Institute for Advanced Study
Kyoto University

Milton Keynes UK
Ingram Content Group UK Ltd.
UKHW041031070124
435586UK00001B/68

9 780691 095868